全国高等中医药院校中药学类专业双语规划教材
Bilingual Planned Textbooks for Chinese Materia Medica Majors in TCM Colleges and Universities

无机化学

Inorganic Chemistry

（供中药学类、药学类及相关专业使用）

(For Chinese Materia Medica, Pharmay and other related majors)

主　审　铁步荣

主　编　关　君　杨爱红

副主编　吴巧凤　杨　婕　徐　飞　姚华刚

编　者（以姓氏笔画为序）

马鸿雁（成都中医药大学）　王　颖（北京中医药大学）
王　熙（天津中医药大学）　王　霞（河南中医药大学）
方德宇（辽宁中医药大学）　吕惠卿（浙江中医药大学）
齐学洁（天津中医药大学）　关　君（北京中医药大学）
杨　婕（江西中医药大学）　杨爱红（天津中医药大学）
李德慧（长春中医药大学）　吴巧凤（浙江中医药大学）
吴品昌（辽宁中医药大学）　张　璐（北京中医药大学）
张凤玲（浙江中医药大学）　张晓青（湖南中医药大学）
阿合买提江·吐尔逊（新疆医科大学）　林　舒（福建中医药大学）
罗　黎（山东中医药大学）　姚　军（新疆医科大学）
姚华刚（广东药科大学）　袁　洁（新疆医科大学厚博学院）
贾力维（黑龙江中医药大学）　徐　飞（南京中医药大学）
徐　暘（黑龙江中医药大学）　郭爱玲（山西中医药大学）
曹　莉（湖北中医药大学）　戴红霞（甘肃中医药大学）

中国健康传媒集团
中国医药科技出版社

内 容 提 要

本教材是《全国高等中医药院校中药学类专业双语规划教材》之一，根据《无机化学》教学大纲的基本要求和课程特点编写而成，内容上涵盖了溶液、化学平衡、酸碱平衡、沉淀－溶解平衡、氧化－还原反应、原子结构与元素周期律、化学键与分子结构、配位化合物、主族元素及相关矿物药、过渡金属元素及其相关矿物药等内容。本教材在内容选择上力争做到少而精，内容编排次序上遵守循序渐进、先易后难的原则。本教材为书网融合教材，即纸质教材有机融合电子教材、教学配套资源(PPT、微课、视频等)、题库系统、数字化教学服务(在线教学、在线作业、在线考试)等，使教学资源更加多样化和立体化。

本教材可供高等医药院校中药学类、药学及相关专业使用。

图书在版编目（CIP）数据

无机化学：汉英对照 / 关君，杨爱红主编．— 北京：中国医药科技出版社，2020.8

全国高等中医药院校中药学类专业双语规划教材

ISBN 978-7-5214-1881-1

Ⅰ.①无…　Ⅱ.①关…②杨…　Ⅲ.①无机化学－双语教学－中医学院－教材－汉、英　Ⅳ.①O61

中国版本图书馆 CIP 数据核字（2020）第 101874 号

美术编辑　陈君杞

版式设计　辰轩文化

出版　**中国健康传媒集团**｜中国医药科技出版社

地址　北京市海淀区文慧园北路甲 22 号

邮编　100082

电话　发行：010-62227427　邮购：010-62236938

网址　www.cmstp.com

规格　889 × 1194 mm $^1/_{16}$

印张　17 $^3/_4$

彩页　1

字数　448 千字

版次　2020 年 8 月第 1 版

印次　2023 年 8 月第 2 次印刷

印刷　三河市万龙印装有限公司

经销　全国各地新华书店

书号　ISBN 978-7-5214-1881-1

定价　59.00 元

出版说明

近些年随着世界范围的中医药热潮的涌动，来中国学习中医药学的留学生逐年增多，走出国门的中医药学人才也在增加。为了适应中医药国际交流与合作的需要，加快中医药国际化进程，提高来中国留学生和国际班学生的教学质量，满足双语教学的需要和中医药对外交流需求，培养优秀的国际化中医药人才，进一步推动中医药国际化进程，根据教育部、国家中医药管理局、国家药品监督管理局等部门的有关精神，在本套教材建设指导委员会主任委员成都中医药大学彭成教授等专家的指导和顶层设计下，中国医药科技出版社组织全国50余所高等中医药院校及附属医疗机构约420名专家、教师精心编撰了全国高等中医药院校中药学类专业双语规划教材，该套教材即将付梓出版。

本套教材共计23门，主要供全国高等中医药院校中药学类专业教学使用。本套教材定位清晰、特色鲜明，主要体现在以下方面。

一、立足双语教学实际，培养复合应用型人才

本套教材以高校双语教学课程建设要求为依据，以满足国内医药院校开展留学生教学和双语教学的需求为目标，突出中医药文化特色鲜明、中医药专业术语规范的特点，注重培养中医药技能、反映中医药传承和现代研究成果，旨在优化教育质量，培养优秀的国际化中医药人才，推进中医药对外交流。

本套教材建设围绕目前中医药院校本科教育教学改革方向对教材体系进行科学规划、合理设计，坚持以培养创新型和复合型人才为宗旨，以社会需求为导向，以培养适应中药开发、利用、管理、服务等各个领域需求的高素质应用型人才为目标的教材建设思路与原则。

二、遵循教材编写规律，整体优化，紧跟学科发展步伐

本套教材的编写遵循“三基、五性、三特定”的教材编写规律；以“必需、够用”为度；坚持与时俱进，注意吸收新技术和新方法，适当拓展知识面，为学生后续发展奠定必要的基础。实验教材密切结合主干教材内容，体现理实一体，注重培养学生实践技能训练的同时，按照教育部相关精神，增加设计性实验部分，以现实问题作为驱动力来培养学生自主获取和应用新知识的能力，从而培养学生独立思考能力、实验设计能力、实践操作能力和可持续发展能力，满足培养应用型和复合型人才的要求。强调全套教材内容的整体优化，并注重不同教材内容的联系与衔接，避免遗漏和不必要的交叉重复。

三、对接职业资格考试，“教考”“理实”密切融合

本套教材的内容和结构设计紧密对接国家执业中药师职业资格考试大纲要求，实现教学与考试、理论与实践的密切融合，并且在教材编写过程中，吸收具有丰富实践经验的企业人员参与教材的编写，确保教材的内容密切结合应用，更加体现高等教育的实践性和开放性，为学生参加考试和实践工作打下坚实基础。

四、创新教材呈现形式，书网融合，使教与学更便捷更轻松

全套教材为书网融合教材，即纸质教材与数字教材、配套教学资源、题库系统、数字化教学服务有机融合。通过“一书一码”的强关联，为读者提供全免费增值服务。按教材封底的提示激活教材后，读者可通过 PC、手机阅读电子教材和配套课程资源（PPT、微课、视频等），并可在线进行同步练习，实时收到答案反馈和解析。同时，读者也可以直接扫描书中二维码，阅读与教材内容关联的课程资源，从而丰富学习体验，使学习更便捷。教师可通过 PC 在线创建课程，与学生互动，开展在线课程内容定制、布置和批改作业、在线组织考试、讨论与答疑等教学活动，学生通过 PC、手机均可实现在线作业、在线考试，提升学习效率，使教与学更轻松。此外，平台尚有数据分析、教学诊断等功能，可为教学研究与管理提供技术和数据支撑。需要特殊说明的是，有些专业基础课程，例如《药理学》等 9 种教材，起源于西方医学，因篇幅所限，在本次双语教材建设中纸质教材以英语为主，仅将专业词汇对照了中文翻译，同时在中国医药科技出版社数字平台“医药大学堂”上配套了中文电子教材供学生学习参考。

编写出版本套高质量教材，得到了全国知名专家的精心指导和各有关院校领导与编者的大力支持，在此一并表示衷心感谢。希望广大师生在教学中积极使用本套教材和提出宝贵意见，以便修订完善，共同打造精品教材，为促进我国高等中医药院校中药学类专业教育教学改革和人才培养做出积极贡献。

全国高等中医药院校中药学类专业双语规划教材
建设指导委员会

数字化教材编委会

主　审　铁步荣

主　编　关　君　杨爱红

副主编　吴巧凤　杨　婕　徐　飞　姚华刚

编　者（以姓氏笔画为序）

马鸿雁（成都中医药大学）
王　颖（北京中医药大学）
王　熙（天津中医药大学）
王　霞（河南中医药大学）
方德宇（辽宁中医药大学）
吕惠卿（浙江中医药大学）
齐学洁（天津中医药大学）
关　君（北京中医药大学）
杨　婕（江西中医药大学）
杨爱红（天津中医药大学）
李德慧（长春中医药大学）
吴巧凤（浙江中医药大学）
吴品昌（辽宁中医药大学）
张　璐（北京中医药大学）
张凤玲（浙江中医药大学）
张晓青（湖南中医药大学）
阿合买提江·吐尔逊（新疆医科大学）
林　舒（福建中医药大学）
罗　黎（山东中医药大学）
姚　军（新疆医科大学）
姚华刚（广东药科大学）
袁　洁（新疆医科大学厚博学院）
贾力维（黑龙江中医药大学）
徐　飞（南京中医药大学）
徐　旸（黑龙江中医药大学）
郭爱玲（山西中医药大学）
曹　莉（湖北中医药大学）
戴红霞（甘肃中医药大学）

前言

为了满足21世纪对国际化高素质人才的需求、推进高等教育的内涵式发展和高等院校的"双一流"建设，根据无机化学教学大纲基本要求和课程特点，我们组织编写了这本供中药学类本科各专业使用的《无机化学》双语教材。

本教材由国内18所中医药院校的29名一线教师通力合作，经过多次集体研究讨论，统筹安排，分工编写，精心修改后完成。在编写过程中，本教材编委会充分调研、参考了大量国内外化学和药学相关教材和资料，吸取了近十年来各个高校在无机化学双语教学课程体系、教学内容和课堂教学方面的众多经验，采用英文为主，专业词汇辅以中文的形式，以利于读者更好掌握英文词汇。

本教材共分为11章，为方便教学，各章都附有练习题（含答案），附录包括了使用本教材需要的无机化学常用数据。我们的目标是秉持少而精、特色鲜明和利于教学的原则，编写一本内容充实、深浅适度、适合教学的高质量教材。为此，我们既考虑了我国中药学及其相关中医药专业对无机化学课程的基本概念、基本理论与基本技能的要求，也努力适应全国众多医药院校药学专业的教学需求，特别重视了在主要内容、名词术语、计量单位等方面的严谨规范，并根据学科发展前沿的特点，适当增加了一些无机化学在药学领域的发展前沿性内容。本教材可作为普通高等院校中药学类、药学类及相关专业无机化学课程的教科书，也可作为化学、医学、环境等相关专业无机化学课程的教学参考书，并可供科研院所、医药企业、药品管理机构从事无机化学工作的科技人员参考。

本教材由关君、杨爱红任主编，负责全书的统稿、定稿工作，铁步荣教授负责全书内容审定工作。各章节编写分工如下：杨婕、马鸿雁、曹莉、吕惠卿编写第一章、第五章；吴巧凤、姚军、林舒、张凤玲编写第二章、第七章；杨爱红、齐学洁、王霞、阿合买提江·吐尔逊、王熙编写第三章、第四章；徐飞、徐暘、戴红霞、方德宇编写第六章、附录；关君、贾力维、王颖、吴品昌、张璐、罗黎、袁洁编写第八章、第十一章；姚华刚、张晓青、郭爱玲、李德慧编写第九章、第十章。

本教材的编写工作获得了全体编者及其所在院校领导的大力支持与帮助，在此一并致谢。

由于受编者学识水平所限，本教材可能存在疏漏与不足之处，敬请广大师生和读者提出宝贵意见，以便再版时加以修改和完善。

编　者

2020年6月

Preface

In order to meet the needs of international high-quality talents in the 21st century, and to promote the connotative development of higher education and the "double first class" construction of colleges and universities, according to the basic requirement and characteristics of the syllabus of *Inorganic chemistry.* we organized and compiled this bilingual textbook of inorganic chemistry for undergraduates majoring in traditional Chinese pharmacology.

This textbook is finished by 29 front-line teachers from 18 medical universities in China. After several collective research discussions, overall arrangement, division of work, and careful revision, the manuscript was unified by the chief editor. In the process of compiling, the editorial board of this textbook fully investigated and summarized the achievements of each university in the course system, teaching content and classroom teaching of inorganic chemistry bilingual teaching in the past ten years, and drew on the experience of compiling chemistry and pharmacy related textbooks at home and abroad.

The textbook is divided into 11 chapters. For the convenience of teaching, each chapter is attached with a certain number of exercises (including answers). The appendixes include the commonly used inorganic chemistry data needed. Our goal is to write a high-quality teaching material with substantial content, appropriate depth and suitable for teaching, adhering to the principle of few but fine, distinctive features and being conducive to teaching. For this reason, we not only consider the basic concepts, basic theories and basic skills of inorganic chemistry courses for Chinese Materia Medica and its related majors, but also strive to adapt to the teaching needs of pharmaceutical majors in many medical colleges and universities across the country. We pay special attention to the strict norms in the main contents, terms, measurement units, etc. And some advanced contents of inorganic chemistry in pharmaceutical field have been added appropriately according to the characteristics of the forefront of discipline development. This textbook can be used for inorganic chemistry courses of Chinese Materia Medica, pharmacy, and other related majors in colleges and universities, as a teaching reference book for inorganic chemistry courses of chemistry, medicine, environment and other related majors, and as a professional reference book for scientific and technological personnel engaged in inorganic chemistry work in scientific research institutes, pharmaceutical enterprises and pharmaceutical management organizations.

The chief editors of this textbook are Jun Guan and Aihong Yang. They were in charge of the final editing, and profession Burong Tie are the chief reviewer. The division of labor of the editorial board is as follows. Jie Yang, Hongyan Ma, Li Cao and Huiqing Lv, Chapter 1 and Chapter 5; Qiaofeng Wu,

Jun Yao, Shu Lin and Fengling Zhang, Chapter 2 and Chapter 7; Aihong Yang, Xuejie Qi, Xia Wang, Ahimatijiang Tursun and Xi Wang, Chapter 3 and Chapter 4; Fei Xu, Min Xu, Hongxia Dai and Deyu Fang, Chapter 6 and Appendix; Jun Guan, Liwei Jia, Ying Wang, Lu Zhang, Pinchang Wu, Li Luo and Jie Yuan, Chapter 8 and Chapter 11; Huagang Yao, Xiaoqing Zhang, Ailing Guo and Dehui Li, Chapter 9 and Chapter 10.

The compilation of this textbook has been greatly supported by the schools of the editorial committee members and their leaders. Thanks very much.

Due to the limited knowledge level of the editors, this textbook may have omissions or inadequacies. We would like to invite readers to provide valuable opinions so that we can revise and perfect this textbook when it is reprinted.

Editors

June 2020

Contents

医药大学堂
WWW.YIYAODXT.COM

医药大学堂
WWW.YIYAODXT.COM

Chapter 1 Introduction
第一章 绪论

学习目标

知识要求

1. **掌握** 无机化学的发展简史。
2. **熟悉** 无机化学的研究内容和新进展。
3. **了解** 无机化学与药学的关系。

能力要求

掌握无机化学课程的学习方法。

微课

The world is composed of matter, chemistry is one of the main methods and means used by human beings to understand and transform the material world, and the achievement is an important symbol of social civilization. With the rapid development of science and technology, people gradually realize that chemistry will become the key science to make human beings survive. Nowadays, chemistry is a central subject to meet the needs of society and it is very important to human water supply, food, energy, materials, resources, environment and health problems. Everyone's life is affected by the scientific achievements with chemistry as the core.

Chemistry is a science which mainly studies the composition, structre, properties and changing law of substance. Inorganic chemistry is the earliest branch of chemistry, and it is the basic subject of chemistry. Modern inorganic chemistry is the experimental test and theoretical exposition of the preparation, composition, structure, properties and reactions of all elements and their compounds (except hydrocarbons and their derivatives).

1. The Brief History of Inorganic Chemistry

PPT

Like the development of other sciences, chemistry is the product of human production and life practice. The origin of chemistry can be traced back to ancient times. In the practice of alchemy and medicine, human beings have obtained preliminary chemical knowledge, and chemistry has formed an inextricable relationship with medicine from the beginning.

Inorganic chemistry is the earliest branch of chemistry. Because most of the initial chemical research is on minerals and other inorganic substances. Therefore, before inorganic chemistry became an independant branch of chemistry, the development history of the chemistry is the development history of inorganic chemistry. According to the characteristics of chemical development, it can be divided into three important stages: ① ancient chemistry (before the 17th century), ② early modern chemistry (from the middle of the 17th century to the end of the 19th century, involving the introduction of the concept of elements, combustion oxidation theory, atomic theory, periodic law of elements, formation of chemical branches of inorganic chemistry, etc.), ③ modern chemistry (from the end of the 19th century, involving the movement law of microscopic particles, the disclosure of the essence of atoms and molecular structures, and the formation of interdisciplines, etc.)

In ancient times, the manufacture of glass and pottery was a major chemical process. It is a symbol of ancient civilization. By reducing or increasing the proportion of air, pottery can be made black or red. By 500 AD, the combination of different colors of glaze and color had reached high degree of professional proficiency. Jingdezhen porcelain in Jiangxi Province is famous at home and abroad in ancient and modern times. Ceramic technology in China has a long history. Pottery is the predecessor of porcelain invention in Tang Dynasty. Porcelain technology in Tang Dynasty has been quite mature. China's porcelain technology began to be introduced into Africa and Europe in Tang Dynasty. Other chemical processes in ancient times included wine making, sugar making, dyestuff, metal smelting, pharmacy and so on.

From the middle of the 17th century to the end of the 19th century, chemistry as an independent discipline began to form and develop in these two hundred years. Modern chemistry was established in the struggle against the traditional alchemy thought and the false theory of burning elements. At this stage, the concepts of acid, alkali, salt, element, compound and chemical reagent were gradually formed, and sulfuric acid, hydrochloric acid, ammonia and alum were found. The scientific concept of elements was put forward, and chemistry was defined as a science for the first time, which laid a solid foundation for the establishment and development of chemistry as a science.

Since the end of the 19th century, with the adoption of new technologies in physics, a series of major scientific discoveries have violently impacted on people's old concept that atoms are inseparable. It promotes the study of chemistry to a deeper level of material structure to explore, but also pregnant with a profound revolution in the development of chemistry, marking the establishment of modern chemistry.

During this period, the establishment of Rutherford's atomic "celestial planet model" revealed the structural mysteries of atoms, and put forward the idea of electron shell configuration, which laid a foundation for the establishment of atomic electronic theory and for people to explore the essential causes of periodic law of elements. In 1930s, the modern atomic structure model and modern chemical bond theory deepened people's understanding of the essence of molecular internal structure and played a powerful role in promoting the development of modern chemistry.

2. The Research Contents and New Progress of Inorganic Chemistry

PPT

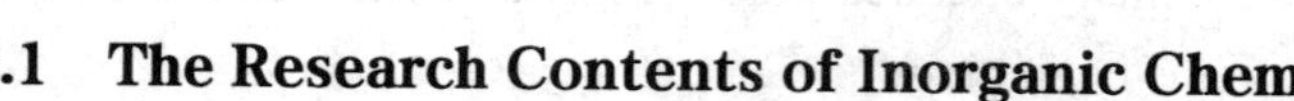

2.1 The Research Contents of Inorganic Chemistry

The research contents of inorganic chemistry include the basic principles of chemistry, the properties of chemical elements and related chemical reactions. By 2019, there are 118 kinds of chemical elements have been found or synthesized. There is an important element, carbon C, which constitutes another important branch of chemistry: organic chemistry. The combination of organic and inorganic chemistry derived organic element chemistry and organometallic chemistry (有机金属化学). In addition, some carbon chemistry, including carbon oxides, oxygen-containing ions or carbides, belongs to the field of inorganic chemistry. Of course, there are no strict boundaries, and no one is trying to strictly define them. And many exciting research results often appear at the intersection of inorganic chemistry and other chemical disciplines. Inorganic chemistry plays an important role in the interlaced part with the subject. These interdisciplines include geochemistry, bioinorganic chemistry, material science and metallurgy.

2.2 New Developments in Contemporary Inorganic Chemistry

In the past 50 years, the pursuit of new methods, new theories, new fields (such as the development of metals in biological systems), new materials, new catalysts, high output and low pollution has strongly promoted the development of inorganic chemistry. Since the 1970s, with the emergence and development of energy, catalysis and biochemistry and other research fields, inorganic chemistry has made new breakthroughs in practice and theory. Nowadays, the most active fields in inorganic chemistry are inorganic material chemistry, bioinorganic chemistry and organometallic chemistry.

The above examples show that the factors that promote the development of disciplines are pure scientific research and applied research. These studies are based on the knowledge of their predecessors. Inorganic chemistry is a systematic chemistry based on the periodic table of elements. Therefore, in order to learn inorganic chemistry, we must firmly master the basic knowledge and understand the development trend of the subject at the same time. This applies to the study of any subject. The application of theory and calculation method in the 21st century will greatly strengthen the closer combination of theory and experiment. At the same time, the in-depth development of various disciplines and the mutual penetration of disciplines form new research fields of many disciplines.

In recent years, remarkable progress has been made in the basic research of inorganic chemistry in China, and many achievements have been made, mainly in the following aspects.

(Ⅰ) On the basis of hydrothermal synthesis, Qian Yitai and Xie Yi of China University of Science and Technology designed and implemented a new inorganic chemical reaction in organic system, and prepared a sequence of nonoxide nanomaterials at relatively low temperature. The principle of solvothermal synthesis is similar to that of hydrothermal synthesis. Organic solvent is used instead of water to realize chemical reaction in sealing system.

(Ⅱ) Shouhua Feng and Ruren Xu of Jilin University successfully synthesized inorganic-organic nanocomposites with spiral structure from simple reaction materials by hydrothermal synthesis technology.

(Ⅲ) Rengen Xiong of Nanjing University designed and synthesized inorganic organic hybrid multidimensional structure with chiral and catalytic function in the assembly of optically active zeolites and their chiral resolution function. They modified the optically active natural organic drug (quinine), and used it as ligand and metal ion self-assembly to form a racemic 2-butanol and 3-methyl-2-butanol. The separation rate of three-dimensional porous zeolites is more than 98%.

(Ⅳ) The work of Maochun Hong and Xintao Wu (Fujian Institute of material structure, Chinese Academy of Sciences) in nanomaterials and inorganic polymers has attracted extensive attention at home and abroad. They successfully synthesized nanometallic molecular cages (nanometer-sized metallomolecular cage), also successfully constructed a new type of molecular sieves with nanometer pores, in which the size of the pores was nearly one nanometer. The synthesis and structure of metal nanowires and metal-organic nanowires had achieved remarkable results. Some metal nanowires, metal-non-metallic nanowires and metal organic nanowires were designed and synthesized.

(Ⅴ) Gaosong Research Group of Peking University has made outstanding achievements in the research of magnetic and molecular materials.

(Ⅵ) Li Yadong of Tsinghua University has made outstanding progress in the preparation and assembly of new one-dimensional nanostructures. Li Yadong research group has discovered a new type of single crystal multi-wall metal nanotube formed by bismuth metal with quasi-layered structure for the first time. This is the first single crystal nanotube formed by metal in the world. The discovery of bismuth nanotube provides a new object and topic for the formation mechanism and application of inorganic nanotube.

In the face of the challenges of the rapid development of life science, material science, information science and other disciplines and the new requirements put forward by human beings to understand and transform nature, chemistry is constantly creating new materials and varieties to meet the material and cultural life of the people, benefit the country and benefit mankind. At present, the effective development and utilization of resources, environmental protection and governance, sustainable development of society and economy, population and health and human safety, and the development and application of high-tech materials have put forward a series of major challenging problems for Chinese scientists. It is urgent for chemists to carry out basic and applied research on chemistry at a higher level, and to discover and create new theories, methods and means.

3. The Relationship Between Inorganic Chemistry and Pharmacy

PPT

Inorganic chemistry and drugs are interrelated, and some inorganic substances can be directly used as drugs. At present, a large number of inorganic preparations also appear in the development of new drugs, which is of course an important task faced by students studying inorganic chemistry. At the end of 1960s, bioinorganic chemistry, a new subject, gradually formed in the intersection of inorganic chemistry and biology. It mainly studies the relationship between the structure, properties and biological activity of bioactive metal ions (including a small number of nonmetals) and their complexes, as well as the mechanism of participating in the reaction in the living environment. Drug inorganic chemistry is a very active aspect in recent ten years. It can be regarded as a branch of bioinorganic chemistry.

There are more than 80 kinds of elements in the human body, which can be divided into constant elements (also known as life structure elements) and trace elements (also known as important elements of life). It is precisely because these elements of life perform their respective duties in the body of organisms, maintain the normal activity of life and promote the development of life. The shortage of

any element (such as a temporary decline in concentrations in body fluids and insufficient reserves in the body) can hinder the normal metabolism of life, leading to malnutrition, stunting, and even disease. As far as we know, many diseases are related to metal ions. As early as 1950s, it was found that metal complexes had antibacterial and antiviral ability, especially the phenanthroline complexes of iron and rhodium, which had a strong inhibitory effect on the division of influenza virus at low concentration. Many cancers are closely related to viruses. Many drug chemists have tried to find effective anticancer drugs.

The most representative inorganic drug used in modern medicine is platinum (Pt) complex with anticancer function. In 1965, when Rosenberg introduced direct current into *E. coli* culture medium containing ammonium chloride with platinum electrode, it was found that the bacteria were no longer divided. After a series of studies, it was confirmed that there were trace platinum complexes in the culture medium, the strongest of which was $PtCl_2(NH_3)_2$ (DDP, cisplatin). It was formed by some interaction between platinum dissolved by electrode and NH_3 and Cl^- in culture medium. Considering that the above phenomena were very similar to those caused by alkylation anticancer drugs, they had done anticancer experiments with these platinum complexes, and the results showed that DDP and its analogues had strong anticancer effect. Since then, a new field of anticancer effect of metal complexes has been opened up. With the continuous development of science, inorganic chemistry and drugs are more and more closely related. Since the earliest pharmaceutical monograph, Shen Nong's Classic of Materia Medica (《神农本草经》), inorganic mineral drugs have been used in the treatment of diseases, a large number of inorganic drugs have also appeared one after another.

Long ago, humans used plants or minerals to treat certain diseases. The era of synthetic drugs can be said to have begun in the 1930s. Pharmaceutical chemists, mainly organic chemists, synthesized thousands of drugs. At present, the total number of drugs has reached tens of thousands, and more than 7,000 kinds of drugs are often used. The study of drugs has entered the stage of drug design or molecular design. Another very important part of drug inorganic chemistry is the inorganic chemical process of organic drugs in organisms, or rather bioinorganic chemical processes. The research in this field is of great significance for exploring the pathogenic factors, expounding the pharmacology and action mechanism of drug molecules, the improvement of drugs and the design of new drugs.

4. The Way to Learn Inorganic Chemistry

PPT

4.1 Purpose and Main Contents of this Course

Inorganic chemistry is an important basic course of pharmacy, traditional Chinese medicine and other majors. Through the study of inorganic chemistry theory course, we can master the basic knowledge of inorganic chemistry, master the general law of chemical reaction and basic chemical calculation method, understand the general method of studying inorganic chemistry, understand the application of basic knowledge of inorganic chemistry, and lay a good theoretical foundation for the study of subsequent specialized courses. Master the basic experimental skills through the experiment. Chemistry is an experimental science in essence. At any time, the development and verification of the new theory must

pass the experiment. Therefore, in the learning process of inorganic chemistry, we should fully realize the importance of chemical experiment in the process of inorganic chemistry learning. In addition, through self-study, we can improve the ability to think independently and solve problems independently.

Inorganic chemistry plays an important role in our knowledge structure, ability training and subject thought. When we acquire chemistry knowledge and training ability, we should also pay attention to the cultivation of our own discipline thought of inorganic chemistry, the formation of the thought will play a very important and extensive role in our future chemistry learning and lifelong development. The contents of inorganic chemistry are very rich (such as the thought of periodic law of elements, the thought of material classification, the thought of dynamic balance, the thought of condition theory and conservation of chemical change, the thought of qualitative and quantitative, the thought of mutual change of mass, the thought of mutual reflection of micro and macro, the thought of energy, etc.). Because the structures and properties of matter are the most basic research objects of chemistry, and the structure relationship of matter is the most basic relationship in chemistry, the idea of "structure determining property" of matter is a very important chemical thought, or one of the core chemical ideas.

The content of this course is divided into two parts: chemical theory and elemental chemistry. The first part of chemical theory includes the basic concepts of chemistry: equilibrium theory (acid-base equilibrium, redox equilibrium, precipitation dissolution equilibrium, coordination equilibrium, etc.); structure theory (atomic structure, molecular structure, crystal structure, complex structure, etc.). The second part is elemental chemistry, focusing on the generality, important reaction laws and applications of non-metallic elements and metal element compounds.

4.2 Methods of Learning This Course

4.2.1 Theoretical Course

(1) The learning mode of middle school　The teaching content of each class is less, the teaching content is repeated more, a large number of classroom exercises and extracurricular exercises, and the content of self-study is less. The learning mode of the university: the teaching content of each class is more, the repetition of the teaching content is small, the amount of work is less, the classroom exercise is less, and the improvement of self-study ability is emphasized. According to the characteristics of university learning, the suggested learning methods are as follows:

(2) Preview before class　Study by yourself before you learn a new class, so that you can understand what you want to learn in this class, and pay special attention to the parts that you don't understand when you listen to the class.

(3) Listen carefully in class　Listen carefully in class, keep up with the teacher's ideas, and put down the problems you don't understand for the time being, and wait future solutions. Otherwise, because of the fast teaching speed, it is easy to accumulate more difficult problems. Make good classroom notes, leave a certain space, mark questions notes or handouts, ask for teachers opinion and discussion in question answering class, and write conclusions.

(4) After class review　review after class is an important process to digest and master what you have learned. This course is characterized by strong theory, and some concepts are more abstract. Only through repeated self-study and thinking, can we gradually deepen understanding and master its essence.

(5) After-school homework　completing a certain amount of exercises after class is helpful to a

deep understanding of the contents of the classroom and to the cultivation of independent thinking and self-study ability. After each class, the teacher will arrange some exercises. We have to answer carefully, and complete it independently.

Reference books: in addition to preview, review, and doing exercises, reading reference books is an important link. It is also an excellent way to cultivate independent thinking and self-study ability.

4.2.2 Experimental Courses

Chemistry is a subject based on experiment, and experiment is very important for the understanding of theory. Therefore, the course arranged nearly 15 related preparation, determination and element properties experiments.

Objective: to master the basic experimental skills, deepen the understanding and memory of theoretical problems through the experiment, and improve the ability to analyze and solve problems.

Requirements: do a good job of preview reports, experimental records and experimental reports.

重点小结

化学主要研究物质的组成、结构和性质，在原子和分子水平上研究物质在变化过程中的变化规律和能量关系。无机化学是化学的最早分支，是化学的基础学科。现代无机化学是对所有元素及其化合物（碳氢化合物及其衍生物除外）的制备、组成、结构、性质和反应的实验测试和理论阐述。

Exercises

1. What is chemistry, what are the categories of chemistry, and what is the work of chemists?
2. Briefly describe the development history of inorganic chemistry.
3. What is the relationship between inorganic chemistry and pharmacy?
4. What are the fields of inorganic medicine research in China from ancient times to the present?

题库

目标检测

1. 什么是化学，化学学科有哪些分类，化学家的工作是什么？
2. 简述无机化学的发展历史。
3. 无机化学跟药学有什么联系？
4. 我国从古至今对无机药物的研究包括哪些领域？

Chapter 2 Solution
第二章 溶液

学习目标

知识要求

1. **掌握** 物质的量浓度和质量摩尔浓度的概念及相关计算；依数性的定义以及利用依数性求物质摩尔质量的方法。

2. **熟悉** 摩尔分数、质量浓度、质量分数及体积分数的概念及有关计算；渗透压在医学中的应用。

3. **了解** 反渗透技术等内容。

能力要求

通过对溶液性质的学习掌握溶液在医、药学中的应用。

A homogeneous and stable dispersion system formed by mixing two or more substances is called solution (溶液). All solutions are composed of solute (溶质) and solvent (溶剂). Molecules, ions and atoms are used as particles of solute, which is evenly distributed in the solvent. Generally, the components those can dissolve other substances are called solvents, and the dissolved substances that are called solutes.

The solution can be liquid, gaseous or solid. And liquid solutions are the most familiar species, especially those with water as the solvent, such as glucose injection and sodium chloride injection. Meanwhile, benzene, gasoline and carbon tetrachloride can be also used as solvent to dissolve organic compounds. Such solution is called non-aqueous solution (非水溶液). Furthermore, the air which is made up of oxygen, nitrogen, carbon dioxide, water vapor and other gases is a gaseous solution. In addition, many alloys are solid solutions (固体溶液). For example, brass is a homogeneous and stable solid solution formed by dissolving zinc in copper. Because the most common solution in chemistry and pharmacy is a water-based solvent, when we refer to the term "solution", unless otherwise stated, we are referring to an aqueous solution.

When a gas or solid is dissolved in a liquid, the liquid component is often used as the solvent, and the gas or solid dissolved in the liquid is called the solute. When a liquid is dissolved in another liquid, a large amount of liquid is usually called the solvent, and a small amount of liquid is called the solute.

The process of solution formation is often accompanied by thermal effects (热效应), volume changes, and sometimes color changes. For example, exothermic (放热) effect occurs when concentrated sulfuric acid is diluted, while endothermic (吸热) effect for ammonium nitrate dilution in aqueous system. The volume of alcohol dissolved in water decreases, and the volume of acetic acid dissolved in benzene

becomes larger. Brown-yellow copper chloride solids show different colors with different concentrations after being dissolved in water. Dilute solution is blue, concentrated solution is green and very concentrated solution is yellow-green. These phenomena indicate that dissolution is not a simple physical mixing, there are also accompanied by a certain degree of chemical change. After the solution is formed, the solute and the solvent can be easily separated by simple physical methods such as evaporation of the solvent. It indicates that the solute and the solvent did not undergo a real chemical reaction during the dissolution process. Therefore, the dissolution process is a physicochemical process.

1. The Concentration of Solution

PPT

1.1 Expression of Solution Concentration

The properties of a solution depend on the relative content of solvent and solute. In order to meet the different needs of research and production, there are many ways to express the concentration of solution. This section describes the most common concentration representations.

1.1.1 Molarity

Molarity (物质的量浓度) of solute B is defined that amount of substance B divided by the volume of the solution, which is concentration for short. It can be represented by symbol c_B.

$$c_B = \frac{n_B}{V} \tag{2-1}$$

In the formula: n_B is the amount of substance of solute B and the SI unit (国际单位制) is mol; V is the volume of the solution and the SI unit is m^3. Since the unit of cubic meter is too large, the common unit of molarity is mol/L or mol/dm^3 in the chemical calculation of concentration in practice.

If there is only one solute, the concentration of the solute can be referred to as the concentration of the solution. If there are several solutes, the total concentration of the solution is the sum of the concentrations of several solutes.

Example 2-1 7.49g of $CuSO_4 \cdot 5H_2O$ crystals were dissolved in water to make a solution with a volume of 150 ml. Find molarity of solution.

Solution $CuSO_4 \cdot 5H_2O$ crystals are dissolved in a solution of water, and the solute is $CuSO_4$. The volume of the solution was 0.150L. The molar mass of $CuSO_4 \cdot 5H_2O$ is 250g/mol

$$n(CuSO_4) = n(CuSO_4 \cdot 5H_2O) = \frac{7.49}{250} = 0.0300\text{mol}$$

The molarity of solute $CuSO_4$ is: $c = \frac{n}{V} = \frac{0.0300}{0.150} = 0.200\text{mol/L}$

1.1.2 Molality

Molality (质量摩尔浓度) is defined that amount of substance of solute B divided by mass of solvent A. It can be represented by symbol b_B.

$$b_B = \frac{n_B}{m_A} \tag{2-2}$$

In the equation: b_B is molality of solute B, and the SI unit is mol/kg; n_B is the amount of substance of solute B and the SI unit is mol; m_A is the mass of solvent A and the SI unit is kg.

In a very dilute aqueous solution, the molarity c_B and the molality b_B of the substance can be approximately considered to be numerically equal. This is because the mass of the solute is negligible when the aqueous solution is very thin, the volume of the solution can be considered to be the same as the volume of water, the density is 1kg/L, and the mass of the solvent water m_A (kg) is approximately equal to the mass of the solution. It is equal to the volume V (L) of the solution.

Example 2-2 5.56g of green vitriol ($FeSO_4 \cdot 7H_2O$) crystals were dissolved in 100g of water. Find the molality of the solution.

Solution The molar mass of $FeSO_4 \cdot 7H_2O$ is 278g/mol

$$n(FeSO_4) = n(FeSO_4 \cdot 7H_2O) = \frac{5.56}{278} = 0.020\text{mol}$$

$$m_A = 100 + 0.020 \times 7 \times 18 = 102.52\text{g} \approx 0.103\text{kg}$$

$$b_B = \frac{n_B}{m_A} = \frac{0.020}{0.103} = 0.194\text{mol/kg}$$

1.1.3 Mole Fraction

The ratio of the amount of the substance B in the mixture to the total amount of the substance in the mixture is called mole fraction (摩尔分数). It can be represented by the symbol x_B.

$$x_B = \frac{n_B}{n_{总}} \tag{2-3}$$

In the equation: x_B is the mole fraction of substance B, and the SI unit is 1; n_B is the amount of substance of solute B, and the SI unit is mol; n_{total} is the total amount of substance in mixture, and the SI unit is mol.

Obviously, the sum of the molar fractions of the components in the solution is equal to 1, that is $\sum_i x_i = 1$

1.1.4 Mass Concentration

The mass of solute B divided by the volume of the solution is called the mass concentration (质量浓度) of solute B, which is denoted by the symbol ρ_B.

$$\rho_B = \frac{m_B}{V} \tag{2-4}$$

In the equation: ρ_B is the mass concentration of solute B, the SI unit is kg/m^3, and the common unit is g/L or g/ml; m_B is the mass of solute B, the SI unit is kg, and the common unit in chemistry is g; V is The volume of the solution, the SI unit is m^3, and the commonly used unit in chemistry is L or ml.

1.1.5 Mass Fraction

The mass of substance B divided by the mass of the mixture is called the mass fraction (质量分数) of substance B. It can be represented by the symbol ω_B.

$$\omega_B = \frac{m_B}{m_{总}} \quad (2\text{-}5)$$

In the equation: ω_B is the mass fraction of substance B, and the SI unit is 1; m_B is the mass of substance B, m_{tolal} is the total mass of the mixture.

1.1.6 Volume Fraction

Under the same temperature and pressure conditions, the volume of component B in the solution before mixing is divided by the total volume of the components before mixing, which is called the volume fraction (体积分数) of component B. It can be represented by the symbol φ_B.

$$\varphi_B = \frac{V_B}{V_{total}} \quad (2\text{-}6)$$

In the equation: V_B is the volume of component B before mixing under the same temperature and pressure conditions as the solution; and is the sum of the volumes of components before mixing under the same temperature and pressure conditions as the solution. The SI units of V_B and V_{total} are both m^3. The common units in chemistry are L or ml.

When the mixed system is in a gaseous phase, its component concentration is usually expressed by a volume fraction.

1.2 Conversion among Different Concentrations

In the above six kinds of solution concentration expressions, c_B, ρ_B, and φ_B are all expressed by the amount of solute contained in a certain volume of solution. The advantage is that it is easy to prepare. However, the disadvantage is that the volume is affected by temperature, so the concentration is also affected by temperature. b_B, x_B and ω_B are expressed by the relative amounts of the solute and the solvent contained in the solution. The advantage is that it is not affected by temperature while the disadvantage is that it is inconvenient to prepare.

In actual work, what kind of concentration representation method should be used depends on actual needs. These kinds of concentration representations can be converted to each other.

Example 2-3 What are the molarity, mass concentration, molar fraction, and molality of a NaOH solution with a mass fraction of 0.200 and a density of $1.22g/cm^3$?

Solution The molar mass of NaOH is 40.0g/mol and the molar mass of water is 18.0g/mol.

Assume 1L solution,

$$m_{tolal} = 1000 \times 1.22 = 1220g$$

$$m_B = 1220 \times 0.200 = 244g$$

$$m_A = 1220 - 244 = 976g = 0.976kg$$

$$n_B = \frac{244}{40.0} = 6.10mol$$

$$n_A = \frac{976}{18.0} = 54.2mol$$

$$c_B = \frac{n_B}{V} = \frac{6.10}{1} = 6.10mol/L$$

$$\rho_B = \frac{m_B}{V} = \frac{244}{1} = 244\text{g / L}$$

$$x_B = \frac{n_B}{n_A + n_B} = \frac{6.10}{6.10 + 54.2} = 0.101$$

$$b_B = \frac{n_B}{m_A} = \frac{6.10}{0.976} = 6.25\text{mol / kg}$$

Example 2-4 The mass concentration of sucrose in the sucrose (蔗糖) solution was 46.8g/L, the solution density was 1.02g/ml, the mass fraction of glucose (葡萄糖) in the glucose aqueous solution was 0.0100, and the solution density was 1.01g/ml. Mix 150ml of this sucrose solution with 250ml of glucose solution, ignore the volume change during the mixing process, and find the molar fraction of glucose, molality, and molarity in the mixed solution.

Solution The molar mass of sucrose, glucose, and water is 342, 180, and 18.0g/mol, respectively.

$$V = 150 + 250 = 400\text{ml}$$

$$m_{(\text{sucrose})} = 150 \times 10^{-3} \times 46.8 = 7.02\text{g}$$

$$m_{(\text{glucose})} = 250 \times 1.01 \times 0.0100 = 2.52\text{g}$$

$$m_{(\text{water})} = (150 \times 1.02 - 7.02) + (250 \times 1.01 - 2.52) = 396\text{g}$$

$$n_{(\text{sucrose})} = \frac{7.02}{342} = 0.0205\text{mol}$$

$$n_{(\text{glucose})} = \frac{2.52}{180} = 0.0140\text{mol}$$

$$n_{(\text{water})} = \frac{396}{18.0} = 22.0\text{mol}$$

$$x_{(\text{glucose})} = \frac{n_{(\text{glucose})}}{n_{\text{total}}} = \frac{0.0140}{0.0205 + 0.0140 + 22.0} = 6.36 \times 10^{-4}$$

$$b_{(\text{glucose})} = \frac{n_{(\text{glucose})}}{m_{(\text{water})}} = \frac{0.0140}{396 \times 10^{-3}} = 0.0354\text{mol/kg}$$

$$c_{(\text{glucose})} = \frac{n_{(\text{glucose})}}{V} = \frac{0.0140}{400 \times 10^{-3}} = 0.0350\text{mol/L}$$

2. The Colligative Properties of Dilute Nonelectrolyte Solutions

PPT

Any substance whose aqueous solution contains ions is called an electrolyte (电解质). Any substance that forms a solution containing no ions is a nonelectrolyte (非电解质). For example, NaCl whose aqueous solutions contain ions is an electrolyte, while $C_{12}H_{22}O_{11}$ that does not form ions in solution is a nonelectrolyte. Electrolytes that are present in solution entirely as ions are strong electrolytes (强电解质), whereas those that are present partly as ions and partly as molecules are weak electrolytes (弱电解质).

Because the composition and nature of every solute are different, and the interaction between the solute and the solvent is different, the solution made up of different solutes not only has different chemical properties, but also has many different physical properties, such as color, density, viscosity, conductivity, and so on. However, in the solution of different solutes, there is also a kind of property that only depends on the concentration of solutes in the solution, but has nothing to do with the type and nature of solutes. This kind of property is called colligative property (依数性). The colligative properties of dilute nonelectrolyte solutions include vapor pressure decreasing (蒸汽压下降), boiling point elevation (沸点升高), freezing point depression (凝固点降低) and osmotic pressure (渗透压). Among them, osmotic pressure is the most closely related to medicine. When the solute is an electrolyte, or a nonelectrolyte but the concentration is high, the above mentioned colligative law of the solution will change greatly. This section only discusses the colligative properties of dilute nonelectrolyte solutions.

2.1 Vapor Pressure Decreasing

When liquid is placed in a sealed container, the molecules of the liquid constantly evaporate and form vapor above the liquid level. At the same time, vapor molecules near the liquid level can condense and return to the liquid. When the evaporation and condensation rates are equal, the gas and liquid phases are in equilibrium, and the vapor pressure is called the saturated vapor pressure (饱和蒸汽压) of the liquid, or vapor pressure. The results show that at the same temperature, the vapor pressure of the solution is lower than that of the pure solvent when the non-volatile nonelectrolyte is dissolved into the solvent to form a dilute solution. This is because the nonvolatile solute molecules in the solution occupy part of the liquid surface of the solution, which reduces the number of solvent molecules per unit area, so that the number of solvent molecules that escape the liquid surface per unit time is less than that of pure solvents. The solvent will reach equilibrium with its vapor at a lower vapor pressure, so the vapor pressure of the solution will drop.

The physical chemist, F. M. Raoult (France, 1832~1901), was the first to make an accurate quantitative study on the vapor pressure decreasing of a solution. According to the experimental results, he published a quantitative relationship in 1887: At a certain temperature, the vapor pressure of a dilute solution is equal to the product of the vapor pressure of the pure solvent and the molar fraction of the solvent. This is Raoult's law (拉乌尔定律).

$$p = p_A^0 x_A \tag{2-7}$$

In the equation: p is the vapor pressure of the solution; p_A^0 is the vapor pressure of the pure solvent; x_A is the molar fraction of the solvent.

Let x_B be the mole fraction of the solute. Since $x_A + x_B = 1$, then:

$$p = p_A^0(1 - x_B)$$

$$p = p_A^0 - p_A^0 x_B$$

$$p_A^0 - p = p_A^0 x_B$$

that is:

$$\Delta p = p_A^0 x_B \tag{2-8}$$

The above equation shows that, at a certain temperature, the decrease value of the vapor pressure of dilute nonvolatile nonelectrolyte solution is directly proportional to the mole fraction of solute, and has nothing to do with the nature of solute. This is another description of Raoult's law.

Let n_A and n_B represent the amount of solute and solvent respectively, because $n_A >> n_B$ in dilute solution, then

$$\Delta p = p_A^0 x_B = p_A^0 \frac{n_B}{n_A + n_B} \approx p_A^0 \frac{n_B}{n_A}$$

In a solution containing 1kg of solvent, $b_B = \frac{n_B}{1} = n_B$

Let M_A (g/mol) be the molar mass of the solvent, then $n_A = \frac{1000}{M_A}$

$$\Delta p = p_A^0 \frac{n_B}{n_A} = p_A^0 n_B \frac{1}{n_A} = p_A^0 \frac{M_A}{1000} n_B = p_A^0 \frac{M_A}{1000} b_B$$

When the temperature is constant, $p_A^0 \frac{M_A}{1000}$ is a constant, Use K instead, then

$$\Delta p = K b_B \tag{2-9}$$

The above equation shows that, for the dilute solutions of nonvolatile nonelectrolyte, the decrease value of vapor pressure just depends on the nature of the solvent and the molality of the solution, being irrelevant with the nature of the solute.

Since we are talking about the dilute solutions of nonvolatile nonelectrolyte, considering that the solutes are essentially nonvolatile, the vapor pressure of the solution is actually the vapor pressure of the solvent in the solution. If both the solute and the solvent are volatile, the total vapor pressure of the solution is equal to the sum of the vapor pressure of the solvent and the solute. For solutions containing volatile solutes, the relevant calculation of vapor pressure is more complicated, which is not discussed in this section.

For the dilute solutions of nonvolatile nonelectrolyte, since the decrease value of vapor pressure is positively proportional to the molality of the solution, the molality of the solute can be calculated according to equation (2-9) as long as the decrease value of vapor pressure of the solution can be determined.

Example 2-5 0.883g of an organic substance was dissolved in 33.9g of benzene. The vapor pressure of the solution was 630 mmHg, and the vapor pressure of pure benzene was 640 mmHg at the same temperature. Find the molar mass of the organic compound.

Solution According to Raoult's law, the decrease of vapor pressure of solution is directly proportional to the mole fraction of the solute.

The molar mass of the solvent benzene is 78.1g/mol, $\Delta p = 640 - 630 = 10\text{mmHg}$

Let the molar mass of the organic compound be, then

$$n_A = \frac{33.9}{78.1} = 0.434\text{mol}$$

$$n_B = \frac{0.883}{M_B}$$

$$\Delta p = p_A^0 \cdot x_B = p_A^0 \cdot \frac{n_B}{n_A + n_B}$$

$$10 = 640 \times \frac{n_B}{0.434 + n_B}$$

$$n_B = 0.00689\text{mol}$$

$$M_B = \frac{0.883}{0.00689} = 128\text{g/mol}$$

2.2 Boiling Point Elevation

Boiling is a violent vaporization that occurs simultaneously on the surface and inside of a liquid. The vapor pressure of a liquid increases with temperature, and when the temperature rises to the point where the vapor pressure is equal to the external pressure, the liquid boils. This temperature is the boiling point of the liquid. The boiling point of the liquid is different under different external pressures. Normally, the boiling point at 101.325kPa is called the normal boiling point (正常沸点).

At the same temperature, the vapor pressure of the solution is lower than that of pure solvent. When the temperature rises to the boiling point of the pure solvent, the vapor pressure of the pure solvent is equal to the external pressure, but the vapor pressure of the solution is still lower than the external pressure. In order to make the solution boil, it must be heated to a higher temperature than the pure solvent to cause the vapor pressure of the solvent to equal atmospheric pressure. It can be seen that the decrease in vapor pressure of the solution causes the boiling point of the solution to be higher than that of the pure solvent.

If T_b represents the boiling point of the solution, T_b^0 represents the boiling point of the pure solvent, the value of boiling point elevation of the solution may be represented by the term $\Delta T_b = T_b - T_b^0$.

A schematic diagram of boiling point elevation and freezing point depression of an aqueous solution is shown in Figure 2-1. The abscissa indicates temperature, and the ordinate indicates vapor pressure. Curves AB and CD represent the relationship between the vapor pressure of pure solvent (water) and the solution as a function of temperature, and T_b is the boiling point of the solution. It can be seen from the figure that at the same temperature (the vertical line is drawn on the same ordinate), the vapor pressure of the pure solvent is higher than the vapor pressure of the solution. At the temperature of 373.15K, the vapor pressure of water is equal to the external pressure of 101.325kPa, and the water begins to boil; at this time, the vapor pressure of the solution is the ordinate corresponding to the point B', which is obviously still less than the external pressure of 101.325kPa. It has not reached the boiling condition. In order to make the solution's vapor pressure reach 101.325kPa, it is necessary to continue heating up to point D (the boiling point of the solution). Obviously, the temperature T_b of point D is higher than the boiling point of 373.15K of the pure solvent, that is to say, the boiling point of the solution is increased.

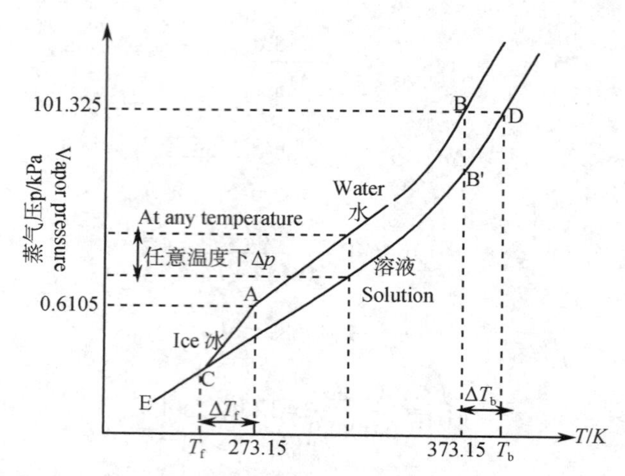

Figure 2-1 Schematic diagram of boiling point elevation and freezing point depression of aqueous solution

The cause of the boiling point elevation of the solution is the decrease of the vapor pressure of the solution. Similar to the drop of the solution's vapor pressure, the increase of the boiling point of the solution is approximately proportional to the mass molar concentration of the solution and irrelevant with the nature of the solute. For the dilute solution of nonvolatile nonelectrolyte, the mathematical equation of boiling point elevation is as follows:

$$\Delta T_b = K_b b_B \quad (2\text{-}10)$$

In the equation: ΔT_b is the value of boiling point elevation of the solution (溶液沸点升高的度数), that is the difference between the boiling point of the solution minus the boiling point of the pure solvent, the unit is K or °C; K_b is the molal boiling point elevation constant (摩尔沸点升高常数) of solvent, which represents the degree of boiling point elevation caused by molarity of solute of 1mol/kg, the unit is (K·kg)/mol or (°C · kg)/mol.

K_b is only related to the nature of the solvent, not to the nature of the solute. Different solvents have different K_b values, and they can be calculated theoretically or measured experimentally. Table 2-1 lists the K_b of several common solvents.

Table 2-1 K_b values of several solvents

solvent	T_b^0/K	K_b/(K · kg/mol)	solvent	T_b^0/K	K_b/(K · kg/mol)
Water	373.1	0.512	Acetic acid	391.0	2.93
Acetone	329.5	1.71	Chloroform	334.2	3.63
Ether	307.7	2.16	Carbon tetrachloride	349.7	5.03
Carbon disulfide	319.1	2.34	Naphthalene	491.0	5.80
Benzene	353.4	2.53	Camphor	481.0	5.95
Nitrobenzene	484.0	5.24	Phenol	454.9	3.56
Ethanol	351.4	1.22	Cyclohexane	354.0	2.79

By measuring the boiling point of the solution, the rising value of the boiling point is calculated. The molality of the solute can be obtained from equation (2-10). The molar mass of the solute can be calculated under the condition that the mass of the solute is known (it can be accurately weighed before measurement). Therefore, the boiling point elevation of solution and the decrease of vapor pressure of solution can be used for the determination of the molar mass of nonvolatile nonelectrolyte solutes.

Example 2-6 The boiling point of pure benzene is 80.10°C. 4.27g of an organic substance was dissolved in 100g of benzene, and the boiling point of the solution was measured to be 80.801°C. Find the molar mass of the organic substance.

Solution The mass m_A of solvent benzene is $100\times10^{-3} = 0.100$kg

Solute B has a mass of 4.27g

$$\Delta T_B = K_b b_B$$

$$b_B = \frac{\Delta T_B}{K_b} = \frac{80.801-80.10}{2.53} = 0.277\text{mol / kg}$$

$$n_B = b_B \cdot m_A = 0.277\times0.100 = 0.0277\text{mol}$$

$$M_B = \frac{m_B}{n_B} = \frac{4.27}{0.0277} = 154\text{g / mol}$$

2.3 Freezing Point Depression

Like liquids, solids also have a certain vapor pressure at a certain temperature. The freezing point of a liquid substance is the temperature at which the liquid and solid phases of the substance have the same vapor pressure and can coexist. If the vapor pressures of the two phases are not equal, the substance will spontaneously change from one phase with the higher vapor pressure to another with the lower vapor pressure. In Figure 2-1, the curve ACE represents the relationship between the vapor pressure of the solid pure solvent and the temperature, and the curve AB represents the relationship between the vapor pressure of the liquid pure solvent and the temperature. As can be seen from Figure 2-1, at 273.15K, the vapor pressure of ice and water is equal, and 273.15K is the freezing point of water. Above 273.15K, the vapor pressure of ice will be greater than that of water, and ice will melt into water. Below 273.15K, the vapor pressure of water is greater than that of ice, and water will solidify into ice.

When the vapor pressure of the solid pure solvent is equal to the vapor pressure of the solution, the solid phase and liquid phase of the solution reach equilibrium, and the temperature at this time is the freezing point of the solution. In Figure 2-1, the curve CD shows the change of the vapor pressure of the solution with temperature. As can be seen from Figure 2-1, at 273.15K, the vapor pressure of pure water is already equal to the vapor pressure of ice, and water begins to solidify. However, since the vapor pressure of the solution is lower than that of the pure solvent, the vapor pressure of the solution is lower than that of the ice, so it cannot solidify and is still liquid. Temperature has a greater effect on the vapor pressure of ice, and ACE is steeper than AB and CD. As the temperature drops, the vapor pressure of the ice drops more than that of the solution. When the temperature drops to point C, the curve ACE intersects with CD, at which point the vapor pressure of the ice is equal to that of the solution, and the temperature T_f is the freezing point of the solution. Obviously, the freezing point T_f of the solution is lower than that of

the pure solvent.

T_f represents the freezing point of the solution, T_f^0 represents the freezing point of the pure solvent, and the value of freezing point depression is equal to $\Delta T_f = T_f^0 - T_f$

The basic reason of the freezing point depression is also the decrease of vapor pressure of the solution. Just as the boiling point elevation of solution, the freezing point depression is in proportion to the molality of the solute, and has nothing to do with the nature of the solute. For the dilute solution of nonvolatile nonelectrolyte, the mathematical equation of freezing point depression is as follows:

$$\Delta T_f = K_f b_B \tag{2-11}$$

In the equation: ΔT_f is the decrease value of freezing point of solution (溶液凝固点下降的度数) and the unit is K or °C; K_f is the molal freezing point depression constant (摩尔凝固点下降常数), which refers to the degree of freezing point reduction caused by the molar concentration of the solute of 1mol/kg, and the unit is K·kg/mol or °C·kg/mol.

K_f is only related to the nature of the solvent, not to the nature of the solute. Different solvents have different K_f values. Table 2-2 lists the K_f of several common solvents.

Table 2-2 K_f values of several solvents

Solvent	T_f^0/K	K_f/(K · kg/mol)	Solvent	T_f^0/K	K_f/(K · kg/mol)
Water	273.0	1.86	Acetic acid	290.0	3.90
Benzene	278.5	5.10	Camphor	451.0	40.0
Phenol	316.2	7.80	Carbon tetrachloride	250.1	32.0
Cyclohexane	279.5	20.2	Ether	156.8	1.80
Naphthalene	353.0	6.90	Nitrobenzene	278.9	7.00

The equation (2-11) can also be used to determine the molar mass of the solute, and its accuracy is higher than that of the vapor pressure method and boiling point method. Both Δp and ΔT_b are not easy to measure, however, with modern experimental technology, it can be measured to 0.0001°C with very small absolute error. For most solvents, $K_f > K_b$, the decrease value of freezing point is greater than the increase value of boiling point for the same solution, so the relative error of the measurement is also small. The boiling point method and the vapor pressure method usually require heating the solutions. Especially the boiling point method needs to be heated to the boiling point of the solution. The solute may be damaged or denatured at higher temperatures. The solvent is volatile, which will cause the concentration of the solution to change. And the freezing point depression method is performed at a low temperature, there is no such problem. In addition, for volatile solutes, the boiling point method or the vapor pressure method cannot be used to determine the molar mass, and only the freezing point depression method can be used. In view of the above reasons, the freezing point depression method is most widely used in determining the molar mass of a solute.

Example 2-7 12.0g of an organic substance was dissolved in 100.0g of water and the freezing point was measured to be −0.653°C. Find the molar mass of the organic substance.

Solution

$$\Delta T_f = K_f b_B$$

$$\Delta T_f = 0 - (-0.653) = 0.653\ ^\circ\text{C}$$

$$K_f = 1.86°C \cdot kg/mol$$

$$b_B = \frac{\Delta T_f}{K_f} = \frac{0.653}{1.86} = 0.351 mol/kg$$

$$m_A = \frac{100.0}{1000} = 0.1000 kg$$

$$b_B = \frac{n_B}{m_A} = \frac{n_B}{0.1000} = \frac{1}{0.1000} \times \frac{12.0}{M_B} = 0.351 mol/kg$$

$$M_B = 342 g/mol$$

The principle of the boiling point elevation and the freezing point depression of a solution have been widely used in production, life, and scientific research. The cooling water of automobile radiators often needs to be added with appropriate amount of glycol, glycerin or methanol in winter, the purpose is to decrease the freezing point of the cooling liquid and prevent freezing. During snow and ice disasters, people sprinkle salt (or directly sprinkle salt water) on frozen roads. There is more or less water on the surface of the ice. The salt dissolves in water to form a solution. The vapor pressure of the solution decreases, which is lower than that of ice. A phase with high vapor pressure spontaneously changes to a phase with low vapor pressure, so the ice melts into solution. From another perspective, it is assumed that the freezing point of the salt solution is reduced to −15°C and the ambient temperature is −10°C. At this temperature, the water has been deeply frozen, while the salt solution has not yet reached the freezing point and will not freeze, thus the smooth road is guaranteed. In addition, for the ice-salt mixture system, the melting of the ice needs to absorb a large amount of heat, and it leads to the ice-salt mixture temperature significantly reduced. Therefore ice-salt mixture is often used as a refrigerant. For example, the refrigerant made of NaCl and ice (about 30g NaCl +100g water) can reduce the temperature to –22.4°C, which is widely used for the preservation and transportation of aquatic products and food.

2.4 Osmotic Pressure

2.4.1 Osmotic Phenomena and Osmotic Pressure

In agricultural production, if the application of fertilizer is too thick, the plants will die. In medicine, patients who lose water often need to intravenous drip sterilization solution of 0.9% NaCl (that is, normal saline, 生理盐水). The concentration of these injections must be strictly controlled at around 0.9%, not too high or too low, or it will cause discomfort to the patient. Why is that? Let's start with the following experiment.

Sucrose solution and pure water were placed on both sides of a connector, which separated by a semi-permeable membrane (半透膜) (Figure 2-2). A semi-permeable membrane is a porous membrane through which a portion of the material can be selectively passed. There are many natural semi-permeable membranes in nature, such as turnip skins, animal casings,

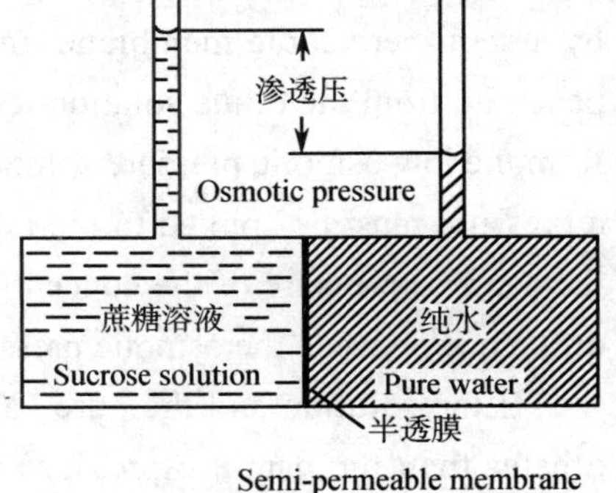

Figure 2-2 Schematic diagram of osmotic pressure generation

bladders, cell membranes of animals and plants, etc. There are also many types of synthetic semi-permeable membranes, such as artificial parchment, synthetic colloidal thin films, cellophane, and solid films of copper ferricyanide ($Cu_2[Fe(CN)_6]$) which is deposited on the surface of plain fired ceramics. In Figure 2-2 the semi-permeable membrane installed in the device just allows solvent (here water molecules is used) to come and go freely, while solute sucrose molecules cannot.

At the beginning, the liquid levels on both sides are equal. Since the number of water molecules per unit volume of sucrose solution is less than the number of water molecules per unit volume of pure water (that is, the concentration of water in pure water is higher than the concentration of water in sucrose solution), water in pure water will spontaneously diffuse from pure water (high concentrations) into the sucrose solution low concentrations through the semi-permeable membrane. Although the concentration of sucrose molecules in the sucrose solution is greater than the concentration of sucrose molecules in pure water, sucrose molecules cannot enter pure water due to the barrier of the semi-permeable membrane. In this way, the water on the pure water side will gradually decrease, and the water in the sucrose solution will gradually increase. After a period of diffusion, the liquid levels in the glass columns on both sides are no longer the same, and the liquid level on the sucrose solution side is higher than the liquid level on the pure water side. The phenomenon in which this solvent diffuses into the solution through a semi-permeable membrane is called osmosis (渗透). It must be pointed out that the migration of solvent molecules is not unidirectional during the process of diffusion. While the water molecules on the side of pure water migrate into the sucrose solution through the semi-permeable membrane, the water molecules in the sucrose solution also migrate into the pure water through the semi-permeable membrane. Just because of the difference in concentration, there are more water molecules entering the sucrose solution through the semi-permeable membrane per unit time than water molecules leaving the sucrose solution and entering pure water. As the liquid level of the sucrose solution rises, the hydrostatic pressure (静水压力) increases, driving water molecules into the pure water through the semi-permeable membrane. When the hydrostatic pressure increases to a certain value, the number of solvent molecules permeating from both sides of the membrane per unit time is equal, and the osmotic equilibrium is reached. At this time, the liquid levels on both sides no longer change, and the static pressure represented by the water level difference between the liquid levels on both sides of the semipermeable membrane is called the osmotic pressure (渗透压) of the solution. Precisely, osmotic pressure refers to the additional pressure that must be applied above the solution just enough to prevent solvent penetration.

It must be pointed out that the osmotic pressure is only exhibited when the solution and solvent are separated by a semi-permeable membrane. When two solutions with different concentrations are separated by a semi-permeable membrane, the phenomenon of penetration will also occur, and the solvent will penetrate from the dilute solution to the concentrated solution. In other words, the solvent will penetrate from the low osmotic pressure solution to the high osmotic pressure solution. In order to prevent osmosis, a pressure must be applied to the liquid surface of the concentrated solution, but this pressure is neither the osmotic pressure of the concentrated solution nor the osmotic pressure of the dilute solution, but the difference between the osmotic pressure of the two solutions.

If the solution and the pure solvent are separated by a semi-permeable membrane and a pressure greater than the osmotic pressure is applied to the solution side, the solvent molecules will be net transferred from the solution side to the pure solvent side, thereby the pure solvent from the solution is extruded. This phenomenon is called reverse osmosis (反渗透). Reverse osmosis can be used for desalination and wastewater purification; some substances that cannot be concentrated through high temperature distillation can be concentrated using normal temperature reverse osmosis technology. The key to the promotion of reverse osmosis technology lies in the development of semi-permeable membranes with stable properties, long-term pressure resistance and affordable prices. In recent years, membrane separation reverse osmosis seawater desalination technology has made great progress in the marine field. For example, the Fuliji seawater desalination unit in the United States (美国富立吉海水淡化装置) has been widely used in ocean fishing vessels. In the Middle East countries, desalination is an important part of their survival. At present, obtaining fresh water through reverse osmosis technology has become the main way of desalination in these countries. In China, the reverse osmosis desalination industry has also developed to a certain extent. For example, in Zhoushan (Zhejiang Province, China), the desalination production capacity in 2013 reached 54,500 tons/day. Among them, the two sets of 10,000 ton reverse osmosis equipment currently available in Liuheng desalination plant (六横海水淡化厂) can output of 20,000 tons freshwater every day, and the plant is currently constructing a reverse osmosis desalination project with a daily output of 100,000 tons.

2.4.2 The Relationship of Osmotic Pressure, Concentration and Temperature

In 1886, physical chemist Van't Hoff (Netherlands, 1852—1911) summarized a large number of experimental data and pointed out that at a certain temperature, the osmotic pressure of a dilute nonelectrolyte solution is proportional to the amount of substances concentration of solute and has nothing to do with the nature of the solute. Its mathematical relationship is:

$$\pi = c_B RT \tag{2-12}$$

In the equation: π is the osmotic pressure of the solution, the unit is kPa; c_B is the amount of substances concentration of solute, the unit is mol/L; R is the molar gas constant (8.314kPa·L/mol·K); T is the absolute temperature; and the unit is K.

When the aqueous solution is very dilute, the molarity is approximately equal to the molar concentration, then:

$$\pi = b_B RT \tag{2-13}$$

In the equation: R is the molar gas constant. Pay attention to the change of unit: 8.314 (kPa · kg)/(mol · K).

Example 2-8 What is the osmotic pressure of a 0.150mol/L dilute solution equivalent to the water column pressure at the temperature of 20°C?

Solution

$$\pi = c_B RT$$

$$\pi = 0.150\times8.314\times(20 + 273.15) \approx 366\text{kPa}$$

$$p = \rho gh = 1.0\times10^3\times9.8h = 3.66\times10^5\text{Pa}$$

$$h = 37.3\text{ m}$$

It can be seen from example 2-8 that even in a dilute solution, the driving force generated by its permeation is still amazing. Generally, the osmotic pressure of plant cell fluid can reach 2000kPa, so water can be transported from the root of the plant to the top of hundreds of meters. There are many towering trees in nature, and osmotic pressure is indispensable.

The osmotic pressure of the solution can also be used to determine the molar mass of the solute. Because it is difficult to directly measure the osmotic pressure, for nonelectrolytes whose molar mass is not too large, the methods of boiling point elevation and freezing point depression are commonly used. The method of freezing point depression is particularly widely used.

However, for macromolecular compounds, it is difficult to accurately determine the boiling point or the freezing point. For example, for a solution containing 25g of a high-molecular substance (such as protein) with a molar mass of 5×10^4 in 1 L of water, the freezing point of the solution will only decrease by 9.3×10^{-4}°C if measured by the freezing point method. It is determined that the osmotic pressure can reach as high as 1240Pa at 25°C. Obviously, it is much better to measure the osmotic pressure than to measure the decrease of freezing point. Therefore, the osmotic pressure of a solution is particularly suitable for determining the molar mass of a polymer compound.

However, for macromolecular compounds, it is difficult to accurately determine the boiling point or the freezing point. For example, for a solution containing 25g of a high-molecular substance (such as protein) with a molar mass of 5×10^4 in 1 L of water, the freezing point of the solution measured by the method of freezing point depression will only decrease by 9.3×10^{-4}°C. It is determined that the osmotic pressure can reach as high as 1240 Pa at 25°C. Obviously, it is much better to measure the osmotic pressure than to measure the decrease of freezing point. Therefore, the osmotic pressure of a solution is particularly suitable for determining the molar mass of a polymer compound.

Example 2-9 The freezing point of human blood is the same as the freezing point of a solution formed by dissolving 54.2g of glucose and 4.00g of a protein with a molar mass of 2.00×10^4 in 1.00kg of water. Find the osmotic pressure of blood at the temperature of 37°C.

Solution For a mixed solution of glucose and protein, the mass of solvent is 1.00kg, then:

$$b_{\mathrm{B}} = b_{(\mathrm{glucose})} + b_{(\mathrm{protein})} = \frac{54.2}{180} + \frac{4.00}{2.00\times10^4} = 0.301\mathrm{mol/kg}$$

$$\pi = b_{\mathrm{B}}\mathrm{R}T = 0.301 \times 8.314 \times (273.15 + 37) = 776\mathrm{kPa}$$

2.4.3 Medical Significance of Osmotic Pressure

Osmosis occurs widely in physiological activities of plants and animals. Plant cell fluids and animal intracellular fluids, plasma, tissue fluids and other body fluids have a certain osmotic pressure. Most of the membranes in the body are semi-permeable membranes (of which the cell membrane is easily permeable to water, and it is almost impervious to dissolving in the cell fluid matter), the phenomenon of penetration runs through life. When a person has severe diarrhea or severe vomiting, a large amount of water will be lost, it results in an increase in the osmotic pressure of the tissue fluid. The water in the blood and intracellular fluid will be transferred to the tissue fluid due to the relatively low osmotic pressure, which results in dehydration of the blood cells and tissue cells. At this time, you should add water in time. Marine fish and freshwater fish both rely on the gill's osmotic function to maintain the osmotic balance between their body fluids and the surrounding water environment. Due to the different gill's osmotic functions of the two types of fish, they cannot exchange living environments. Synthetic osmotic membranes have been used in the treatment of diseases. For example, the kidney in the human body is a special permeator. When it fails, metabolic waste cannot be eliminated. At this time, artificial kidneys are used instead of kidneys to perform osmotic functions. The key component of artificial kidneys is a synthetic permeable membrane. The protoplast layer of plant cells is equivalent to a semi-permeable membrane. When the concentration of the cell fluid is lower than that of the external solution, the water in the cell fluid enters the external solution through the protoplast layer, and the plasma wall separation occurs after constant water loss. The above-mentioned situation of excessively fertilizing plants will cause the plant to die, which is actually because when the concentration of plant cell fluid is lower than the concentration of external fertilizer, water penetrates from the plant to the outside.

Human body fluids (such as plasma, interstitial fluid, lymph fluid, intracellular fluid, etc.) are complex dispersion systems with water as a dispersion medium, which contains a variety of inorganic ions (such as Na^+, Ca^{2+}, Cl^-, etc.), gas molecules (mainly O_2 and CO_2), small and medium molecular organic substances (MMS, 中分子物质, such as glucose, urea, amino acids, etc.) and high molecular substances (HMS, 高分子物质, such as proteins). The osmotic pressure is determined by the concentration of various particles (molecules and ions) dissolved in it. The total molarity of solute particle in the solution that can produce osmosis is called the osmotic concentration, and the osmotic concentration of normal human plasma is about 304 mmol/L. Because many electrolytes and small molecular substances can form crystals, high molecular substances usually have some properties of colloids when dispersed in water. Therefore, in medicine, the osmotic pressure generated by electrolytes and small molecular substances is called crystal osmotic pressure (晶体渗透压). The osmotic pressure generated by high molecular substances is called colloid osmotic pressure (胶体渗透压).. The osmotic pressure of plasma is the sum of these two osmotic pressures.

In plasma, the content of high molecular substances such as proteins is as high as 7%, which is about 9 to 10 times that of electrolytes and low molecule substances (LMS, 低分子物质). However, due to the disparity in molecular weights, the osmotic pressure of crystals accounts for 99.5% in the total plasma osmotic pressure, and the colloidal osmotic pressure only accounts for about 0.5%. Water can permeate the cell membrane freely, and many electrolytes and small molecular substances cannot pass through the cell membrane freely, but can pass through pore capillaries. Therefore, the osmotic pressure of the crystal plays an important role in maintaining the water balance inside and outside the cell. Doctors demand that patients with edema nephropathy eat as little salt as possible, the purpose of which is to prevent excessive levels of salt in the plasma and interstitial fluid, and to attract more water from the cells to the interstitial fluid, thereby the edema is aggravated. The capillary wall is also a semi-permeable membrane. Water and low molecular substances can come in and out freely, but high molecular protein substances cannot pass through. Therefore, colloid osmotic pressure plays a major role in maintaining blood volume and water and salt balance inside and outside blood vessels. If the protein content in the plasma decreases, the colloid osmotic pressure in the plasma will decrease, and the water in the plasma will enter the tissue fluid through the capillary wall, which results in a decrease in blood volume and an increase in tissue fluid, and forms edema.

For intravenous infusion, a solution with the same osmotic pressure as the total plasma osmotic pressure must be used, which is medically called an isotonic solution (等渗溶液)[a solution with an osmotic pressure lower than the total plasma osmotic pressure is a hypotonic solution (低渗溶液), and a solution higher than the total plasma osmotic pressure is hypertonic solution (高渗溶液)]. Isotonic solutions commonly used in clinical practice include 0.9% physiological saline and 5% glucose solutions. If a non-isotonic solution is used for intravenous infusion, adverse consequences can occur. If a large amount of hypotonic solution with an osmotic pressure lower than that of plasma is input, water will penetrate into the red blood cells through the cell membrane, which will cause the red blood cells to swell or even rupture. This phenomenon is medically called hemolysis (溶血). Normal people's red blood cells begin to hemolyze in 0.42% ~ 0.46% NaCl solution, and it may be completely hemolyzed in NaCl solution below 0.34%. If a large amount of hypertonic solution with an osmotic pressure higher than the osmotic pressure of plasma is input, the water in the red blood cells will gradually leak out, which will

cause cell shrinkage, not only losing the function of transporting oxygen, but the shrunken red blood cells will also easily stick together to form clumps. Small blood vessels are blocked to form blood clots, which will endanger human health.

3. Theory of Strong Electrolyte Solution

In a molten state or aqueous solution, a compound that conducts electricity is called an electrolyte. According to the structure of electrolyte and their ionization behavior in aqueous solution, they are divided into strong electrolyte（强电解质）and weak electrolyte（弱电解质）.

Most of the strong electrolytes are ionic bond compounds or strongly polar covalent compounds, which are completely ionized in water, so they exist in solution as hydrated ions with strong conductivity, such as HCl, NaCl, NaOH and so on.

3.1 The Colligative Properties of Electrolyte Solutions

In the discussion of colligative properties of the dilute nonelectrolyte solutions in this chapter, it has been known that the colligative properties of nonelectrolyte are only proportional to the number of solute particles in the solution, and is irrelevant with the nature of the solute. But for electrolyte solutions, due to ionization, colligative properties are abnormal when the value of freezing point elevation is measured. The decrease value measured is larger than the corresponding value of the nonelectrolyte solution with the same concentration, as shown in Table 2-3.

Table 2-3 The decreased values of freezing point of several inorganic salt aqueous solutions

Electrolyte	Concentration (mol/kg)	Calculated value ΔT_f/K	Experimental value $\Delta T_f'$/K	$i=\Delta T_f'/\Delta T_f$
KCl	0.2	0.372	0.673	1.81
KNO_3	0.2	0.372	0.664	1.78
$MgCl_2$	0.1	0.186	0.519	2.79
$Ca(NO_3)_2$	0.1	0.186	0.461	2.48
NaCl	0.1	0.186	0.347	1.87

In addition, when the osmotic pressure of dilute solution of electrolyte is measured, it is also found that the measured value is higher than that obtained by the osmotic pressure calculation equation. For the occurrence of such deviation, Van't Hoff first suggested introducing the correction coefficient i into the equation, namely, $\pi = ic_BRT$

In this equation, symbol i represents the number of particles dissociated from a strong electrolyte.

But through the experimental data, it is found that, even though the correction equation, there is still some error between the experimental value and the theoretical value. In general, the experimental measured value is always less than the value calculated by the correction equation. The ionization degree

of the following electrolytes has been measured experimentally. The results are shown in Table 2-4.

Table 2-4 The ionization degrees of the strong electrolytes

b_B (mol/Kg)	Ionization degree α/ 100%		
	NaCl	HCl	HAc
0.10	0.87	0.91	0.011
0.050	0.89	0.92	0.020
0.010	0.93	0.97	0.048
0.0050	0.94	0.99	0.060

The ionization degree of the strong electrolyte measured experimentally is not actual ionization degree of the strong electrolyte in solution. However, it only reflects the strength of ion interaction in strong electrolyte solution, so it is called apparent degree of ionization (表观电离度). Experimental results show that the apparent ionization degree of strong electrolysis is less than 100%. Is the strong electrolyte not 100% ionization?

3.2 Ionic Atmosphere

Why is the apparent degree of ionization of strong electrolyte not 100% but less than 100%? These difficulties can be resolved with a theory of electrolyte solutions (电解质溶液理论) proposed by Peter Debye (Netherlands, 1884~1966) and Erich Huckel (Germany, 1896~1980) in 1923. This theory continues to view strong electrolytes as existing only in ionic form in aqueous solution, but the ions in solution do not behave independently of one another. Instead, each cation is surrounded by a cluster of ions in which anions predominate, and each anion is surrounded by a cluster in which cations predominate. In short, each ion is enveloped by an ionic atmosphere (离子氛) with a net charge opposite its own (Figure 2-3).

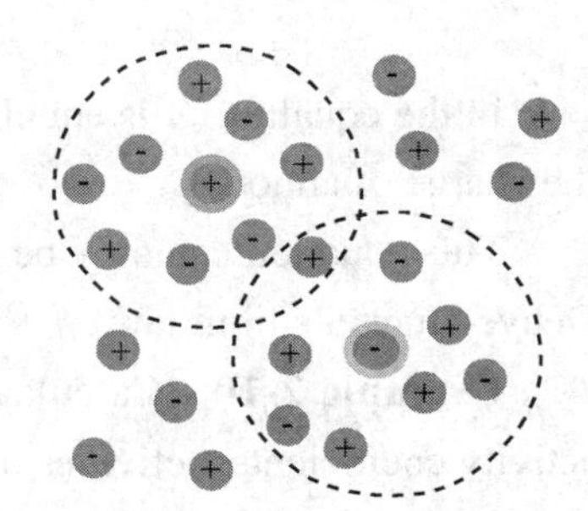

Figure 2-3 Schematic diagram of ionosphere model

In an electric field, the mobility of each ion is reduced because of the attraction or drag exerted by its ionic atmosphere. Similarly, the magnitudes of colligative properties are reduced. This explains why, for example, the value of *i* for NaCl (0.10mol/kg) is 1.87 rather than 2.00 (shown in Table 2-4). What we can say is that each type of ion in an aqueous solution has two “concentrations”. One is called the stoichiometric concentration (化学计量浓度) and is based on the amount of solute dissolved. The other is the “effective” concentration (有效浓度), called the activity (活度), which takes into account interionic attractions.

3.3 Activity

In a strong electrolyte solution, the ion concentration that really plays its role is somewhat smaller than the ion concentration that should be achieved when the electrolyte is completely ionized due to the ion trapping effect. In an electrolyte solution, the ion concentration that actually plays its role is called the

effective concentration or activity.

Qualitative relationship between activity (a) and stoichiometric concentration (c) is:

$$a_i = \gamma_i \cdot c_i \tag{2-14}$$

In the equation: i represents the ith ion in solution; γ_i represents activity coefficient (活度系数) of the ith ion and is usually a number less than 1.

3.4 Ionic Strength

The activity coefficient of a certain ion is affected not only by its own concentration and charge, but also by the concentration and charge of other ions in the solution. In order to measure the strength of the interaction between ions in the solution and its ionic atmosphere, the concept of ionic strength (离子强度) is introduced.

$$I = \frac{1}{2}(b_1Z_1^2 + b_2Z_2^2 + b_3Z_3^2 + \ldots) = \frac{1}{2}\sum_{i=1}^{n} b_iZ_i^2 \tag{2-14}$$

In the equation: symbol I represents ionic strength and the unit is mol/kg; b_i represents the molality of the ith ion in solution; Z_i represents the charge of the ith ion in the solution.

When the concentration of the solution is very dilute, the approximate calculation is $c_i = b_i$.

In 1923, Debye and Hueckel derived the approximate relationship between the activity coefficient of an ion and the ionic strength of the solution from the theory of electricity and molecular motion:

$$\lg\gamma_i = -Az_i^2\sqrt{I} \tag{2-15}$$

In the equation: A is equal to 0.509 in an aqueous solution at the temperature 298.15K; Z_i represents the charge of an ion.

This equation can only be established when the solution is extremely dilute (I<0.01). It is also called Debye-Huckel's limit law (德拜–休克尔极限公式).

Example 2-10 Calculate the ionic strength of the 0.01mol/L NaCl solution at 298K, and the activity coefficients, activities of Na^+ and Cl^-, and osmotic pressure of solution.

Solution $I = \frac{1}{2}\sum(b_iz_i^2) = \frac{1}{2}[(0.01\text{mol/L})\times(1)^2 + (0.01\text{mol/L})\times(-1)^2 = 0.010\text{mol/L}$

$$\lg\gamma_i = -Az_i^2\sqrt{I}$$

$$\lg\gamma_{Na^+} = \lg\gamma_{Cl^-} = -Az_i^2\sqrt{I} = -0.509\times1^2\times\sqrt{0.010} = -0.051$$

$$\gamma_{Na^+} = \gamma_{Cl^-} = 0.89$$

$$a_{Na^+} = a_{Cl^-} = \gamma_{Na^+}\times c_{Na^+} = \gamma_{Cl^-}\times c_{Cl^-} = 0.89\times0.010\text{mol/L} = 0.0089\text{mol/L}$$

$$\pi = [a_{Na^+} + a_{Cl^-}]RT = 2\times0.0089\times8.314298 = 44.1\text{kPa}$$

Classic Reading

Applying the Like-Dissolves-Like Rule

The like-dissolves-like rule (the rule of the likes dissolve each other, 相似相容原理) means that when the forces within the solute are similar to those within the solvent, the forces replace each other and a solution forms. For example, salts are soluble in water because the strong ion-dipole (离子偶极) attractions between ion and water are similar to the strong attractions between the ions and the strong hydrogen bonds (氢键) between water molecules, so they can replace each other. Salts are insoluble in hexane (C_6H_{14}) because the weak ion-induced dipole forces between ion and nonpolar hexane cannot replace attractions between the ions.

Oil is insoluble in water because the weak dipole-induced dipole forces between oil and water molecules cannot replace the strong hydrogen bonds within water molecules and the extensive dispersion forces within the oil. Oil is soluble in hexane because dispersion forces in one readily replace dispersion forces in the other.

重点小结

1. 溶液的浓度 一种物质以分子、原子或离子状态分散于另一种物质中所构成的均匀而又稳定的分散体系叫做溶液。能溶解其它物质的化合物叫做溶剂。被溶解的物质叫做溶质。溶液中溶剂和溶质的相对含量称为浓度。常用溶液浓度的表示方法见下表：

名称	定义	数学表达式	单位
质量分数	溶质B的质量 m_B 与溶液质量 m 之比值	$w_B = m_B/m$	1
摩尔分数	物质B的物质的量 n_B 与混合物的物质的量 $\sum ini$ 之比	$x_B = n_B/\sum ini$	1
质量摩尔溶液	溶质B的物质的量 n_B 除以溶剂的质量 m_A	$b_B = n_B/m_A$	mol/kg
质量浓度	溶质B的质量 m_B 除以溶液的体积V	$\rho_B = m_B/V$	g/L 或 g/ml
物质的量浓度（简称浓度）	溶质B的物质的量 n_B 除以混合物的体积V	$c_B = n_B/V$	mol/L 或 mmol/L

2. 溶液的依数性 难挥发性非电解质稀溶液的依数性包括蒸气压下降、沸点升高、凝固点降低、渗透压，这些性质与溶液中所含溶质粒子的浓度有关，而与溶质的本性无关。

同一温度下，难挥发性非电解质稀溶液的蒸气压下降（Δp）与溶质粒子的摩尔分数（x_B）成正比，而与溶质的本性无关。这一规律称为拉乌尔定律。其表达式为：$\Delta p = p_A^0 x_B$

溶液的沸点（T_b）总是高于相应纯溶剂的沸点（T_b^0）。难挥发性非电解质稀溶液沸点升高（ΔT_b）与溶液的质量摩尔浓度（b_B）成正比，而与溶质的本性无关。其表达式为 $\Delta T_b = K_b b_B$。

溶液的凝固点（T_f^0）总是低于相应纯溶剂的凝固点（T_f^0）。难挥发性非电解质稀溶液凝固点下降（ΔT_f）与溶液的质量摩尔浓度（b_B）成正比，而与溶质的本性无关。其表达式为：$\Delta T_f = K_f b_B$。

溶剂透过半透膜进入而使溶液一侧的液面升高的自发过程称为渗透。为维持只允许溶剂通过的膜所隔开的溶液与溶剂之间的渗透平衡，在溶液一侧需加的额外压力称为渗透压。产生渗透现象的两个必要条件包括：半透膜的存在及其膜两侧存在浓度差。

在一定温度下，稀溶液渗透压的大小仅与单位体积溶液中溶质的物质的量有关，而与溶质的

本性无关。其表达式为：$\pi = c_B RT$

3. 离子的活度 电解质溶液由于溶质发生电离，使溶液中溶质粒子数增加，计算时应考虑其电离的因素，否则会使计算得到的 Δp、ΔT_b、ΔT_f、π 值比实验测得值小；另一方面，电解质溶液由于离子间的静电引力比非电解质之间的作用力大得多，因此用离子浓度来计算强电解质溶液的 Δp、ΔT_b、ΔT_f、π 时，其计算结果与实际值偏离较大，应该用活度代替浓度进行计算。

题库

Exercises

1. Choice question

(1) The unit of mass concentration is

A. g/L B. mol/L C. g/mol D. g/g E. L/mol

(2) Given that the number of moles of solute B is n_B and the number of moles of solvent is n_A, then the mole fraction of solute B in this solution x_B is

A. $\frac{n_B}{n_A + n_B}$ B. $\frac{n_A}{n_A + n_B}$ C. $1 - x_B$

D. $x_B - x_A = 1$ E. None of the above is true

(3) Given that the molar mass of glucose $C_6H_{12}O_6$ is 180g/mol, and the solution of 1 L water contains 18g of glucose, then the molarity of glucose in this solution is

A. 0.05mol/L B. 0.10mol/L C. 0.20mol/L D. 0.30mol/L E. 0.15mol/L

(4) Which of the following concentration representations is temperature dependent

A. Mass fraction B. Molality C. Molarity

D. Mole fraction E. none of the above is correct

(5) The high content of NaCl in the soil makes it difficult for plants to survive, which is related to the following properties of dilute solutions

A. Vapor pressure drop B. Boiling point rise C. Freezing point drop

D. Osmotic pressure E. Boiling point lowering

(6) The essence of the colligative property of dilute solutions is

A. Permeability B. Boiling point elevation C. Vapor pressure decreasing

D. Freezing point depression E. Vapor pressure increasing

(7) When the molecular weight of glucose is measured by freezing point depression method, if the glucose sample contains insoluble impurities, the molecular weight of glucose is measured

A. Low B. High C. No effect

D. No measurement E. None of the above is correct

(8) There are sucrose ($C_{12}H_{22}O_{11}$), sodium chloride (NaCl), calcium chloride ($CaCl_2$), three solutions, their concentrations are all 0.1mol/L, the order of osmotic pressure from low to high is

A. $CaCl_2 < NaCl < C_{12}H_{22}O_{11}$ B. $C_{12}H_{22}O_{11} < NaCl < CaCl_2$

C. $NaCl < C_{12}H_{22}O_{11} < CaCl_2$ D. $C_{12}H_{22}O_{11} < CaCl_2 < NaCl$

E. $NaCl = C_{12}H_{22}O_{11} = CaCl_2$

(9) The solution freezing point depression value is ΔT_f, the solute mass is g grams, the solvent mass is G grams, the molecular weight of the solute is

A. $\dfrac{G\times g\times 1000}{K_f\times \Delta T_f}$ B. $\dfrac{K_f\times g\times 1000}{G\times \Delta T_f}$ C. $\dfrac{G\times 1000}{K_f\times g\times \Delta T_f}$

D. $\dfrac{K_f\times g\times \Delta T_f}{G\times 1000}$ E. $\dfrac{g\times 1000\times \Delta T_f}{K_f\times G}$

(10) When a 1.0g/L glucose solution and a 1.0g/L sucrose solution are separated by a semi-permeable membrane, which of the following phenomena occurs?

A. No penetration between the two solutions

B. Water molecules in the glucose solution penetrate the semi-permeable membrane into the sucrose solution

C. Water molecules in the sucrose solution penetrate the semi-permeable membrane into the glucose solution

D. Glucose and sucrose solutions are isotonic

E. Glucose molecules enter the sucrose solution through a semi-permeable membrane

(11) The concentration of concentrated sulfuric acid is ω = 98% and the density is 1.84g/ml. The concentration of concentrated sulfuric acid is

A.18.4mol/L B. 1.84mol/L C. 18.4g/L

D. 184mol/L E. 18.4mol/kg

(12) The activity coefficient γ of Ba^{2+} is 0.24, the activity a of 0.050mol/L Ba^{2+} is

A. 0.012mol/L B. 0.014mol/L C. 0.016mol/L

D. 0.050mol/L E. 0.013mol/L

(13) In a 0.020mol/L $NaNO_3$ solution, the ionic strength I is

A. 0.10mol/L B. 0.010mol/L C. 0.020mol/L

D. 0.040mol/L E. 0.050mol/L

(14) What is incorrect about the ionic strength (I) is

A. The greater the ionic strength of the solution, the greater the interaction between the ions.

B. The greater the ionic strength, the smaller the ion activity coefficient γ

C. Ionic strength is related to the charge and concentration of ions

D. The greater the ionic strength, the greater the ion activity a

E. Ionic strength has nothing to do with the nature of ions

2. Judgment Question

(1) The sum of the molar fractions of each solute and the solvent in the solution is 1.

(2) Strong electrolytes are completely ionized in solution.

(3) The greater the ionic strength of the solution, the greater the activity of the ions.

(4) Any two solutions are separated by a semi-permeable membrane, and a penetrating phenomenon occurs.

(5) The boiling point of a solution is the temperature at which the boiling temperature of the solution does not change.

3. Simple answer question

(1) What are the advantages and disadvantages of using the molar mass concentration and the mass concentration of a substance to indicate the concentration of a substance?

(2) Considering that the cell membrane is a permeable membrane, try to explain why the lettuce salad containing salt and vinegar can be softened within hours.

(3) The same mass of ethanol, glycerol, glucose and sucrose were dissolved in 100 ml of water, respectively, and these solutions were cooled at the same speed and frozen. What are the first freeze and the last freeze? What is the result if the same amount of ethanol, glycerol, glucose and sucrose were replaced? Please explain.

(4) What is the connection between the four colligative properties of a dilute solution? Please explain.

目标检测

1. 选择题

（1）质量浓度的单位是

A. g/L　B. mol/L　C. g/mol
D. g/g　E. L/mol

（2）已知溶质 B 的摩尔数为 n_B，溶剂的摩尔数为 n_A，则溶质 B 在此溶液中的摩尔分数 X_B 为

A. $\frac{n_B}{n_A+n_B}$　B. $\frac{n_A}{n_A+n_B}$　C. $1-x_B$
D. $x_B-x_A=1$　E. 以上均不对

（3）已知葡萄糖 $C_6H_{12}O_6$ 的摩尔质量是 180g/mol，1 升水溶液中含葡萄糖 18g，则此溶液中葡萄糖的物质的量浓度为

A. 0.05mol/L　B. 0.10mol/L　C. 0.20mol/L
D. 0.30mol/L　E. 0.15mol/L

（4）下列那种浓度表示方法与温度有关

A. 质量分数　B. 质量摩尔浓度　C. 物质的量浓度
D. 摩尔分数　E. 以上均不对

（5）土壤中 NaCl 含量高使植物难以生存，这与下列稀溶液的性质有关

A. 蒸汽压下降　B. 沸点升高　C. 凝固点下降
D.　渗透压　E. 沸点降低

（6）稀溶液依数性的本质是

A. 渗透性　B. 沸点升高　C. 蒸汽压下降
D. 凝固点降低　E. 蒸汽压升高

（7）用冰点降低法测定葡萄糖分子量时，如果葡萄糖样品中含有不溶性杂质，则测的分子量

A. 偏低　B. 偏高　C. 无影响
D. 无法测定　E. 以上都不对

（8）有蔗糖（$C_{12}H_{22}O_{11}$），氯化钠（NaCl），氯化钙（$CaCl_2$），三种溶液，它们的浓度均为 0.1mol/L, 按渗透压由低到高的排列顺序是

A. $CaCl_2 < NaCl < C_{12}H_{22}O_{11}$　B. $C_{12}H_{22}O_{11} < NaCl < CaCl_2$
C. $NaCl < C_{12}H_{22}O_{11} < CaCl_2$　D. $C_{12}H_{22}O_{11} < CaCl_2 < NaCl$
E. $NaCl = C_{12}H_{22}O_{11} = CaCl_2$

（9）溶液凝固点降低值为 ΔT_f，溶质为 g 克，溶剂为 G 克，溶质的分子量是

A. $\frac{G\times g\times 1000}{K_f\times \Delta T_f}$　B. $\frac{K_f\times g\times 1000}{G\times \Delta T_f}$　C. $\frac{G\times 1000}{K_f\times g\times \Delta T_f}$

D. $\dfrac{K_f \times g \times \Delta T_f}{G \times 1000}$　　　　E. $\dfrac{g \times 1000 \times \Delta T_f}{K_f \times G}$

（10）1.0g/L 的葡萄糖溶液和 1.0g/L 的蔗糖溶液用半透膜隔开后，会发生以下哪种现象

A. 两个溶液之间不会发生渗透

B. 葡萄糖溶液中的水分子透过半透膜进入蔗糖溶液中

C. 蔗糖溶液中的水分子透过半透膜进入葡萄糖溶液中

D. 葡萄糖溶液和蔗糖溶液是等渗溶液

E. 葡萄糖分子透过半透膜进入蔗糖溶液中

（11）浓硫酸质量分数 $\omega = 98\%$，密度 1.84g/ml，则浓硫酸物质的量浓度为

A.18.4mol/L　　B. 1.84mol/L　　C. 18.4g/L

D. 184mol/L　　E. 18.4mol/Kg

（12）已知 Ba^{2+} 的活度系数 $\gamma = 0.24$，则 0.050mol/L Ba^{2+} 的活度 α 为

A. 0.012mol/L　　B. 0.014mol/L　　C. 0.016mol/L

D. 0.050mol/L　　E. 0.013mol/L

（13）0.020mol/L $NaNO_3$ 溶液中，离子强度 I 为

A. 0.10mol/L　　B. 0.010mol/L　　C. 0.020mol/L

D. 0.040mol/L　　E. 0.050mol/L

（14）有关离子强度 (I) 的说法不正确的是

A. 溶液的离子强度越大，离子间相互牵制作用越大

B. 离子强度越大，离子的活度系数 γ 越小

C. 离子强度与离子的电荷及浓度有关

D. 离子强度越大，离子的活度 α 也越大

E. 离子强度与离子的本性无关

2. 判断题

（1）溶液中各溶质与溶剂的摩尔分数之和为 1。

（2）强电解质在溶液中是完全电离的。

（3）溶液的离子强度越大，离子的活度也越大。

（4）任何两种溶液用半透膜隔开，都有渗透现象发生。

（5）溶液的沸点是指溶液沸腾温度不变时的温度。

3. 问答题

（1）用质量摩尔浓度和物质的量浓度表示物质的浓度时，各有何优缺点？

（2）考虑到细胞膜为一渗透膜，试解释为何含盐与醋的莴苣沙拉于数小时内即行软化？

（3）相同质量的乙醇、甘油、葡萄糖和蔗糖分别溶于 100 ml 水中，将这几种溶液以相同速度降温冷冻，最先结冰和最后结冰的分别是？如果换成相同物质的量的乙醇、甘油、葡萄糖和蔗糖，结果又怎样？请加以说明。

（4）稀溶液的四种依数性之间的联系是什么？请加以说明。

PPT

Chapter 3 Chemical Equilibrium
第三章 化学平衡

学习目标

知识要求

1. 掌握 标准平衡常数和反应商的概念、标准平衡常数表达式的书写及其有关计算。

2. 熟悉 浓度、压力、温度对化学平衡移动的影响；多重平衡规则。

3. 了解 催化剂与化学平衡的关系；Le Châtelier 原理。

能力要求

熟练掌握平衡常数的计算并判定反应进行的方向；改变影响平衡的条件对化学平衡的影响，为学习四大平衡等内容打下理论基础。

1. Chemical Equilibrium

In this chapter, we will discuss the reversible reaction (可逆反应) when forward and reverse reactions (正向和逆向反应) proceed at the same rate on certain conditions. Our main tool in dealing with equilibrium will be the equilibrium constant (平衡常数). We will begin with equilibrium constant expression (平衡常数表达式); then we will make qualitative predictions about the condition of equilibrium; and finally we will do some equilibrium calculations.

Reversible reaction is a chemical reaction that results in an equilibrium mixture of reactants and products. All reactions are reversible. When the forward and reverse reaction rates are equal, the system has reached equilibrium (平衡). After this point, there is no further observable change. Sometimes we can consider some reactions as nonreversible when they go right almost completely. Nonreversible reaction is a chemical reaction that goes to the (large) full extent and almost no reactant is left in the system, such as:

$$Mg + HCl \rightarrow MgCl_2 + H_2 \uparrow$$

Only very few reactions can be considered as nonreversible.

Chemical equilibrium is a state when no net change of both the concentration of reactant(s) and product(s) occur, because the forward and reverse reactions proceed at the same rate. The reaction rates of the forward and backward reactions are not zero, but equal. Such a state is known as dynamic equilibrium (动态平衡). We show equilibrium reactions by the sign $\rightleftharpoons$.

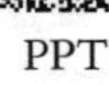

For example, consider the reaction between hydrogen and iodine carried out in a sealed glass tube at 400°C:

$$H_2(g) + I_2(g) \rightleftharpoons 2HI(g)$$

The reversible reaction will inevitably lead to the realization of chemical equilibrium.

Characters of chemical equilibrium:

(a) No change of the concentration of both reactant(s) and product(s) can be observed.

(b) It is a dynamic equilibrium. Reactions in both directions are not stopped.

(c) Equilibrium applies to the extent (or yield) of a reaction. At the time of equilibrium, the reaction has proceeded to the largest extent.

(d) Chemical equilibrium is established with certain conditions. When reaction condition is changed, the equilibrium will be disturbed and after some time a new equilibrium will be established.

1.1 Equilibrium Constant

Equilibrium constant is an important parameter to show the extent of equilibrium reaction (平衡常数是衡量平衡反应进行限度的一个重要参数). It is a quantitative view of equilibrium: a constant ratio of constants. Let's see how reactant and product concentrations affect this process. At a particular temperature, when the system reaches equilibrium, we have

$$\text{rate}_{\text{fwd}} = \text{rate}_{\text{rev}}$$

We take the following reaction as example:

$$N_2O_4\ (\text{g, colorless}) \rightleftharpoons 2NO_2\ (\text{g, brown})$$

In this reaction system, both forward and reverse reactions are elementary steps (基元反应) , so we can write their rate laws directly from the balanced equation:

$$k_{\text{fwd}}\ [N_2O_4]_{\text{eq}} = k_{\text{rev}}\ [NO_2]^2_{\text{eq}}$$

Here we use [ ] to show the equilibrium concentrations (用中括号表示平衡浓度) of the reactant and products. k_{fwd} and k_{rev} are the forward and reverse rate constants (速率常数), respectively, and the subscript "eq" refers to concentrations at equilibrium. By rearranging, we set the ratio of the rate constants equal to the ratio of the concentration terms:

$$\frac{k_{\text{fwd}}}{k_{\text{rev}}} = \frac{[NO_2]^2_{\text{eq}}}{[N_2O_4]_{\text{eq}}}$$

The ratio of constants creates a new constant called the equilibrium constant (K):

$$K = \frac{k_{\text{fwd}}}{k_{\text{rev}}} = \frac{[NO_2]^2_{\text{eq}}}{[N_2O_4]_{\text{eq}}} \tag{3-1}$$

Equation (3-1) is the equilibrium constant expression, it shows that at a given temperature, a chemical system reaches a state in which a particular ratio of product to reactant concentrations has a constant value. This is a statement of the law of chemical equilibrium, or the law of mass action (质量作用定律).

The equilibrium constant K is a number equal to a particular ratio of equilibrium concentrations of product(s) to reactant(s) at a particular temperature.

Equilibrium constant (K) being as a measure of reaction extent, is an indication of how far a reaction proceeds toward product at a given temperature. Different reactions, even at the same temperature, have a wide range of concentrations at equilibrium. Large K indicates that most of the reactants are converted to products. Small K indicates that only small amounts of products are formed.

K does not vary with the concentration or partial pressure of each substance. K is the function of the temperature. When the temperature is different, the K value is different (K 不随各物质的浓度或分压的变化而变化，但随温度的变化而变化，当温度不同时，K 值不同).

1.2 Standard Equilibrium Constant and Calculation

For an equilibrium reaction, there is a relationship between the concentrations of the reactants and products which is given by the equilibrium constant. In the expression of the equilibrium constant, if each substance uses its own standard state as the reference state, the obtained equilibrium constant is the standard equilibrium constant (标准平衡常数).

1.2.1 Standard Equilibrium Constant of Dilute Solution

For the following reaction:

$$a\mathrm{A(aq)}+d\mathrm{D(aq)} \rightleftharpoons e\mathrm{E(aq)}+f\mathrm{F(aq)}$$

At a certain temperature, the standard equilibrium constant is expressed as

$$K^{\ominus}=\frac{\left([\mathrm{E}]/c^{\ominus}\right)^{e}\left([\mathrm{F}]/c^{\ominus}\right)^{f}}{\left([\mathrm{A}]/c^{\ominus}\right)^{a}\left([\mathrm{D}]/c^{\ominus}\right)^{d}} \tag{3-2}$$

In equation (3-2), [A], [D], [E] and [F] represent the equilibrium molarity (物质的量浓度) of A, D, E and F, respectively. $[\mathrm{A}]/c^{\ominus}$, $[\mathrm{D}]/c^{\ominus}$, $[\mathrm{E}]/c^{\ominus}$ and $[\mathrm{F}]/c^{\ominus}$ are the relative equilibrium concentrations (相对平衡浓度)of substances A, D, E and F, respectively. And $c^{\ominus}$ = 1mol/L, the standard equilibrium constant can also be abbreviated as:

$$K^{\ominus}=\frac{[\mathrm{E}]^{e}[\mathrm{F}]^{f}}{[\mathrm{A}]^{a}[\mathrm{D}]^{d}} \tag{3-3}$$

1.2.2 Standard Equilibrium Constant for Gaseous Reactions

Similarly, a reversible reaction for any ideal gas:

$$a\mathrm{A(g)}+d\mathrm{D(g)} \rightleftharpoons e\mathrm{E(g)}+f\mathrm{F(g)}$$

$$K^{\ominus}=\frac{[p_{\mathrm{eq}}(\mathrm{E})/p^{\ominus}]^{e}[p_{\mathrm{eq}}(\mathrm{F})/p^{\ominus}]^{f}}{[p_{\mathrm{eq}}(\mathrm{A})/p^{\ominus}]^{a}[p_{\mathrm{eq}}(\mathrm{D})/p^{\ominus}]^{d}} \tag{3-4}$$

In equation (3-4), $p_{\mathrm{eq}}(\mathrm{A})$, $p_{\mathrm{eq}}(\mathrm{D})$, $p_{\mathrm{eq}}(\mathrm{E})$ and $p_{\mathrm{eq}}(\mathrm{F})$ represent the equilibrium partial pressure (平衡分压) of A, D, E and F, usually with unit of kPa or atm. The ratio of equilibrium partial pressure and $p^{\ominus}$ is the relative equilibrium partial pressure (相对平衡分压). $p^{\ominus}$ is standard pressure which is 100kPa or 1 atm. $p^{\ominus}$ cannot be omitted in (3-4).

1.2.3 Standard Equilibrium Constant of Heterogeneous System (多相体系)

For the following heterogeneous reversible reaction

$$a\mathrm{A(s)} + d\mathrm{D(aq)} \rightleftharpoons e\mathrm{E(g)} + f\,\mathrm{F(l)}$$

$$K^{\ominus}=\frac{[p_{\mathrm{eq}}(\mathrm{E})/p^{\ominus}]^{e}}{[\mathrm{D}]^{d}}$$

In the equation, only species in gas or aqueous phases are included because the concentration (in solution) of pure liquid or solid will not change. And they are considered as 1.

1.2.4 Important Points about Writing $K^{\ominus}$

(1) $K^{\ominus}$ is unitless, so in the expression of the standard equilibrium constant, each species takes its own standard state as the reference state. So, if a species is a gas, the partial pressure is divided by $p^{\ominus}$(100kPa); If in aqueous solution, the concentration is divided by $c^{\ominus}$ (1mol·L^{-1}).

(2) For pure liquid or solid, their concentration is 1 and should not be listed in the equation. For aqueous reaction, water is considered as pure liquid; for reactions in other solvent, water is considered as solute and should be included in equation (纯固体或纯液体，浓度看成 1 不必列入平衡常数表达式。水溶液中反应,水被认为纯液体,对于非水溶剂中的反应,水被看成溶质要列入平衡常数表达式中).

(3) Same reaction with different presenting ways will have different $K^{\ominus}$. For example:

$$2SO_2(g) + O_2(g) \rightleftharpoons 2SO_3(g) \qquad K_1^{\ominus} = \frac{[p_{eq}(SO_3)/p^{\ominus}]^2}{[p_{eq}(SO_2)/p^{\ominus}]^2[p_{eq}(O_2)/p^{\ominus}]}$$

$$SO_2(g) + \frac{1}{2}O_2(g) \rightleftharpoons SO_3(g) \qquad K_2^{\ominus} = \frac{[p_{eq}(SO_3)/p^{\ominus}]}{[p_{eq}(SO_2)/p^{\ominus}][p_{eq}(O_2)/p^{\ominus}]^{1/2}}$$

$$2SO_3(g) \rightleftharpoons 2SO_2(g) + O_2(g) \qquad K_3^{\ominus} = \frac{[p_{eq}(SO_2)/p^{\ominus}]^2[p_{eq}(O_2)/p^{\ominus}]}{[p_{eq}(SO_3)/p^{\ominus}]^2}$$

Here, $K_1^{\ominus} \neq K_2^{\ominus} \neq K_3^{\ominus}$, the relationship is: $K_1^{\ominus} = (K_2^{\ominus})^2 = \frac{1}{K_3^{\ominus}}$

1.2.5 Multiple Equilibrium Rule

In a multi-reaction equilibrium system, the equilibrium concentration or partial pressure of any substance must simultaneously satisfy multiple equilibria in the system. Therefore, there is a certain relationship between the standard equilibrium constants. This is called multiple equilibrium rule (多重平衡规则). Such as:

$$(1)\ SO_2(g)+\frac{1}{2}O_2(g) \rightleftharpoons SQ_3(g) \qquad K_1^{\ominus}$$

$$(2)\ NO_2(g) \rightleftharpoons NO(g)+\frac{1}{2}O_2(g) \qquad K_2^{\ominus}$$

(1) + (2) = (3): (3) $SO_2(g)+NO_2(g) \rightleftharpoons SO_3(g)+NO(g)$ $\qquad K_3^{\ominus}$

So $\qquad K_3^{\ominus} = K_1^{\ominus} \cdot K_2^{\ominus}$

Example 3-1 The equilibrium constants for the following reactions at 298K:

$$(1)\ N_2O_4(g)) \rightleftharpoons 2NO_2(g) \qquad K_1^{\ominus}$$

$$(2)\ \frac{1}{2}N_2(g) + O_2(g) \rightleftharpoons NO_2(g) \qquad K_2^{\ominus}$$

Calculate $K_3^{\ominus}$ for reaction (3): $N_2(g) + 2O_2(g) \rightleftharpoons N_2O_4(g)$ at 298K.

Solution The relationship between the three reactions is as follows: (3) = (2) × 2 – (1), so based on multiple equilibrium rule:

$$K_3^{\ominus} = \frac{(K_2^{\ominus})^2}{K_1^{\ominus}}$$

When two independent reactions are added or subtracted together, the standard equilibrium constant of the total reaction obtained is equal to the product or quotient of the standard equilibrium constants of the two reactions. This rule is called multiple equilibrium rule (在多重平衡系统中,如果某一平衡反应可以由几个平衡反应相加（或相减）得到,则该平衡反应的标准平衡常数等于几个平衡反应的标准平衡常数的乘积（或商），这种关系称为多重平衡规则).

The standard equilibrium constants can be calculated indirectly according to this rule for the reactions of some equilibrium constants that are difficult to directly measure or not easy to check from the literature. It is worth noting that the multi- equilibrium rule requires that all chemical reactions take place at the same temperature.

1.2.6 Calculation of Chemical Equilibrium

The standard equilibrium constant can be used to measure the degree of a reaction and calculate the equilibrium concentrations of the reactants and products. The degree to which the reaction is progressing is also often expressed as the equilibrium conversion rate (平衡转化率). The conversion rate of a reaction is the percentage of the initial amount of the reactant that has been converted at equilibrium.

$$\text{Conversion rate } (\alpha)=\frac{\text{conversion amount of reactant at equilibrium}}{\text{initial amount of reactant}}\times 100\%$$

If the volume is unchanged before and after the reaction, it can be expressed as:

$$\text{Conversion rate } (\alpha)=\frac{\text{(initial concentration- equilibrium concentration) of reactant}}{\text{initial concentration of the reactant}}\times 100\%$$

The chemical equilibrium state is the maximum extent of the reaction (化学平衡状态是一个反应进行的最大限度的状态). The equilibrium conversion rate is the maximum conversion rate under certain conditions. The higher the conversion rate, the more the reaction goes right.

Example 3-2 At a certain temperature, consider the following equilibrium:

$$CO(g) + H_2O(g) \rightleftharpoons H_2(g) + CO_2(g) \quad K^{\ominus} = 9$$

0.020mol of CO(g) and 0.020mol of H_2O(g) are injected into a closed 1L container, what is the maximum conversion rate of CO under such conditions?

Solution When equilibrium is established, suppose that x mol CO_2 and H_2 formed

	CO(g)	+	H_2O(g)	⇌	H_2(g)	+ CO_2(g)
Initial (mol)	0.020		0.020		0	0
Equilibrium (mol)	$0.020-x$		$0.020-x$		x	x

$$K^{\ominus}=\frac{\left[\frac{p_{eq}(H_2)}{p^{\ominus}}\right]\cdot\left[\frac{p_{eq}(CO_2)}{p^{\ominus}}\right]}{\left[\frac{p_{eq}(CO)}{p^{\ominus}}\right]\cdot\left[\frac{p_{eq}(H_2O)}{p^{\ominus}}\right]}=\frac{n_{eq}(H_2)\cdot n_{eq}(CO_2)}{n_{eq}(CO)\cdot n_{eq}(H_2O)}$$

$$\frac{(x\text{mol})^2}{(0.020\text{mol}-x\text{mol})^2}=9$$

$$x=0.015$$

$$\text{Conversion of CO}=\frac{0.015\text{mol}}{0.020\text{mol}}\times 100\%=75\%$$

Example 3-3 Some nitrogen and hydrogen are placed in an empty 5L container at 500°C. When equilibrium is established, 3.01mol of N_2, 2.10mol of H_2, and 0.565mol of NH_3 are present. Evaluate $K_c^\ominus$ for the following reaction at 500°C.

Solution

$$N_2(g)+3H_2(g) \rightleftharpoons 2NH_3(g)$$

$[N_2]$: 3.01mol/5L = 0.602mol/L
$[H_2]$: 2.10mol/5L = 0.420mol/L
$[NH_3]$: 0.565mol/5L = 0.113mol/L

$$K_c^\ominus = \frac{[NH_3]^2}{[N_2][H_2]^3} = \frac{(0.113)^2}{(0.602)(0.420)^3} = 0.29$$

Example 3-4 Myoglobin (Mb) is present in muscle tissue and has the ability to carry O_2. The reaction between oxygen and myoglobin can be expressed as:

$$Mb(aq) + O_2(g) \rightleftharpoons MbO_2(aq)$$

At 310K, $K^\ominus = 1.20\times10^2$. When the partial pressure of O_2 was 5.15kPa, calculate the ratio of equilibrium concentrations of oxygenated-myoglobin (MbO_2) and myoglobin (Mb).

Solution The equilibrium constant expression:

$$K^\ominus = \frac{[MbO_2]}{[Mb][p_{O_2}/p^\ominus]}$$

The ratio of the equilibrium concentrations is:

$$\frac{[MbO_2]}{[Mb]} = [p_{O_2}/p^\ominus]\cdot K^\ominus = \frac{5.15\text{kPa}}{100\text{kPa}} \times 1.20\times10^2 = 6.18$$

2. Shifts of Chemical Equilibrium

Chemical equilibrium exists when two opposing reactions occur simultaneously at the same rate. Changes in experimental conditions such as concentration, pressure, volume, and temperature disturb the balance and shift the equilibrium position.

Le Châtelier's principle predicts the direction in which the equilibrium will shift. It states that if a system at equilibrium experiences a change, the system will shift its equilibrium to try to compensate for the change.

Le Châtelier's principle describes what happens to a system when something momentarily takes it away from equilibrium. This section focuses on three ways in which we can change the conditions of a chemical reaction at equilibrium:

(a) changing the concentration of one of the components of the reaction

(b) changing the pressure on the system

(c) changing the temperature at which the reaction is run.

2.1 Effect of Concentration on Chemical Equilibrium

The reaction quotient (反应商) helps us to predict the direction in which the equilibrium is moving. The effect of concentration on chemical equilibrium can be determined by the relative size of reaction quotient Q and $K^{\ominus}$.

For a reaction at any state (equilibrium has been reached or not):

$$aA(aq) + dD(aq) \rightleftharpoons eE(aq) + fF(aq)$$

We can obtain the reaction quotient (Q_c, also named as concentration quotient, 又叫浓度商) by the formula:

$$Q_c = \frac{\left(c_E / c^{\ominus}\right)^e \left(c_F / c^{\ominus}\right)^f}{\left(c_A / c^{\ominus}\right)^a \left(c_D / c^{\ominus}\right)^d} \quad (3\text{-}5)$$

In equation (3-5), c_A, c_D, c_E, and c_F represent the molarity of A, D, E and F at A certain time. $c_A/c^{\ominus}$, $c_D/c^{\ominus}$, $c_E/c^{\ominus}$ and $c_F/c^{\ominus}$ are the relative concentrations of A, D, E and F at a certain time.

The Qp (also named pressure quotient, 又叫压力商)expression for the gas reaction is:

$$Q_p = \frac{\left(p_E / p^{\ominus}\right)^e \left(p_F / p^{\ominus}\right)^f}{\left(p_A / p^{\ominus}\right)^a \left(p_D / p^{\ominus}\right)^d} \quad (3\text{-}6)$$

In equation (3-6), p_A, p_D, p_E, and p_F represent the partial pressures of A, D, E, and F at a certain time. $p_A/p^{\ominus}$, $p_B/p^{\ominus}$, $p_D/p^{\ominus}$and $p_E/p^{\ominus}$ are the relative partial pressures of A, D, E and F at a certain time.

Keep in mind that the method for calculating Q is the same as that for $K^{\ominus}$. $K^{\ominus}$ is a constant at a given temperature, but Q varies according to the relative amounts of reactants and products present.

To predict and determine the direction in which the net reaction will proceed to achieve equilibrium, we compare the values of Q and $K^{\ominus}$, the three possible cases are as follows:

$Q < K^{\ominus}$ The ratio of initial concentrations of products to reactants is too small. To reach equilibrium, reactants must be converted to products. The system proceeds from left to right to reach equilibrium.

$Q = K^{\ominus}$ The initial concentrations are equilibrium concentrations. The system is at equilibrium.

$Q > K^{\ominus}$ The ratio of initial concentrations of products to reactants is too large. To reach equilibrium, products must be converted to reactants. The system proceeds left to reach equilibrium.

Example 3-5 At the start of a reaction, there are 0.249mol N_2, 3.21×10^{-2}mol H_2, and 6.42×10^{-4}mol NH_3 in a 3.50L reaction vessel at 375°C. For the reaction

$$N_2(g) + 3H_2(g) \rightleftharpoons 2NH_3(g)$$

If the equilibrium constant $K_c^{\ominus} = 1.2$ at this temperature, is the system at equilibrium? If it is not, which way the net reaction will shift?

Solution Calculate the initial concentration of the reacting species

$$c(N_2) = \frac{0.249\text{mol}}{3.50\text{L}} = 0.0711\text{mol/L}$$

$$c(H_2) = \frac{3.21\times10^{-2}\text{ mol}}{3.50\text{L}} = 9.17\times10^{-3}\text{ mol/L}$$

$$c(NH_3) = \frac{6.42\times10^{-4}\text{ mol}}{3.50\text{L}} = 1.83\times10^{-4}\text{ mol/L}$$

Calculate Q_c

$$Q_c = \frac{(1.83\times10^{-4})^2}{(0.0711)(9.17\times10^{-3})^3} = 0.611$$

Compare: 0.611<1.2, so $Q_c < K_c^\ominus$

The system is not at equilibrium, and the net reaction will proceed from left to right until equilibrium is reached.

Example 3-6 At 298K, the following reaction happens in a solution with $K^\ominus = 3.2$,

$$Ag^+(aq) + Fe^{2+}(aq) \rightleftharpoons Ag(s) + Fe^{3+}(aq)$$

If the initial concentration: $c(Ag^+) = c(Fe^{2+}) = 0.10$mol/L.

Calculate:

1) What is the equilibrium concentration of Ag^+, Fe^{2+} and Fe^{3+}?

2) What is the percent conversion of Ag^+?

3) If $c(Ag^+)$ does not change, $c(Fe^{2+}) = 0.300$mol/L, what is percent conversion of Ag^+?

Solution

(1) We assume the equilibrium concentration: $c(Fe^{3+}) = x$ mol/L,

	$Ag^+(aq)$ +	$Fe^{2+}(aq) \rightleftharpoons Ag(s)$ +	$Fe^{3+}(aq)$
Initial concentration /(mol/L)	0.10	0.10	0
Equilibrium concentration /(mol/L)	$0.10 - x$	$0.10 - x$	x

$$K^\ominus = \frac{[Fe^{3+}]}{[Ag^+][Fe^{2+}]} = \frac{x}{(0.10 - x)^2} = 3.2$$

$$x = 0.020$$

So
$$c(Fe^{3+}) = 0.020\text{mol/L}$$
$$c(Fe^{2+}) = 0.080\text{mol/L}$$
$$c(Ag^+) = 0.080\text{mol/L}$$

(2) the percent conversion α of Ag^+

$$\alpha = \frac{0.020}{0.10}\times100\% = 20.00\%$$

(3) We assume the α of Ag^+ at equilibrium

$$Ag^+(aq) + Fe^{2+}(aq) \rightleftharpoons Ag(s) + Fe^{3+}(aq)$$

	Ag^+	Fe^{2+}	Fe^{3+}
Equilibrium concentration /(mol/L)	$0.10(1 - \alpha)$	$0.30 - 0.10\alpha$	0.10α

$$K^\ominus = \frac{[Fe^{3+}]}{[Ag^+][Fe^{2+}]} = \frac{0.10\alpha}{0.10(1-\alpha)(0.30-0.10\alpha)} = 3.2$$

So
$$\alpha = 45.33\%$$

For a reversible reaction, if the concentration of one reactant is increased or the concentration of the product is reduced, $Q_c < K^\ominus$, it can increase the conversion rate of another reactant. In production, this principle is often used to improve the conversion rate of reactants.

2.2 Effect of Pressure on Chemical Equilibrium

Pressure change has a negligible effect on liquids and solids, because they are almost impressible. Here we mainly discuss the effect of pressure on gas reaction. Pressure changes can occur in three ways:

2.2.1 Changing of the Partial Pressure of a Gaseous Component

Suppose that the equilibrium system:

$$N_2O_4(g) \rightleftharpoons 2NO_2(g)$$

Effect of changing partial pressure is same as that of concentration. The equilibrium position shifts to the right if the partial pressure of the reactant (N_2O_4) is increased or that of product (NO_2) is decreased. The equilibrium position shifts to the left if the partial pressure of the reactant (N_2O_4) is decreased or that of product (NO_2) is increased.

2.2.2 Changing of the System Pressure (or the Vessel Volume)

For the reaction of N_2O_4-NO_2, we assume it is in a cylinder fitted with a movable piston. What happens if we increase the pressure on the gases by pushing down on the piston at constant temperature?

Because the volume decreases, the concentrations of both NO_2 and N_2O_4 increase.

$$Q_c = \frac{[NO_2]^2}{[N_2O_4]}$$

Thus, $Q_c > K^{\ominus}$ and the net reaction will shift to the left until $Q_c = K^{\ominus}$.

In general, an increase in pressure (decrease in volume) favors the net reaction that decreases the total number of moles of gases (the reserve reaction, in this case), and a decrease in pressure (increase in a volume) favors the net reaction that increases the total number of moles of gases (here, the forward reaction). For reactions in which there is no change in the number of moles of gases, a pressure (or volume) change has no effect on the position of equilibrium.

2.2.3 Adding an Inert Gas to the System

Inert gas means that it does not take part in the reaction.

(1) Constant-volume system (恒容体系). It is possible to change the pressure of a system without changing its volume. If the NO_2-N_2O_4 system has a constant volume, we can increase the total pressure in the vessel by adding an inert gas (helium, for example) to the equilibrium system. Adding helium to the equilibrium mixture at constant volume, the total gas pressure increases, but all partial pressures of reactants and products do not change, so the equilibrium position will not change.

(2) Constant-pressure system (恒压体系). If the system has a constant-pressure, the volume of the system increases when adding an inert gas. All partial pressures of reactants and products decrease. The equilibrium position will shift to the side which has more more gas molecules.

Example 3-7 At 310K and 100kPa, when the reaction $N_2O_4(g) \rightleftharpoons 2NO_2(g)$ reached equilibrium, the partial pressures: $p_{N_2O_4} = 58\text{kPa}$, $p_{NO_2} = 42\text{kPa}$,and $K^{\ominus} = 0.30$, calculate:

(1) The volume of the system is compressed to increase the pressure to 200kPa, which direction will the reaction proceed?

(2) At 310K and 200kPa, the initial amount: $n(N_2O_4) = 1.0\text{mol}$, $n(NO_2) = 0.10\text{mol}$, when the reaction reached equilibrium, 0.15mol N_2O_4 has been transformed, then what is the partial pressure of N_2O_4 and NO_2?

Solution

(1) When the total pressure of the system is doubled, the partial pressure of the components in the system is doubled.

$$p(N_2O_4) = 2 \times 58\text{kPa} = 116\text{kPa},\ p(NO_2) = 2{\times}42\text{kPa} = 84\text{kPa}$$

$$Q_p = \frac{(p_{NO_2}/p^{\ominus})^2}{(p_{N_2O_4}/p^{\ominus})} = \frac{(84\text{kPa}/100\text{kPa})^2}{116\text{kPa}/100\text{kPa}} = 0.61$$

$Q_p > K^{\ominus}$, the equilibrium shifts left. System pressure is increased, the reaction will move to the side with fewer number of gas.

(2) $N_2O_4(g) \rightleftharpoons 2NO_2(g)$

	$N_2O_4(g)$	$2NO_2(g)$
Initial amount/(mol)	1.0	0.10
Equilibrium amount/(mol)	1.0-0.15	0.10+2×0.15

The total amount at equilibrium: $n = (1.0 - 0.15)\text{mol} + (0.10 + 2 \times 0.15)\text{mol} = 1.25\text{mol}$

$$p_{N_2O_4} = x_{N_2O_4} p_{\text{total}} = \frac{n(N_2O_4)}{n} p_{\text{total}} = \frac{(1.0-0.15)\text{mol}}{1.25\text{mol}} \times 200\text{kPa} = 136\text{kPa}$$

$$p_{NO_2} = p_{\text{tota}} - p_{N_2O_4} = 200\text{kPa} - 136\text{kPa} = 64\text{kPa}$$

2.3 Effect of Temperature on Chemical Equilibrium

Changes in concentration and pressure shift the chemical equilibrium by changing the reaction quotient Q to make $Q \neq K^{\ominus}$. Then equilibrium shifts until $Q = K^{\ominus}$. Changes in temperature shift the chemical equilibrium by changing $K^{\ominus}$. The effect of temperature on the equilibrium constant is related to the change in enthalpy (焓变) of the chemical reaction.

The effect of temperature on standard equilibrium constant can be derived through chemical thermodynamics (化学热力学)

$$\ln\frac{K_2^{\ominus}}{K_1^{\ominus}} = \frac{\Delta_r H_m^{\ominus}}{R}\left(\frac{1}{T_1} - \frac{1}{T_2}\right) = \frac{\Delta_r H_m^{\ominus}}{R}\left(\frac{T_2 - T_1}{T_1 T_2}\right) \tag{3-7}$$

(3-7) is called the Van't Hoff equation. $K_1^{\ominus}$ and $K_2^{\ominus}$ represent the standard equilibrium constants at T_1 and T_2, respectively. R is gas constant, R=8.314J/(mol · K). $\Delta_r H_m^{\ominus}$ is the average molar enthalpy change (平均摩尔焓变) in T_1~T_2 temperature scope. In fact, according to the thermodynamic derivation, when the temperature changes little, that is, the average molar enthalpy change of the reaction can be regarded as a constant.

It can be clearly seen from the above equation that the relationship between the temperature change and the standard equilibrium constant and the reaction heat is as follows:

$\Delta_r H_m^{\ominus} < 0$, is exothermic reaction (放热反应), when $T_2 > T_1$, $K_2^{\ominus} < K_1^{\ominus}$, the equilibrium moves in the reverse direction; when $T_2 < T_1$, $K_2^{\ominus} < K_1^{\ominus}$, the equilibrium moves in the forward direction.

$\Delta_r H_m^{\ominus} > 0$, is an endothermic reaction (吸热反应), when $T_2 > T_1$, $K_2^{\ominus} > K_1^{\ominus}$, the equilibrium moves in the forward direction; when $T_2 < T_1$, $K_2^{\ominus} < K_1^{\ominus}$, the equilibrium moves in the reverse direction.

In short, if you raise the temperature, the equilibrium moves in the direction of absorption; Lower the temperature, and the equilibrium moves in the exothermic direction.

With above formula, the change of enthalpy $\Delta_r H_m^{\ominus}$ of a reaction can be obtained by the standard equilibrium constants at two different temperatures. The change in enthalpy of the chemical reaction and the standard equilibrium constant at known temperature can also be used to obtain the standard equilibrium constant at another temperature.

Example 3-8 At 800K, for the reaction: $2SO_2(g)+O_2(g) \rightleftharpoons 2SO_3(g)$, $K^{\ominus} = 910$, $\Delta_r H_m^{\ominus} = -197.8$kJ/mol. Calculate the $K^{\ominus}$ at 900K.

We assume that the effect of temperature on $\Delta_r H_m^{\ominus}$ of this reaction is negligible.

Soluiton

$$\ln\frac{K^{\ominus}(900\text{K})}{K^{\ominus}(800\text{K})}=\ln\frac{K^{\ominus}(900\text{K})}{910}=\frac{\Delta_r H^{\ominus}_{m,298.15\text{K}}}{8.314}\left(\frac{900-800}{800\times 900}\right)$$

$$\ln K^{\ominus}(900\text{K})=\ln 910+\frac{(-197.78\times 1000)}{8.314}\times\frac{100}{800\times 900}$$

$$K^{\ominus}=33.4$$

2.4 Effect of Catalyst

Adding a catalyst to a system changes the rate of the reaction, but this does not shift the equilibrium in favor of either products or reactants. Because a catalyst affects the activation energy of both forward and reverse reactions equally, it changes both rate constants by the same factor, so their ratio, $K^{\ominus}$, does not change.

So, adding a catalyst to a reaction at equilibrium has no effect on the amounts of reactants or products; it changes neither Q nor $K^{\ominus}$.

The same equilibrium mixture is achieved with or without the catalyst, but the equilibrium is established more quickly in the presence of a catalyst.

重点小节

1. 标准平衡常数 反应达到平衡时，系统中生成物浓度（以反应方程式中的化学计量数为指数的幂）的乘积与反应物浓度（以反应方程式中的化学计量数为指数的幂）的乘积之比为一常数，此常数称为化学平衡常数，用 K 表示，这一规律称为化学平衡定律。若把平衡浓度除以标准浓度 $c^{\ominus}(c^{\ominus}=1\text{mol/L})$，得到的比值称为相对平衡浓度（$c_{\text{eq,B}}/c^{\ominus}$），但为了书写方便，习惯上用 [B] 表示相对平衡浓度。若将平衡浓度用相对平衡浓度或相对平衡分压（$c_{\text{eq,B}}/p^{\ominus}$，$p^{\ominus}$ 为标准压力，$p^{\ominus}=100\text{kPa}$）表示，则该常数称作标准平衡常数，用 $K^{\ominus}$ 表示，它的量纲为 1。

标准平衡常数的应用：①判断反应的进行程度；②预测反应的进行方向。

2. 多重平衡规则 在同一温度下，将几个独立反应相加，所得总反应的标准平衡常数等于相加反应的标准平衡常数之乘积。将几个独立反应相减，所得反应的标准平衡常数等于这几个反应标准平衡常数之商。

3. 化学平衡的移动 因外界条件改变，使可逆反应从一种平衡状态向另一种平衡状态转变的过程，称为化学平衡的移动，浓度、压力、温度等外界因素都会使化学平衡发生移动，但催化剂不能使平衡移动，它通过改变反应机理、降低活化能而加快反应速率。根据反应商 Q 与 $K^{\ominus}$ 大小的比较，可以推测化学平衡移动的方向。

（1）当 $Q=K^{\ominus}$，则化学反应达到平衡。

（2）当 $Q<K^{\ominus}$，化学反应未达到平衡，平衡应正向自发进行，直至 $Q=K^{\ominus}$。

（3）当 $Q>K^{\ominus}$，化学反应未达到平衡，平衡应逆向自发进行，直至 $Q=K^{\ominus}$。

4. Le Châtelier 原理 改变平衡体系的条件，如浓度、压力、温度，体系将向削弱这个改变的方向移动。

题库

Exercises

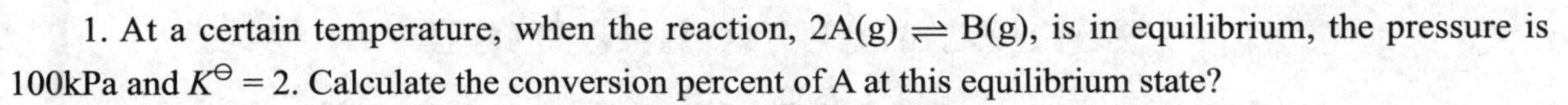

1. At a certain temperature, when the reaction, $2A(g) \rightleftharpoons B(g)$, is in equilibrium, the pressure is 100kPa and $K^{\ominus} = 2$. Calculate the conversion percent of A at this equilibrium state?

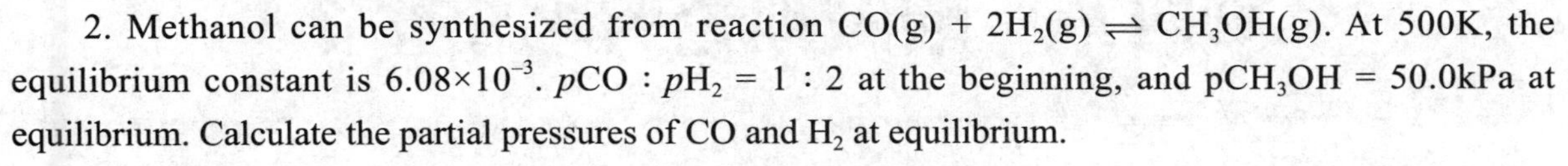

2. Methanol can be synthesized from reaction $CO(g) + 2H_2(g) \rightleftharpoons CH_3OH(g)$. At 500K, the equilibrium constant is 6.08×10^{-3}. pCO : pH$_2$ = 1 : 2 at the beginning, and pCH$_3$OH = 50.0kPa at equilibrium. Calculate the partial pressures of CO and H_2 at equilibrium.

3. 1mol CO(g) and 1mol $H_2O(g)$ are filled in a 2L container. When reaction $CO(g) + H_2O(g) \rightleftharpoons CO_2(g) + H_2(g)$ get equilibrium at 800℃, the concentration of CO is 0.25mol/L. ① Calculate the equilibrium constant at this temperature. ② Keep the temperature constant, if 1mol CO(g) and 4mol $H_2O(g)$ are filled in this container, calculate the concentrations and conversion percent of CO at equilibrium.

4. At a particular temperature, 2mol $PCl_5(g)$ and 1mol $PCl_3(g)$ are filled in a sealed vessel. When reaction PCl_5 (g) $\rightleftharpoons$ PCl_3 (g) + Cl_2 (g) get equilibrium, the total pressure is 202kPa and the conversion percent of PCl_5 is 91%. Calculate the equilibrium constant at this temperature.

5. The equilibrium constants of reaction $CaCO_3(s) \rightleftharpoons CaO(s) + CO_2(g)$ are 3.0×10^{-2} and 1.0 at 973 K and 1173K, respectively. Calculate the equilibrium constant at 1073K.

目 标 检 测

1. 某特定温度下，100kPa 时某反应 $2A(g) \rightleftharpoons B(g)$ 达到平衡，其标准平衡常数 $K^{\ominus} = 2$。求 A 的平衡转化率是多少？

2. 甲醇可以通过反应 $CO(g) + 2H_2(g) \rightleftharpoons CH_3OH(g)$ 合成，500K 时 反应的标准平衡常数 $K^{\ominus} = 6.08\times10^{-3}$。初始时 CO 和 H_2 的分压比为 1 : 2，平衡时 CH_3OH 的分压为 50.0kPa，计算平衡时 CO 和 H_2 的分压各是多少。

3. 将 1mol CO(g) 和 1mol $H_2O(g)$ 一并引入到体积为 2L 的密闭容器中，当反应 $CO(g) + H_2O(g) \rightleftharpoons CO_2(g) + H_2(g)$ 在 800 度达到平衡时，CO 的浓度为 0.25mol/L。①求此温度下的平衡常数；②若保持体系温度不变，将 1mol CO(g) 和 4mol $H_2O(g)$ 引入上述容器，求平衡时的 CO 浓度和转化率。

4. 在某一特定温度下，2mol $PCl_5(g)$ 和 1mol $PCl_3(g)$ 同时被引入到一恒容密封体系，发生反应 $PCl_5(g) \rightleftharpoons PCl_3(g) + Cl_2(g)$，反应平衡时体系的总压为 202kPa，$PCl_5$ 的转化率为 91%，求该温度下反应的标准平衡常数 $K^{\ominus}$。

5. 反应 $CaCO_3(s) \rightleftharpoons CaO(s) + CO_2(g)$ 在 973 和 1173K 的平衡常数分别为 3.0×10^{-2} 和 1.0，求 1073K 时该反应的平衡常数。

Chapter 4 Theories of Acids Bases and Ionization Equilibrium
第四章 酸碱理论与电离平衡

学习目标

知识要求

1. **掌握** 电离平衡常数的意义和一元弱电解质的电离平衡及有关计算；缓冲溶液的作用原理及有关计算；同离子效应和盐效应的概念、共轭酸碱对的概念

2. **熟悉** 多元弱酸分步电离及近似计算

3. **了解** 酸碱质子论和电子论

能力要求

灵活运用稀释定律及简化公式对一元弱酸弱碱电离平衡进行相关计算；选择合适的缓冲对配制缓冲溶液。

1. Theories of Acids and Bases

PPT

In this content, we develop three definitions of acids and bases that allow us to understand the theories of acids and bases. Acids and bases have been used as laboratory chemicals for centuries, as well as in the home. Common household acids include acetic acid (CH_3COOH, vinegar), citric acid (枸橼酸) ($H_3C_6H_5O_7$, in citrus fruits), and phosphoric acid (H_3PO_4, a flavoring in carbonated beverages). Sodium hydroxide (NaOH, drain cleaner) and ammonia (NH_3, glass cleaner), are household bases. The original distinction between acids and bases was based on criteria of taste and feel: acids were sour and bases felt soapy, but this approach is dangerous and unregulated.

1.1 Ionic Acid-Base Theory—Arrhenius Theory

Svante Arrhenius, Swedish chemist, proposed the definition of acids and bases firstly in 1884. An acid can be defined as a substance that gives hydrogen ions (proton) in aqueous solution and no other cations. A base gives hydroxide ions in aqueous solution and no other anions. This theory first clarified

the concept of acid and base. However, it has its limitations: one of the problems is in its treatment of the weak base ammonia, no OH^- can be found. Another limitation is that the reaction must involve water.

$$HCl(aq) \rightarrow H^+(aq) + Cl^-(aq)$$

$$NaOH(aq) \rightarrow Na^+(aq) + OH^-(aq)$$

1.2 Protonic acid-base theory—Brønsted-Lowry theory

In 1923, Johannes Brønsted in Denmark and Thomas Lowry in Great Britain independently proposed that the key process responsible for the properties of acids (and bases) was the transfer of an ion (a proton) from one substance to another. For example, when acids dissolve in water, ions are transferred from acid molecules to water molecules, as shown below for HCl and CH_3COOH.

$$HCl(aq) + H_2O(l) \rightarrow H_3O^+(aq) + Cl^-(aq)$$

$$CH_3COOH(aq) + H_2O(l) \rightleftharpoons H_3O^+(aq) + CH_3COO^-(aq)$$

In the above two equations, the acid molecules are acting as proton donors and the water molecules are acting as proton acceptors. According to the Brønsted-Lowry theory, an acid is a proton donor (质子给予体). And a base is a proton acceptor (质子接受体).

The definitions are sufficiently broad that any hydrogen-containing molecule or ion capable of releasing a proton, H^+, is an acid, whereas any molecule or ion that can accept a proton is a base.

Conjugate acid-base pairs (共轭酸碱对):

$$\text{Conjugate acid} \rightleftharpoons \text{proton} + \text{conjugate base}$$

$$HCl \rightleftharpoons H^+ + Cl^-$$

$$NH_4^+ \rightleftharpoons H^+ + NH_3$$

$$HCO_3^- \rightleftharpoons H^+ + CO_3^{2-}$$

$$H_2CO_3 \rightleftharpoons H^+ + HCO_3^-$$

$$H_2O \rightleftharpoons H^+ + OH^-$$

$$H_3O^+ \rightleftharpoons H^+ + H_2O$$

$$[Cr(H_2O)_6]^{3+} \rightleftharpoons H^+ + [Cr(H_2O)_5(OH)]^{2+}$$

In the above examples, each acid has a conjugate base (共轭碱), and each base has a conjugate acid (共轭酸). These conjugate pairs only differ by a proton (共轭酸碱对之间只相差一个质子). For example, NH_4^+ and NH_3 are referred to as a conjugate pair. NH_3 is considered as a base, NH_4^+ is the conjugate acid. Thus, a conjugate acid can lose an H^+ ion to form a base, and a conjugate base can gain an H^+ ion to form an acid. Water can be an acid or a base (amphiprotic, 两性的). It can gain a proton to become a hydronium ion (H_3O^+), its conjugate acid, or lose a proton to become the hydroxide ion (OH^-), its conjugate base. According to the Brønsted–Lowry definitions, any species that contains hydrogen can potentially act as an acid, and any compound that contains a lone pair of electrons can act as a base.

A Brønsted -Lowry acid-base reaction occurs when an acid and a base react to form their conjugate base and conjugate acid. The nature of an acid-base reaction is the transfer of a proton (H^+) between conjugate pairs (酸碱反应的实质是共轭酸碱对间质子的传递反应).

$$\underset{\text{Acid}_1}{HAc(aq)} \rightleftharpoons H^+(aq) + \underset{\text{Base}_1}{Ac^-(aq)}$$

$$H^+(aq) + \underset{\text{Base}_2}{H_2O(l)} \rightleftharpoons \underset{\text{Acid}_2}{H_3O^+(aq)}$$

$$HAc(aq) + H_2O(l) \xrightarrow{H^+} Ac^-(aq) + H_3O^+(aq)$$

$$\text{acid}_1 + \text{base}_2 \rightleftharpoons \text{base}_1 + \text{acid}_2$$

Relative acid-base strength determines the net direction of reaction (酸碱的相对强弱决定反应的方向)

The net direction of an acid-base reaction depends on relative acid and base strengths. A reaction proceeds to the greater extent in the direction in which a stronger acid and stronger base form a weaker acid and weaker base.

The net direction of the reaction of H_2S and NH_3 is to the right because H_2S is a stronger acid than NH_4^+, and NH_3 is a stronger base than HS^-.

$$H_2S + NH_3 \rightleftharpoons HS^- + NH_4^+$$

stronger acid + stronger base ⟶ weaker base + weaker acid

The process is a competition for the proton between the two bases, NH_3 and HS^-, in which NH_3 wins.

Based on evidence from many reactions, we can rank conjugate pairs in terms of the ability of the acid to transfer its proton. Note that a weaker acid has a stronger conjugate base (酸越弱，它的共轭碱的碱性越强): the acid can't give up its proton very readily because its conjugate base is holding it too strongly. For example, H_2O is the conjugate acid of OH^-. H_2O is a very weak acid. It's hard to give up a proton, because its conjugate base OH^- is a very strong base and hold the proton tightly. So the stronger the acid is, the weaker its conjugate base.

1.3 Electronic Acid-Base—Lewis Acids and Bases

In 1923, Professor G. N. Lewis (1875–1946) presented the most comprehensive acid–base theory. The Lewis acid base theory is not limited to reactions involving proton and it extends acid base concepts to reactions in gases and in solids. It is especially important in describing certain reactions between organic molecules.

A Lewis acid is a species (an atom, ion, or molecule) that is an electron-pair acceptor, and a Lewis base is a species that is an electron-pair donor (路易斯酸是电子对的接受体，它可能是一个原子、离子或者分子)；路易斯碱是地电子对的给予体. A reaction between a Lewis acid (A) and a Lewis base results in the formation of a covalent bond (共价键) between them. The product of a Lewis acid base reaction is called adduct (or addition compound). The reaction can be represented as:

$$B + A \rightarrow B-A$$

Here, B-A is the adduct (加合物). The formation of a covalent chemical bond by one species donating a pair of electrons to another is called coordination, and the bond joining the Lewis acid and Lewis base is called a coordinate covalent bond (配位共价键). Lewis acids are species with vacant orbitals (空轨道) that can accept electron pairs; Lewis bases are species with lone-pair electrons (孤对电子) that can share the electron pairs.

By these definitions, OH^-, a Brønsted Lowry base, is also a Lewis base because lone-pair electrons are present on the O atom. So too is NH_3 a Lewis base.

Species with an incomplete valence shell are Lewis acids. When the Lewis acid forms a coordinate covalent bond with a Lewis base, the octet is completed. A good example of octet completion is the reaction of BF_3 and NH_3.

$$H_3N: + BF_3 \longrightarrow H_3N{-}BF_3$$

The nature of acid-base reaction is the process of forming acid-base complex by coordination bond (路易斯酸碱反应的实质就是酸碱通过配位共价键形成配合物的过程). For example:

$$HNO_3 + :NH_3 \longrightarrow [H \leftarrow NH_3]^+ + NO_3^-$$

$$Zn^{2+} + 4:NH_3 \longrightarrow [Zn(\leftarrow NH_3)_4]^{2+}$$

$$H^+ + :OH^- \longrightarrow H \leftarrow OH$$

$$CH_3CO^+ + :OC_2H_5^- \longrightarrow CH_3CO \leftarrow OC_2H_5$$

Complex ions can also form between transition metal ions and other Lewis bases, such as NH_3. For instance, Zn^{2+} combines with NH_3 to form the complex ion $[Zn(NH_3)_4]^{2+}$. The central Zn^{2+} ion accepts electrons from the Lewis base NH_3 to form coordinate covalent bonds. It is a Lewis acid.

PPT

2. Proton Transfer Equilibrium of Weak Electrolyte

The weak electrolytes dissociate slightly into ions in the aqueous solution, and there is dissociation equilibrium between the dissociated ions and the undissociated molecules. Water is a very weak electrolyte.

2.1 Autoionization of Water and the pH Scale

Careful measurements show that pure water has a weak ability to conduct electricity. It shows that pure water can ionize very slightly into ions in an equilibrium process known as autoionization (or self-ionization) (自电离):

$$2H_2O(l) \rightleftharpoons H_3O^+(aq) + OH^-(aq)$$

We will often show it in the simplified notation:

$$H_2O(l) \rightleftharpoons H^+(aq) + OH^-(aq)$$

This autoionization (self-ionization) of water is an acid–base reaction according to the Brønsted-Lowry theory. Water is amphiprotic (两性的); That is, H_2O molecules can both donate and accept protons (水分子既可以作为质子的给予体，也可以作为质子的接受体).

The equilibrium constant (平衡常数) of water is called the ion product of water (水的离子积). It is symbolized as $K_W^\ominus$ at 25°C:

$$K_W^\ominus = [H^+][OH^-] = 1.00 \times 10^{-14}$$

$K_W^\ominus$ is just a function of temperature (温度的函数), so if the temperature is constant (恒定的), the value of $K_W^\ominus$ is constant. Since $K_W^\ominus$ is equilibrium constant, the product of the concentrations of the hydrogen ion (H^+, 氢离子) and hydroxide (氢氧根) ions must always equal $K_W^\ominus$. If the concentration of H^+ is increased by the addition of an acid, then the concentration of OH^- must decrease to maintain the value of $K_W^\ominus$, vice versa.

Both H^+ and OH^- ions are present in all aqueous systems (水溶液体系). Thus, all acidic solutions contain a low $[OH^-]$, and all basic solutions contain a low $[H^+]$. The equilibrium nature of autoionization allows us to define "acidic" and "basic" solutions in terms of relative magnitudes of $[H^+]$ and $[OH^-]$:

In an acidic solution, $[H^+] > [OH^-]$

In a neutral solution, $[H^+] = [OH^-]$

In a basic solution, $[H^+] < [OH^-]$

Expressing the hydronium ion concentration: the pH scale (氢离子浓度表示法— pH值)

Because $K_W^\ominus$ in an aqueous solution is only 10^{-14}, so the $[H^+]$ and $[OH^-]$ are very small. Now we want to consider a more convenient way to describe hydrogen ion and hydroxide ion concentrations. In 1909, the Danish biochemist Søren Sørensen proposed the term pH to refer to the potential of hydrogen ion. He defined pH as the negative of the logarithm of $[H^+]$:

$$pH = -\log[H^+]$$

In the same way, the pOH can be defined as: $pOH = -\log[OH^-]$

Another useful expression can be derived by taking the negative logarithm of the $K_W^\ominus$ expression (written for 25°C) and introducing the symbol $pK_W^\ominus$.

$$K_W^\ominus = [H_3O^+][OH^-] = 1.0 \times 10^{-14}$$

$$pK_W^\ominus = pH + pOH = 14$$

There are two ways to show the pH, one is the color of an acid base indicator; the other, the response of a pH meter (有两种方式表示 pH，一种是通过酸碱指示剂的颜色，另一种是通过 pH 计测量).

2.2 Proton Transfer Equilibrium for Weak Monoprotic Acid/Base

Acids and bases are categorized in terms of their strength, the degree to which they dissociate into ions in water. Strong acids and strong bases dissociate completely into ions. Weak acids and weak bases dissociate very little into ions, remaining mostly as initial molecules. Therefore, weak electrolytes mean that they conduct a small current.

In fact, weak acids are much more numerous than strong acids. Several weak acids are familiar to us. Vinegar is about 5% solution of acetic acid (CH_3COOH). Carbonated drinks (碳酸饮料) are saturated solutions of carbon dioxide in water, which produces carbonic acid. Citrus fruits contain citric acid, $C_3H_5O(COOH)_3$ (柠檬酸). Some ointments and powders used for medicinal purposes contain, H_3BO_3 boric acid (硼酸). Another important constituent of citrus fruit is ascorbic acid (抗坏血酸), or vitamin C, a dietary requirement to prevent scurvy.

Weak monoprotic acid can donate only one proton per dissociated molecule, and weak monobasic can accept only one proton per dissociated molecule.

In a dilute solution (稀溶液) of a weak acid, the great majority of HA molecules are undissociated. Thus, $[H^+] << c(HAc)_{init}$ and $[HA] \approx c(HAc)_{init}$, so K_a is very small.

Note: brackets with no subscript mean relative molar concentration at equilibrium; that is, $[H^+]$ means $[H^+]_{eq}$, [HA] means the concentration of HA at equilibrium.

2.2.1 Dissociation Constant, $K_a^\ominus$

We take the weak monoprotic acids HAc and NH_4^+ as the examples to show the proton transfer equilibrium in aqueous solution:

$$HAc+H_2O \rightleftharpoons H_3O^+ +Ac^- \qquad NH_4^+ +H_2O \rightleftharpoons H_3O^+ +NH_3$$

Simplified:

$$HAc \rightleftharpoons H^+ +Ac^- \qquad NH_4^+ +H_2O \rightleftharpoons H^+ +NH_3 \cdot H_2O$$

Expression of equilibrium constant:

$$K_a^\ominus = \frac{[H^+][Ac^-]}{[HAc]} \qquad K_a^\ominus = \frac{[H^+][NH_3 \cdot H_2O]}{[NH_4^+]}$$

$K_a^\ominus$ is the acid-dissociation constant or acid-ionization constant.

Similarly, there is proton transfer equilibrium of a weak base, $NH_3 \cdot H_2O$、Ac^- in solution:

$$NH_3+H_2O \rightleftharpoons OH^- +NH_4^+ \qquad Ac^- +H_2O \rightleftharpoons HAc+OH^-$$

Expression of equilibrium constant:

$$K_b^\ominus = \frac{[NH_4^+][OH^-]}{[NH_3]} \qquad K_b^\ominus = \frac{[HAc][OH^-]}{[Ac^-]}$$

$K_b^\ominus$ is the base-dissociation constant or base-ionization constant.

Like any equilibrium constant, $K_a^\ominus$ or $K_b^\ominus$ is a number whose magnitude tells how far to the right the reaction has proceeded to reach equilibrium. At equilibrium, for example, the stronger the acid, the higher $[H^+]$, and the larger the $K_a^\ominus$:

$$\text{Stronger acid} \Rightarrow \text{higher } [H_3O^+] \Rightarrow \text{larger } K_a^\ominus$$

The value of the $K_a^\ominus$ or $K_b^\ominus$ is only related to the temperature, not the concentration. However, the

effect of temperature on them is not significant. When the temperature changes little, the values at room temperature are usually adopted (当温度变化不大时，我们往往采用室温下的值).

Table 4-1 contains names, formulas, ionization constants, and $pK_a^{\ominus}$ and $pK_b^{\ominus}$ values for a few common weak acids and bases; Appendix 3 contains a longer list of $K_a^{\ominus}$ and $K_b^{\ominus}$ values.

Table 4-1 Ionization constants and $pK_a^{\ominus}$ and $pK_b^{\ominus}$ values for some weak monoprotic acids/bases at 25℃.

Acid	Ionization Rection	$K_a^{\ominus}$ (or $K_b^{\ominus}$)	$pK_a^{\ominus}$ (or $pK_b^{\ominus}$)
hydrofluoric acid	$HF + H_2O \rightleftharpoons H_3O^+ + F^-$	6.31×10^{-4}	3.20
nitrous acid	$HNO_2 + H_2O \rightleftharpoons H_3O^+ + NO_2^-$	5.62×10^{-4}	3.25
acetic acid	$CH_3COOH + H_2O \rightleftharpoons H_3O^+ + CH_3COO^-$	1.75×10^{-5}	4.765
hypochlorous acid	$HClO + H_2O \rightleftharpoons H_3O^+ + ClO^-$	3.98×10^{-8}	7.40
hydrocyanic acid	$HCN + H_2O \rightleftharpoons H_3O^+ + CN^-$	6.17×10^{-10}	9.21
ammonia	$NH_3 + H_2O \rightleftharpoons OH^- + NH_4^+$	1.74×10^{-5}	4.76
hydroxylamine	$NH_2OH + H_2O \rightleftharpoons OH^- + NH_3OH^+$	9.12×10^{-9}	8.04

Then what's the relation between $K_a^{\ominus}$ and $K_b^{\ominus}$ of a conjugate acid-base pair?

A key relationship exists between the $K_a^{\ominus}$ of HA and the $K_b^{\ominus}$ of A^-, which we can see by writing the two reactions as a reaction sequence and adding them:

$$\cancel{HA} + H_2O \rightleftharpoons H_3O^+ + \cancel{A^-}$$

$$\cancel{A^-} + H_2O \rightleftharpoons \cancel{HA} + OH^-$$

$$2H_2O \rightleftharpoons H_3O^+ + OH^-$$

The sum of the two dissociation reactions is the autoionization of water. Recall from Chapter 2 that, for a reaction that is the sum of two or more reactions, the overall equilibrium constant is the product of the individual equilibrium constants (如果一个反应是两个或多个反应的加和，那么这个反应的平衡常数就是这两个或多个反应平衡常数的乘积). Therefore, writing the equilibrium expressions for each reaction gives

$$\frac{[H_3O^+][\cancel{A^-}]}{[\cancel{HA}]} \times \frac{[\cancel{HA}][OH^-]}{[\cancel{A^-}]} = [H_3O^+][OH^-]$$

$$\text{or} \qquad K_a^{\ominus} \times K_b^{\ominus} = K_w^{\ominus} \tag{4-1}$$

This relationship allows us to find $K_a^{\ominus}$ of the acid in a conjugate pair from the $K_b^{\ominus}$ of the base, and vice versa. So you can calculate them by looking up the value for the molecular conjugate species and relating it to $K_W^{\ominus}$. To find the $K_b^{\ominus}$ value for Ac^-, for instance, we look up the $K_a^{\ominus}$ value for HAc and apply (4-1):

$$K_a^{\ominus} \text{ of HAc} = 1.75\times10^{-5} \quad \text{(from Appendix 3)}$$

$$\text{So, we have } K_a^{\ominus} \text{ of HAc} \times K_b^{\ominus} \text{ of } Ac^- = K_W^{\ominus}$$

$$\text{or} \quad K_b^{\ominus} \text{ of } Ac^- = \frac{K_w^{\ominus}}{K_a^{\ominus} \text{ of HAc}} = \frac{1.0\times10^{-14}}{1.75\times10^{-5}} = 5.71\times10^{-10}$$

The similar result can be obtained by multiplying together the expressions for the acidity constant of NH_4^+ and the basicity constant of NH_3. The implication of the (4-1) is that the larger the value of $K_b^\ominus$, the smaller the value of $K_a^\ominus$(式4-1说明共轭碱的碱常数越大，其共轭酸的酸常数越小). That is, the stronger the base, the weaker is its conjugate acid. A further implication of (4-1) is that the $K_b^\ominus$ of base can be obtained in terms of the $K_a^\ominus$ of its conjugate acid.

Acidity or basicity constants are always very small, it is convenient to show them as their negative of the logarithms like pH.

$$pK = -\log K$$

It follows from this definition and the relation in (4-1) that

$$pK_a^\ominus + pK_b^\ominus = pK_w^\ominus$$

2.2.2 The Extent of Dissociation for Weak Acid/Base

The extent of dissociation of weak acid and base in solution can be expressed by the degree of dissociation (解离度)(the symbol of α).

$$\alpha = \frac{\text{mole of dissociated electrolyte}}{\text{mole of the initial electrolyte}} \times 100\%$$

Because the dissociated and initial electrolytes are in the same solution with the same volume, so the above formula can also be written as following:

$$\alpha = \frac{\text{concentration of dissociated electrolyte}}{\text{concentration of the initial electrolyte}} \times 100\%$$

In fact, in 0.1mol/L CH_3COOH at 25°C, only about 1.3×10^{-3} mol/L of the acid molecules dissociates into ions, so the extent of dissociation is only 1.3%. A similarly small percentage of ammonia molecules form ions when 0.1mol/L NH_3 reacts with water.

α is used to show the extent of the dissociation of electrolytes. α depends not only on the nature of the electrolyte but also on external factors such as the solvent, temperature, and concentration of the solution. With the same temperature and same concentration, the smaller is α, the weaker is the electrolyte. For the same weak electrolyte solute, more dilute solution has a greater degree of dissociation (对于相同的弱电解质溶液，越稀越电离).

The range of $K_a^\ominus$ values are as follows.

Acid-dissociation constants of weak acids range over many orders of magnitude. Most weak acids have $K_a^\ominus$ values ranging from about 10^{-2} to 10^{-10}. According to the $K_a^\ominus$ values, we can classify weak acids into the following cases.

(1) For a weak acid with a relatively high $K_a^\ominus$ (~10^{-2}), a 1mol/L solution has ~10% of the HA molecules dissociated. The $K_a^\ominus$ of chlorous acid ($HClO_2$) is 1.1×10^{-2}, and 1mol · L^{-1} $HClO_2$ is 10% dissociated.

(2) For a weak acid with a moderate $K_a^\ominus$(~10^{-5}), a 1mol/L solution has ~0.3% of the HA molecules dissociated. The $K_a^\ominus$ of acetic acid (CH_3COOH) is 1.8×10^{-5}, and 1mol/L CH_3COOH is 0.42% dissociated.

(3) For a weak acid with a relatively low $K_a^\ominus$(~10^{-10}), a 1mol/L solution has ~0.001% of the HA molecules dissociated. The $K_a^\ominus$ of HCN is 6.2×10^{-10}, and 1mol·L^{-1} HCN is ~0.0025% dissociated.

Thus, for solutions of the same initial HA concentration, the smaller the $K_a^\ominus$, the lower the percent dissociation of HA:

$$\text{Weaker acid} \Rightarrow \text{lower \% dissociation of HA} \Rightarrow \text{smaller } K_a^\ominus$$

2.2.3 Rule of Dilution (稀释定律)

Both dissociation constant and dissociation degree can reflect the extent of dissociation of weak electrolytes, but there are connection and difference between them. The dissociation constant belongs to the chemical equilibrium constants that depend only on temperature and not on concentration; Dissociation degree is a form of conversion rate that represents the dissociation percentage of a weak electrolyte under certain conditions, depending on both temperature and concentration.

We often use HA as a general representation for a monoprotic acid and A^- for its conjugate base. The relationship between dissociation degree and dissociation constant can be deduced as follows:

We assume the initial concentration of weak monoprotic acid HA as *c* mol/L, the dissociation constant as $K_a^\omines$, and the dissociation degree as α.

For the aqueous solution of weak monoprotic acid, HA, there are the following two proton transfer equilibriums:

$$HA \rightleftharpoons H^+ + A^- \qquad K_a^\ominus = \frac{[H^+][A^-]}{[HA]}$$

$$H_2O \rightleftharpoons H^+ + OH^- \qquad K_W^\ominus = [H^+][OH^-]$$

Here the total $[H^+]$ in the solution comes from ionization of water and HA, and as the $[H^+]$ from water is very small. So if $c \cdot K_a^\ominus \geqslant 20K_W^\ominus$, the $[H^+]_{water}$ can be neglected. If the initial concentration of HA is *c* mol/L, the degree of dissociation is α:

	$HA \rightleftharpoons H^+ + A^-$		
Relative initial concentration	c	0	0
Relative equlibrium concentration	$c\text{-}c\alpha$	$c\alpha$	$c\alpha$

$$K_a^\ominus = \frac{[H^+][A^-]}{[HA]} = \frac{c\alpha^2}{1-\alpha}$$

$$\text{If } \alpha < 5\%, 1-\alpha \approx 1,\ K_a^\ominus = c\alpha^2 \quad \alpha = \sqrt{\frac{K_a^\ominus}{c}} \tag{4-2}$$

If $\alpha > 5\%$, we need to solve the quadratic equation:

$$c\alpha^2 + K_a^\ominus\alpha - K_a^\ominus = 0$$

$$\alpha = \frac{-K_a^\ominus \pm \sqrt{K_a^\ominus + 4cK_a^\ominus}}{2c}$$

Similarly, for weak monoprotic base, if $\alpha < 5\%$, it can be deduced that:

$$K_b^\ominus = c\alpha^2 \qquad \alpha = \sqrt{\frac{K_b^\ominus}{c}} \tag{4-3}$$

(4-2) and (4-3) are named as the rule of dilution: the extent of dissociation of a weak electrolyte is inversely proportion to the square root of the concentration (稀释定律：弱电解质的解离度与弱电解质浓度的平方根成反比).

In many cases, and of course include the above, we can use chemical "common sense" to make an assumption that simplifies the math by avoiding the need to use the quadratic formula to find *x* (here, $x=\alpha$). In general, if a reaction has a relatively small *K* and a relatively large initial reactant concentration, the degree of dissociation will be very small. If x is so neglected can be ignored, then the calculation is very simple. To justify the assumption that x is negligible, we make sure the error introduced is not significant. One common criterion for "significant" is the 5% rule: if the assumption results in a change that is less than 5% of the initial concentration, the error is not significant, and the assumption is justified.

So, if $\alpha < 5\%$, the concentration change (x) can often be neglected, i.e. $1 - \alpha \approx 1$.

Example 4-1 what are the pH and the initial concentration for the HAc solution with $\alpha = 1.00\%$? $K_a^\ominus = 1.75 \times 10^{-5}$.

Solution For $\alpha = 1.00\% < 5\%$, we can use equation (4-2) to calculate:

$$c = \frac{K_a^\ominus}{\alpha^2} = \frac{1.75\times10^{-5}}{0.0100^2} = 0.175$$

$$[H^+] = c\alpha = 0.175 \times 1.00\% = 1.75\times10^{-3}$$

$$pH = -\lg[H^+] = -\lg 1.75\times10^{-3} = 2.76$$

$$c(HAc)_{init} = 0.175 mol/L,\ pH = 2.76.$$

2.2.4 Simplifying Calculation about Weak Monoprotic Acid/Base

The simplifying assumption can also be used to determine the pH from initial concentrations and K_c.

For a weak monoprotic acid HA with the initial concentration of c mol/L, and if $c \cdot K_a^\ominus \geqslant 20K_W^\ominus$, the $[H^+]$ from autoionization of water is negligible and neglect it, so:

	HA	⇌ H⁺	+ A⁻
The initial concentration:	c	0	0
The equilibrium concentration:	c-[H⁺]	[H⁺]	[H⁺]

$$K_a^\ominus = \frac{[H^+][A^-]}{[HA]} = \frac{[H^+]^2}{c-[H^+]} \tag{4-4}$$

We need to solve the quadratic equation:

$$[H^+]^2 + K_a^\ominus \cdot [H^+] - c \cdot K_a^\ominus = 0$$

$$[H^+] = -\frac{K_a^\ominus}{2} + \sqrt{\frac{K_a^{\ominus 2}}{4} + c \cdot K_a^\ominus}$$

If $\alpha < 5\%$, i.e., $\frac{c}{K_a^\ominus} < 400$, then $c - [H^+] \approx c$, (4-4) can be simplified as the following:

$$K_a^\ominus = \frac{[H^+]^2}{c} \tag{4-5}$$

$$[H^+] = \sqrt{c \cdot K_a^\ominus} \tag{4-6}$$

(4-6) is an approximate formula for the calculation of $[H^+]$ of weak monoprotic acid, and it is named as the simplest formula.

In the same way, the approximate formula for calculating the $[OH^-]$ of weak monoprotic base can be deduced as:

$$[OH^-] = \sqrt{c \cdot K_b^\ominus} \tag{4-7}$$

Here, if a reaction starts with a high $[reactant]_{init}$ and proceeds very little to reach equilibrium (small

K), the reactant concentration at equilibrium, [reactant]$_{eq}$, will be nearly the same as [reactant]$_{init}$. This assumption is reasonable.

Example 4-2 What are the pH and the degree of dissociation of the following solutions? ($K_b^\ominus = 1.74 \times 10^{-5}$)

(1) 0.100mol/L (2) 1.00×10^{-3}mol/L

Solution (1) $c = 0.100$mol/L, $K_b^\ominus = 1.74 \times 10^{-5}$

$$\because\ c \cdot K_b^\ominus = 1.74 \times 10^{-6} > 20K_W^\ominus,\ \frac{c}{K_b^\ominus} = \frac{0.100}{1.74 \times 10^{-5}} > 400$$

$$\therefore\ [OH^-] = \sqrt{c \cdot K_b^\ominus} = \sqrt{0.100 \times 1.74 \times 10^{-5}} = 1.32 \times 10^{-3}$$

$$pOH = -\lg[OH^-] = -\lg 1.32 \times 10^{-3} = 2.88$$

$$pH = 14 - p\,OH = 11.12$$

$$\alpha = \frac{[OH^-]}{c} \times 100\% = \frac{1.32 \times 10^{-3}}{0.100} \times 100\% = 1.32\%$$

(2) 1.00×10^{-3}mol/L, $c \cdot K_b^\ominus = 1.74 \times 10^{-8} > 20K_W^\ominus,\ \frac{c}{K_b^\ominus} = \frac{1.00 \times 10^{-3}}{1.74 \times 10^{-5}} < 400$

We need to solve the quadratic equation:

$$[OH^-] = -\frac{K_b^\ominus}{2} + \sqrt{\frac{K_b^{\ominus 2}}{4} + c \cdot K_b^\ominus}$$

$$= -\frac{1.74 \times 10^{-5}}{2} + \sqrt{\frac{(1.74 \times 10^{-5})^2}{4} + 1.00 \times 10^{-3} \times 1.74 \times 10^{-5}}$$

$$= 1.23 \times 10^{-4}\ \text{mol/L}$$

$$pOH = -\lg[OH^-] = -\lg 1.23 \times 10^{-4} = 3.91$$

$$pH = 14 - pOH = 10.09$$

$$\alpha = \frac{[OH^-]}{c} \times 100\% = \frac{1.23 \times 10^{-4}}{1.00 \times 10^{-3}} \times 100\% = 12.3\%$$

Example 4-3 What is the pH of the mixed solution of equal volume 0.200mol/L HAc and 0.200mol/L NaOH?

Solution Because the mixed solution is from HAc and NaOH with equal molarity and equal volume, so the HAc and NaOH can exactly react to form NaAc. Ac^- is the conjugated base of HAc, which is a weak monoprotic base.

$$K_b^\ominus(Ac^-) = \frac{K_W^\ominus}{K_a^\ominus(HAc)} = \frac{1.00 \times 10^{-14}}{1.75 \times 10^{-5}} = 5.71 \times 10^{-10}$$

$$\because\ c = \frac{0.200\text{mol/L}}{2} = 0.100\text{mol/L},\ c \cdot K_b^\ominus = 5.71 \times 10^{-11} > 20K_W^\ominus,\ \frac{c}{K_b^\ominus} = \frac{0.100}{5.71 \times 10^{-10}} > 400$$

$$[OH^-] = \sqrt{c \cdot K_b^\ominus} = \sqrt{0.100 \times 5.71 \times 10^{-10}} = 7.56 \times 10^{-6}$$

$$pOH = -\lg[OH^-] = -\lg 7.56 \times 10^{-6} = 5.12$$

$$pH = 14 - pOH = 8.88$$

Example 4-4 what is the pH of the solution of 0.100mol/L NH_4Cl and the degree of dissociation, α of NH_4^+?

Solution NH_4Cl is a strong electrolyte, and can ionize completely to NH_4^+ and Cl^-. NH_4^+ is the conjugated acid of $NH_3 \cdot H_2O$. Its acid-ionization constant is:

$$K_a^\ominus = \frac{K_W^\ominus}{K_b^\ominus} = \frac{1.00\times10^{-14}}{1.74\times10^{-5}} = 5.75\times10^{-10}$$

$\because c = 0.100\text{mol/L}$, $c\cdot K_a^\ominus = 5.75\times10^{-11} > 20K_W^\ominus$,

$\frac{c}{K_a^\ominus} = \frac{0.100}{5.75\times10^{-10}} > 400$,

$[H^+] = \sqrt{c\cdot K_a^\ominus} = \sqrt{0.100\times5.75\times10^{-10}} = 7.58\times10^{-6}$

$pH = 5.12$

$\alpha = \frac{[H^+]}{c}\times100\% = \frac{7.58\times10^{-6}}{0.100}\times100\% = 7.58\times10^{-3}\%$

Hydrolysis (水解): In fact, the other large group of Brønsted-Lowry bases consists of anions of weak acids. And Brønsted-Lowry acids also consist of cations of weak bases. Here, AC^- is the conjugate base of HAc and NH_4^+ is the conjugate acid of $NH_3 \cdot H_2O$ in Brønsted-Lowry theory. However, in the Arrhenius theory, AC^- and NH_4^+ belong to salts. The above two equilibrium reactions are called hydrolysis reactions (水解反应).

In pure water at 25°C, $[H^+] = [OH^-] = 1\times10^{-7}$mol/L and pH=7. Pure water is pH neutral. When NaCl dissolves in water at 25°C, complete dissociation into Na^+ and Cl^- ions occurs, and the pH of the solution remains 7. We can represent this fact with the equation:

$$Na^+ + Cl^- + H_2O \rightarrow \text{no reaction}$$

As shown in example 4-3, when Ac^- is added to water, the pH rises above 7. This means that $[OH^-] > [H^+]$ in the solution. A reaction producing OH^- must occur.

$$Na^+ + H_2O \rightarrow \text{no reaction}$$

$$CH_3COO^- + H_2O \rightleftharpoons CH_3COOH + OH^-$$

The reaction between CH_3COO^- and H_2O is fundamentally no different from other acid base reactions. A reaction between an ion and water is often called a hydrolysis reaction. We say that acetate ion hydrolyzes (水解) (and sodium ion does not).

When NH_4Cl is dissolved in water, the pH falls below 7 (see example 4-4). This means $[H^+] > [OH^-]$ in the solution. Here, ammonium ion hydrolyzes.

$$Cl^- + H_2O \rightarrow \text{no reaction}$$

$$NH_4^- + H_2O \rightleftharpoons NH_3 + H_3O^+$$

Salts of strong bases and strong acids do not hydrolyze (for example, NaCl): for the solution, pH = 7.

Salts of strong bases and weak acids hydrolyze (for example, CH_3COONa). pH >7 (The anion acts as a base).

Salts of weak bases and strong acids hydrolyze (for example, NH_4Cl): pH < 7 (The cation acts as an acid).

Salts of weak bases and weak acids hydrolyze (for example, CH_3COONH_4). (The cations are acids, and the anions are bases. Whether the solution is acidic or basic, however, depends on the relative values of $K_a^\ominus$ and $K_b^\ominus$ for the ions.)

2.3 Proton Transfer Equilibrium of Weak polyprotic Acid/Base

Acids with more than one ionizable proton are polyprotic acids. Table 4-2 lists ionization constants for several polyprotic acids. Additional listings can be found in Appendix 3.

Table 4-2 Equilibria constants of weak polyprotic acids (298K)

	Ionization equations	$K_a^\ominus$
H_2S	$H_2S \rightleftharpoons H^+ + HS^-$	$K_{a1}^\ominus = 8.91\times10^{-8}$
	$HS^- \rightleftharpoons H^+ + S^{2-}$	$K_{a2}^\ominus = 1.00\times10^{-19}$
H_2SO_3	$H_2SO_3 \rightleftharpoons H^+ + HSO_3^-$	$K_{a1}^\ominus = 1.41\times10^{-2}$
	$HSO_3^- \rightleftharpoons H^+ + SO_3^{2-}$	$K_{a2}^\ominus = 6.31\times10^{-8}$
H_2CO_3	$H_2CO_3 \rightleftharpoons H^+ + HCO_3^-$	$K_{a1}^\ominus = 4.47\times10^{-7}$
	$HCO_3^- \rightleftharpoons H^+ + CO_3^{2-}$	$K_{a2}^\ominus = 4.68\times10^{-11}$
H_3PO_4	$H_3PO_4 \rightleftharpoons H^+ + H_2PO_4^-$	$K_{a1}^\ominus = 6.92\times10^{-3}$
	$H_2PO_4^- \rightleftharpoons H^+ + HPO_4^{2-}$	$K_{a2}^\ominus = 6.17\times10^{-8}$
	$HPO_4^{2-} \rightleftharpoons H^+ + PO_4^{3-}$	$K_{a3}^\ominus = 4.79\times10^{-13}$

In solution, each dissociation step has a different $K_a^\ominus$. For example, H_2S is a biprotic acid (two ionizable protons, 二元酸), so it has two $K_a^\ominus$ values:

Step 1: $H_2S + H_2O \rightleftharpoons H_3O^+ + HS^-$

Simplified: $H_2S \rightleftharpoons H^+ + HS^-$ $\quad K_{a1}^\ominus = \dfrac{[H^+][HS^-]}{[H_2S]} = 8.91\times10^{-8}$

Step 2: $HS^- + H_2O \rightleftharpoons H_3O^+ + S^{2-}$

Simplified: $HS^- \rightleftharpoons H^+ + S^{2-}$ $\quad K_{a2}^\ominus = \dfrac{[H^+][S^{2-}]}{[HS^-]} = 1.00\times10^{-19}$

Another example, phosphoric acid is a triprotic acid (three ionizable protons, 三元酸), so it has three K_a values:

Step 1: $PO_4^{3-} + H_2O \rightleftharpoons HPO_4^{2-} + OH^-$

$$K_{b1}^\ominus = \frac{[HPO_4^{2-}][OH^-]}{[PO_4^{3-}]} = \frac{K_W^\ominus}{K_{a3}^\ominus} = \frac{1.00\times10^{-14}}{4.79\times10^{-13}} = 2.09\times10^{-2}$$

Step 2: $HPO_4^{2-} + H_2O \rightleftharpoons H_2PO_4^- + OH^-$

$$K_{b2}^\ominus = \frac{[H_2PO_4^-][OH^-]}{[HPO_4^{2-}]} = \frac{K_W^\ominus}{K_{a2}^\ominus} = \frac{1.00\times10^{-14}}{6.17\times10^{-8}} = 1.62\times10^{-7}$$

Step 3: $H_2PO_4^- + H_2O \rightleftharpoons H_3PO_4 + OH^-$

$$K_{b3}^{\ominus}=\frac{[H_3PO_4][OH^-]}{[H_2PO_4^-]}=\frac{K_W^{\ominus}}{K_{a1}^{\ominus}}=\frac{1.00\times10^{-14}}{6.92\times10^{-3}}=1.45\times10^{-12}$$

For H_3PO_4, the relative K_a values show that H_3PO_4 is a much stronger acid than $H_2PO_4^-$, which is much stronger than HPO_4^{2-}. So does H_2S. Note that the general pattern seen for H_3PO_4 occurs for all polyprotic acids (see the list in Appendix 3):

$$K_{a1}^{\ominus} >> K_{a2}^{\ominus} >> K_{a3}^{\ominus}$$

This trend occurs because it is more difficult for an H^+ ion to leave a singly charged anion (such as $H_2PO_4^-$) than to leave a neutral molecule (such as H_3PO_4), and more difficult still for it to leave a doubly charged anion (such as HPO_4^{2-}). Successive $K_{a1}^{\ominus}$ values typically differ by several orders of magnitude.

In addition to the multi-step proton transfer equilibrium, the solution of multiple weak acids (bases) also has proton self-transfer equilibrium of water, which belongs to the system of multiple equilibrium reactions and obeys the rules of multiple equilibrium rules. The concentration of H^+ or OH^- in solution is the sum of the concentration of H^+ and OH^- provided by all equilibrium reactions. However, since the proton transfer equilibrium reactions in the second and third steps as well as water self-ionization are very weak, only the first step of proton transfer equilibrium needs to be considered.

Example 4-5 At room temperature, solubility of is H_2S 0.10mol/L. What is $[H^+]$, $[HS^-]$ and $[S^{2-}]$ in concentrated H_2S solution?

$$K_{a1}^{\ominus}=8.91\times10^{-8}\text{ , }K_{a2}^{\ominus}=1.00\times10^{-19}$$

Solution $\because\ c\cdot K_{a1}^{\ominus}>20K_W^{\ominus}$, the water dissociation equilibrium can be ignored.

$K_{a1}^{\ominus} >> K_{a2}^{\ominus}$, $\therefore$ the step 2 dissociation equilibrium can be ignored, it can be calculated as a monotonic weak acid.

$$\because\ \frac{c}{K_{a1}^{\ominus}}\geqslant 400\text{ , }\therefore\ [H^+]=\sqrt{cK_{a1}^{\ominus}}=\sqrt{0.100\times8.91\times10^{-8}}=9.44\times10^{-5}$$

Since the second step of H_2S is very weak, so $[HS^-]\approx[H^+]=9.44\times10^{-5}$。

$[S^{2-}]$ can be obtained from the step 2 dissociation equilibrium:

$$HS^- \rightleftharpoons H^+ + S^{2-}\qquad K_{a2}^{\ominus}=\frac{[H^+][S^{2-}]}{[HS^-]}=1.00\times10^{-19}$$

Because $[HS^-]\approx[H^+]$, so $[S^{2-}]\approx\ _{a2}=1.00\times10^{-19}$

We can make three key statements about the ionization of H_2S, as illustrated in Example 4-5.

(1) For diprotic weak acid H_2A, if $K_{a1}^{\ominus}/K_{a2}^{\ominus}\geqslant10^4$, $c(H^+)$ of solution is decided mainly by first ionic equilibrium. And if $c_{acid}/K_{a1}^{\ominus}\geqslant400$, $c(H^+)$ can be calculated with the simple equation.

(2) For diprotic weak acid H_2A, $c(A^{2-})\approx K_{a2}^{\ominus}$ regardless of the molarity of the acid.

(3) For diprotic weak acid H_2A, the concentration of acidic group $c(A^{2-})$ is normally very low, so if a large amount of A^- is needed, the salt but not the acid is used.

2.4 Proton Transfer Equilibrium of Amphoteric Materials (两性物质)

In the Protonic acid-base theory, substances that can both donate and accept protons are called amphoteric substances, such as NH_4Ac, NH_4CN, glycine (NH_2CH_2COOH，甘氨酸), etc.

HCO_3^- is taken as an example to show that amphoteric substances in solution have two proton transfer equilibrium:

HCO_3^- as an acid: $HCO_3^- \rightleftharpoons H^+ + CO_3^{2-}$ $K_a^\ominus(HCO_3^-) = K_{a2}^\ominus(H_2CO_3) = 4.68\times10^{-11}$

HCO_3^- as a base: $HCO_3^- + H_2O \rightleftharpoons H_2CO_3 + OH^-$

$$K_b^\ominus(HCO_3^-) = \frac{K_W^\ominus}{K_{a1}^\ominus(H_2CO_3)} = \frac{1.00\times10^{-14}}{4.47\times10^{-7}} = 2.24\times10^{-8}$$

Because $K_a^\ominus(HCO_3^-) < K_b^\ominus(HCO_3^-)$, so HCO_3^- shows basic.

It can be seen that the acid and base of aqueous solution of amphoteric substances can be judged according to the relative size of $K_a^\ominus$ and $K_b^\ominus$. If, then its ability to give protons is greater than its ability to accept protons, the aqueous solution becomes acidic, such as NH_4F; If, then its ability to give protons is less than the ability to accept protons, the solution is alkaline, such as NH_4CN; If its ability to give protons is equal to its ability to accept protons, the solution is neutral, such as NH_4Ac.

HCO_3^- is taken as the example to show the approximate calculation formula of pH of amphoteric substances:

$$HCO_3^- \rightleftharpoons H^+ + CO_3^{2-} \qquad K_a^\ominus(HCO_3^-) = K_{a2}^\ominus(H_2CO_3) = 4.68\times10^{-11}$$

$$HCO_3^- + H_2O \rightleftharpoons H_2CO_3 + OH^- \qquad K_b^\ominus(HCO_3^-) = \frac{K_W^\ominus}{K_{a1}^\ominus(H_2CO_3)} = \frac{1.00\times10^{-14}}{4.47\times10^{-7}} = 2.24\times10^{-8}$$

because $K_a^\ominus < K_b^\ominus$, so the solution show basic, and $[OH^-] > [H^+]$,

$$[OH^-]=[H_2CO_3]-[CO_3^{2-}] \tag{1}$$

From the above two equilibrium constant expressions, it can be obtained:

$$K_a^\ominus = K_{a2}^\ominus = \frac{[H^+][CO_3^{2-}]}{[HCO_3^-]} \qquad [CO_3^{2-}] = \frac{K_{a2}^\ominus[HCO_3^-]}{[H^+]} \tag{2}$$

$$K_b^\ominus = \frac{K_W^\ominus}{K_{a1}^\ominus} = \frac{[H_2CO_3][OH^-]}{[HCO_3^-]} \qquad [H_2CO_3] = \frac{K_W^\ominus[HCO_3^-]}{K_{a1}^\ominus[OH^-]} = \frac{[H^+][HCO_3^-]}{K_{a1}^\ominus} \tag{1}$$

formulas (2) and (3) substitute into (1):

$$[OH^-] = \frac{[H^+][HCO_3^-]}{K_{a1}^\ominus} - \frac{K_{a2}^\ominus[HCO_3^-]}{[H^+]}$$

Multiply both sides by $K_{a1}^\ominus[H^+]$:

$$[H^+] = \sqrt{\frac{K_{a1}^\ominus(K_W^\ominus + K_{a2}^\ominus[HCO_3^-])}{[HCO_3^-]}} \tag{4}$$

Because $K_a^\ominus$ and $K_b^\ominus$ are very small, so $[HCO_3^-] \approx c$,

(4) is approximately as:

$$[H^+] = \sqrt{\frac{K_{a1}^\ominus(K_W^\ominus + cK_{a2}^\ominus)}{c}} \tag{5}$$

Because $c\cdot K_{a2}^\ominus \geqslant 20K_W^\ominus$, so $K_W^\ominus + cK_{a2}^\ominus \approx cK_{a2}^\ominus$

(5) is approximately as:

$$[H^+]=\sqrt{K_{a1}^{\ominus}\cdot K_{a2}^{\ominus}} \tag{4-8}$$

When applied to other amphoteric substances, the approximate formula of H^+ concentration is:

$$[H^+]=\sqrt{K_a^{\ominus}\cdot K_a^{\ominus *}} \tag{4-9}$$

$$pH=\frac{1}{2}pK_a^{\ominus}+\frac{1}{2}pK_a^{\ominus *} \tag{4-10}$$

$K_a^{\ominus}$ is the acid constant of conjugated acid as basic constituent in amphoteric substance, and $K_a^{\ominus *}$ is the acid constant of the acidic constituent in amphoteric substance.

Example 4-6 Calculate the pH of the following solutions: ①0.100mol/L NaH_2PO_4; ②0.100mol/L Na_2HPO_4.

For H_3PO_4: $pK_{a1}^{\ominus}=2.16, pK_{a2}^{\ominus}=7.21, pK_{a3}^{\ominus}=12.32$

Solution (1) for 0.100mol/L NaH_2PO_4

As an acid, $H_2PO_4^-$ ionizes to produce HPO_4^{2-} ($K_a^{\ominus}=K_{a2}^{\ominus}$). And as a base, it hydrolyzes to produce H_3PO_4, so H_3PO_4 is the conjugated acid of $H_2PO_4^-$ ($K_a^{\ominus *}=K_{a1}^{\ominus}$).

According to (4-10):

$$pH=\frac{1}{2}pK_{a2}^{\ominus}+\frac{1}{2}pK_{a1}^{\ominus}$$

$$=\frac{1}{2}(2.16+7.21)=4.69$$

(2) for 0.100mol/L Na_2HPO_4

As an acid, HPO_4^{2-} ionizes to produce PO_4^{3-} ($K_a^{\ominus}=K_{a3}^{\ominus}$). And as a base, it hydrolyzes to produce $H_2PO_4^-$, so $H_2PO_4^-$ is the conjugated acid of HPO_4^{2-} $K_a^{\ominus *}=K_{a2}^{\ominus}$).

According to (4-10):

$$pH=\frac{1}{2}pK_{a3}^{\ominus}+\frac{1}{2}pK_{a2}^{\ominus}$$

$$=\frac{1}{2}(7.21+12.32)=9.77$$

Example 4-7 Calculate the pH of 0.100mol/L NH_4CN.

For $NH_3 \cdot H_2O$, $K_b^{\ominus}=1.74\times10^{-5}$, for HCN, $K_a^{\ominus}=6.17\times10^{-10}$

Solution NH_4CN can completely dissociate into NH_4^+ and CN^-, NH_4^+ as the acidic constituent and CN^- as the basic constituent.

$$NH_4^+ + H_2O \rightleftharpoons H^+ + NH_3\cdot H_2O \qquad K_a^{\ominus}=\frac{K_w^{\ominus}}{K_b^{\ominus}}=\frac{1.00\times10^{-14}}{1.74\times10^{-5}}=5.75\times10^{-10}$$

$$CN^- + H_2O \rightleftharpoons HCN + OH^- \qquad K_a^{\ominus}(HCN)=6.17\times10^{-10}$$

According to (4-9): $[H^+]=\sqrt{K_a^{\ominus}\cdot K_a^{\ominus *}}$

$$=\sqrt{K_a^{\ominus}(NH_4^+)\cdot K_a^{\ominus}(HCN)}$$

$$= \sqrt{5.75 \times 10^{-10} \times 6.17 \times 10^{-10}}$$
$$= 5.96 \times 10^{-10}$$
$$pH = -\lg 5.96 \times 10^{-10} = 9.22$$

微课

2.5 The Shift of Proton Transfer Equilibrium between Acid and Base

The acid-base proton transfer equilibrium is a dynamic equilibrium. When the external conditions such as concentration, temperature change, the equilibrium will shift. And the dissociation degree of weak acid/base will also change. The effect of concentration on proton transfer equilibrium of weak acid/base will be discussed.

2.5.1 Common-Ion Effect in Acid Base Equilibria (酸碱平衡中的同离子效应)

The questions answered in the above contents were mostly of the type, "What is the pH of 0.10 mol/L $NH_3 \cdot H_2O$, of 0.10 mol/L HAc, of 0.10 mol/L NaAc, of 0.10 mol/L $NaHCO_3$?" In each of these cases, we think of dissolving a single substance in aqueous solution and determining the concentrations of the species present at equilibrium. In most situations in this content, a solution of a weak acid or weak base initially contains a second source of one of the ions produced in the ionization of the acid or base. The added ions are said to be common to the weak acid or weak base. The presence of a common ion can have some important consequences.

For example, when NaAc is added to the solution of HAc, HAc is a weak monoacid, it has the proton transfer equilibrium in the solution, while NaAc is a strong electrolyte in the solution and can dissociate completely:

$$HAc \rightleftharpoons \boxed{Ac^-} + H^+$$
$$NaAc \rightarrow \boxed{Ac^-} + Na^+$$
$$\xleftarrow{\hspace{3cm}}$$
shift left

With the addition of NaAc, the concentration of Ac^- ion in the solution increases, causing the proton transfer balance of HAc to shift left and the degree of dissociation of HAc to decrease.

Example 4-8 Calculate the $[H^+]$ and α of the following solutions: ① 0.100mol/L HAc; ② add some crystal of NaAc into 0.100mol/L HAc solution, and make [NaAc]=0.200mol/L.

$K_a^\ominus = 1.75 \times 10^{-5}$, the change in volume caused by the addition of solids is ignored.

Solution (1) $\because$ $c \cdot K_a^\ominus = 1.75 \times 10^{-6} > 20K_W^\ominus$, the proton transfer equilibrium in water is ignored.

$$\because \ \frac{c}{K_a^\ominus} = \frac{0.100}{1.75 \times 10^{-5}} > 400$$

$$[H^+] = \sqrt{c \cdot K_a^\ominus} = \sqrt{0.100 \times 1.75 \times 10^{-5}} = 1.32 \times 10^{-3}$$

$$\alpha = \frac{[H^+]}{c} \times 100\% = \frac{1.32 \times 10^{-3}}{0.100} \times 100\% = 1.32\%$$

(2)	HAc	$\rightleftharpoons$	H^+ +	Ac^-
$c_{initial}$	0.100		0	0.200
c_{equal}	$0.100-[H^+]$		$[H^+]$	$0.200+[H^+]$

$$K_a^{\ominus}=\frac{[H^+][A^-]}{[HA]}=\frac{[H^+](0.200+[H^+])}{0.100-[H^+]}$$

$$\frac{c}{K_a^{\ominus}}=\frac{0.100}{1.75\times10^{-5}}>400,\ 0.100-[H^+]\approx0.100,\ 0.200+[H^+]\approx0.200$$

$$[H^+]=\frac{0.100\times1.75\times10^{-5}}{0.200}=8.75\times10^{-6}$$

$$\alpha=\frac{[H^+]}{c}\times100\%=\frac{8.75\times10^{-6}}{0.100}\times100\%=8.75\times10^{-3}\%$$

Now we see the consequence of adding a strong electrolyte acid (NaAc) to a weak acid (HAc): The concentration of the H^+ ion is greatly reduced. Between solutions ① and ② of Example 4-8, $[H^+]$ is lowered from 1.32×10^{-3}mol/L to 8.75×10^{-6}mol/L over two orders of magnitude decrease. The phenomenon that the dissociation degree of the weak electrolyte is reduced by adding a strong electrolyte containing the same ions to the weak electrolyte solution is called the common ion effect (在弱电解质溶液中，加入与该弱电解质含有相同离子的强电解质，使弱电解质解离度降低的现象称为同离子效应).

Example 4-9 H_2S is pumping into 0.200mol/L HCl to be saturated, calculate $[S^{2-}]$. $K_{a1}^{\ominus}=8.91\times10^{-8}$, $K_{a2}^{\ominus}=1.00\times10^{-19}$.

Solution The ionization of the weak acid, H_2S, is suppressed when a strong acid, HCl, is added. Here, H^+ is the common ion, and its increased concentration shifts the equilibrium of H_2S to the left greatly. HCl has the common ion effect on the proton transfer of H_2S, making the H^+ transferred from H_2S very little. So in the solution, $[H^+]\approx0.200$mol/L

$$H_2S \rightleftharpoons 2H^+ + S^{2-}$$

$$c_{equal}\quad 0.100\quad 0.200\quad [S^{2-}]$$

$$K_{a1}^{\ominus}K_{a2}^{\ominus}=\frac{[H^+]^2[S^{2-}]}{[H_2S]}=\frac{0.200^2[S^{2-}]}{0.100}=8.91\times10^{-8}\times1.00\times10^{-19}$$

$$[S^{2-}]=2.23\times10^{-26}$$

The concentration of the S^{2-} ion is greatly reduced. Compared with Example 3-5, $[S^{2-}]$ is lowered from 1.00×10^{-19}mol/L to 2.23×10^{-26}mol/L.

It can be seen that by adjusting the concentration of H^+ in saturated H_2S solution, the concentration of S^{2-} can be controlled. Therefore, in the precipitation-Dissolution equilibrium, the purpose of separating different metal by producing metal sulfides is achieved by adjusting the pH value of the solution (在沉淀–溶解平衡一章中，通过调节饱和H_2S溶液中的H^+浓度，可以控制S^{2-}浓度，从而使有的金属硫化物沉淀，有的不沉淀，达到分离的目的).

In practice, the pH of the solution can inhibit or promote the proton transfer reaction of some protonic acids/bases (salts in ionization theory), which is called salt hydrolysis reaction in Arrhenius theory. For example, in salt solutions (Sn^{2+}, Sb^{3+}, Bi^{3+}, Fe^{3+}, Pb^{2+} and Hg^{2+} etc.), it is easy to precipitate because of hydrolysis if the pH is not properly controlled.

$$SnCl_2 + H_2O \rightleftharpoons Sn(OH)Cl\downarrow + HCl$$

$$SbCl_3 + H_2O \rightleftharpoons SbOCl\downarrow + 2HCl$$

$$Pb(NO_3)_2 + H_2O \rightleftharpoons Pb(OH)NO_3\downarrow + HNO_3$$

$$Bi(NO_3)_3 + H_2O \rightleftharpoons BiONO_3\downarrow + 2HNO_3$$

When preparing these salt solutions, we usually dissolve the salt into small amount of the corresponding concentrated acid to inhibit hydrolysis and then dilute the required concentration with water.

2.5.2 Salt Effect (盐效应) in Acid-Base Equilibria

We have explored the effect of common ions on a ionization equilibrium, but what effect do ions different from those involved in the equilibrium have on extent of dissociation? The effect of diverse ions is not as striking as the common ion effect. Moreover, diverse ions tend to increase rather than decrease extent of dissociation (不同的离子使电离度增加，而不是减少). As the total ionic concentration of a solution increases, interionic attractions become more important. Activities (活度) (effective concentrations (有效浓度) become smaller than the stoichiometric concentration. For the ions involved in the ionization process, this means that higher concentrations must appear in solution before equilibrium is established—the extent of dissociation increases. The diverse ion effect is more commonly called the salt effect (盐效应).

When NaAc is added into the HAc solution, the salt effect will be accompanied by the common ion effect. However, the effect of the salt effect is much weaker than that of the common ion effect in dilute solution. Therefore, the effect of the salt effect is generally ignored and only the common ion effect is considered.

微课

3. The Buffer Solution

PPT

Many chemical reactions, especially those in organisms, need to be carried out under certain pH conditions. For example, the enzyme activity is reduced or even lost if the pH is slightly deviated in an enzyme catalytic reaction in an organism. Many drugs are acid or base, and their preparations, analytical conditions and pharmacological effects are closely related to the control of a certain pH. How to control the pH of the solution? The concept of buffer solution (缓冲溶液) is proposed.

A buffer solution is a solution in which the pH does not change significantly when small amounts of acids, bases or water are added (能抵抗外来少量强酸、强碱或水的稀释而保持本身 pH 基本不变的溶液称为缓冲溶液). The anti-acid, anti-base and anti-dilution effects of a buffer solution are called buffer effects (缓冲作用).

3.1 Composition and Action Principle of Buffer Solution

3.1.1 Composition of Buffer Solution

Buffer solution consists of a conjugate acid-base pair, such as HAc-Ac^-, NH_4^+-NH_3, etc. A conjugate acid-base pair which makes up a buffer solution is called a buffer pair (缓冲对) or a buffer system (缓冲系).

Common buffer pairs have the following categories:

a weak acid and its conjugate base (共轭碱): HAc-$NaAc$;

a weak base and its conjugate acid (共轭酸): NH_3-NH_4Cl, CH_3NH_2-CH_3NH_3Cl;

Amphoteric substances and their conjugate acids or bases: H_2CO_3-$NaHCO_3$, H_3PO_4-NaH_2PO_4, NaH_2PO_4-Na_2HPO_4, $C_6H_4(COOH)_2$-$C_6H_4(COOH)COOK$, NaH_2PO_4-Na_2HPO_4, Na_2HPO_4-Na_3PO_4, $NaHCO_3$-Na_2CO_3.

3.1.2 Principle of a Buffer

The action principle of a buffer is closely related to the common ion effect (同离子效应) mentioned earlier. For example, in the HAc-NaAc buffer system, HAc is a weak electrolyte with proton transfer equilibrium, and NaAc is a strong electrolyte which is completely dissociated.

$$\begin{array}{l} HAc \rightleftharpoons \boxed{H^+} + Ac^- \\ \xleftarrow{\qquad\qquad\qquad} \\ NaAc \rightarrow \boxed{Na^+} + Ac^- \end{array}$$

Obviously, because the high concentration of Ac^- ion in the solution produces the common ion effect, the proton transfer equilibrium of HAc moves to the left. The decrease of the dissociation degree of HAc results in the higher concentration of HAc, Ac^- and the smaller concentration of H^+ in the solution.

When a small amount of strong acid is added, the conjugate base Ac^- interacts with H^+ and the equilibrium shifts to the left. Because the solution contains a large amount of HAc and Ac^-, the added H^+ almost becomes HAc with a slight increase in the concentration of HAc and a slight decrease in the concentration of Ac^-. Therefore, the pH of the solution dose not change significantly. The conjugate base Ac^- plays a role in resisting a small amount of strong acids, so Ac^- is the anti-acid component (抗酸成分) of the buffer solution.

When a small amount of strong base is added, OH^- interacts with H^+ and the equilibrium moves to the right. HAc dissociation supplements the H^+ consumed. Because the solution contains a large amount of HAc and Ac^-, the consumed H^+ is almost completely replenished with a slight decrease in the concentration of HAc and a slight increase in the concentration of Ac^-. Therefore, the pH of the solution is almost unchanged. The conjugate acid HAc plays a role in resisting a small amount of strong base, so HAc is an anti-base component (抗碱成分) of the buffer solution.

When a small amount of water is added for a moderate dilution, buffer solutions resist pH changes. Diluting a buffer solution means increasing its volume V by adding water. This action produces the same change of the ratio [conjugate base]/ [acid]. The ratio itself remains unchanged, as does the pH, which is not difficult to see from equation (4-12).

In summary, because the buffer solution contains a large amount of anti-base and anti-acid components at the same time, the proton transfer equilibrium of the weak acid or base can resist and consume a small amount of strong acids and bases from outside so that the pH of the solution does not change significantly. This is the principle of buffering.

It should be noted that if a large amount of strong acid or base is added to the buffer solution, the buffering effect will be lost after the anti-acid or anti-base component in the buffer solution is consumed.

In addition to the above buffer system, strong acid solution or strong base solution with higher concentration also has a certain buffer capacity. Because a small amount of acid or base has little effect on strong acid and strong base concentration, the pH is basically unchanged.

3.2 Approximate Calculation of a Buffer Solution

3.2.1 Approximate Calculation

The approximate calculation of pH in buffer solution is equivalent to that of the common ion effect.

For example, the following proton transfer equilibrium exists in the buffer solution composed of weak acid HA and its conjugate base A^-:

	HA	$\rightleftharpoons$	H^+ +	A^-
Relative initial concentration	$c(HA)$		0	$c(A^-)$
Relative equilibrium concentration	$c(HA)-[H^+]$		$[H^+]$	$c(A^-)+[H^+]$

Proton transfer of weak acids is suppressed due to the common ion effect

$$c(HA)-[H^+]\approx c(HA),\ c(A^-)+[H^+]\approx c(A^-)$$

Substitution equilibrium constant expression: $K_a^\ominus=\dfrac{[H^+]\cdot c(A^-)}{c(HA)}$, $[H^+]=\dfrac{K_a^\ominus\cdot c(HA)}{c(A^-)}$

Take the negative logarithm of each side: $pH=pK_a^\ominus-\lg\dfrac{c(HA)}{c(A^-)}$

$$\text{or}\quad pH=pK_a^\ominus+\lg\frac{c(A^-)}{c(HA)}\tag{4-11}$$

Generalized to all buffer systems consisting of conjugate acid-base pairs:

$$pH=pK_a^\ominus+\lg\frac{c(\text{conjugate base})}{c(\text{conjugate acid})}\tag{4-12}$$

(4-12) is the equation known as the Henderson-Hasselbalch equation. Biochemists and molecular biologists commonly use this equation. Note that in the calculation formula of the pH of the buffer solution, the relative concentration is taken in and the SI unit is 1.

Example 4-10 Calculate the pH of a buffer solution consisting of 0.10mol/L $NH_3\cdot H_2O$ and 0.10mol/L NH_4Cl. ($NH_3\cdot H_2O$ $pK_b^\ominus=4.76$)

Solution Because $pK_a^\ominus=14-pK_b^\ominus=14-4.76=9.24$ (NH_4^+),

$c(NH_3\cdot H_2O)=0.10$mol/L, $c(NH_4^+)=0.10$mol/L,

Substituting into (4-12):

$$pH=pK_a^\ominus+\lg\frac{c(\text{conjugate base})}{c(\text{conjugate acid})}=9.24+\lg\frac{c(NH_3\cdot H_2O)}{c(NH_4^+)}=9.24+\lg\frac{0.10}{0.10}=9.24$$

Example 4-11 Take 90ml of the buffer solution from the above example, ① add 10ml 0.010mol/L HCl, ② add 10mL 0.010mol/L NaOH, and ③ add 10ml water. Calculate the pH of the above three solutions?

Solution (1) H^+ and $NH_3\cdot H_2O$ combine to form NH_4^+ when HCl is added in the buffer solution composed of NH_4^+ and $NH_3\cdot H_2O$, which causes $c(NH_3\cdot H_2O)$ to decrease and $c(NH_4^+)$ to increase. The calculations are as follows.

$$c(NH_4^+)=0.10\text{mol/L}\times\frac{90\text{ml}}{90\text{ml}+10\text{ml}}+0.010\text{mol/L}\times\frac{10\text{ml}}{90\text{ml}+10\text{ml}}=0.091\text{mol/L}$$

$$c(NH_3 \cdot H_2O) = 0.10\text{mol/L} \times \frac{90\text{ml}}{90\text{ml} + 10\text{ml}} - 0.010\text{mol/L} \times \frac{10\text{ml}}{90\text{ml} + 10\text{ml}} = 0.089\text{mol/L}$$

Substituting into (4-12):

$$\text{pH} = \text{p}K_a^{\ominus} + \lg\frac{c(\text{conjugate base})}{c(\text{conjugate acid})} = 9.24 + \lg\frac{c(NH_3 \cdot H_2O)}{c(NH_4^+)} = 9.24 + \lg\frac{0.089}{0.091} = 9.23$$

(2) OH^- and NH_4^+ combine to form $NH_3 \cdot H_2O$ when NaOH is added in the buffer solution composed of NH_4^+ and $NH_3 \cdot H_2O$, which causes $c(NH_4^+)$ to decrease and $c(NH_3 \cdot H_2O)$ to increase. The calculations are as follows:

$$c(NH_3 \cdot H_2O) = 0.10\text{mol/ L} \times \frac{90\text{ml}}{90\text{ml} + 10\text{ml}} + 0.010\text{mol/ L} \times \frac{10\text{ml}}{90\text{ml} + 10\text{ml}} = 0.091\text{mol/ L}$$

$$c(NH_4^+) = 0.10\text{mol/ L} \times \frac{90\text{ml}}{90\text{ml} + 10\text{ml}} - 0.010\text{mol/ L} \times \frac{10\text{ml}}{90\text{ml} + 10\text{ml}} = 0.089\text{mol/ L}$$

Substituting into (4-12):

$$\text{pH} = \text{p}K_a^{\ominus} + \lg\frac{c(\text{conjugate base})}{c(\text{conjugate acid})} = 9.24 + \lg\frac{c(NH_3 \cdot H_2O)}{c(NH_4^+)} = 9.24 + \lg\frac{0.091}{0.089} = 9.25$$

(3) Adding 10ml of water to a buffer solution composed of NH_4^+ and $NH_3 \cdot H_2O$ caused $c(NH_4^+)$ and $c(NH_3 \cdot H_2O)$ to decrease to the same extent.

$$c(NH_4^+) = c(NH_3 \cdot H_2O) = 0.10\text{mol/ L} \times \frac{90\text{ml}}{90\text{ml} + 10\text{ml}} = 0.090\text{mol/ L}$$

Substituting into (4-12):

$$\text{pH} = \text{p}K_a^{\ominus} + \lg\frac{c(\text{conjugate base})}{c(\text{conjugate acid})} = 9.24 + \lg\frac{c(NH_3 \cdot H_2O)}{c(NH_4^+)} = 9.24 + \lg\frac{0.090}{0.090} = 9.24$$

The calculation results show that when a small amount of strong acid, strong base and water are added to the buffer solution, the pH of the buffer solution is almost unchanged.

Example 4-12 20ml of 0.10mol/L HAc and 10mL of 0.10mol/L NaOH were mixed. Calculate the pH of the mixed solution. (HAc $\text{p}K_a^{\ominus} = 4.76$)

Solution After mixing, HAc and NaOH undergo acid-base neutralization reaction. Because the HAc is excessive, the mixed solution is a buffer solution composed of excess HAc and generated NaAc.

The concentration of excess HAc is:

$$c(\text{HAc}) = \frac{0.10\text{mol/ L} \times 20\text{ml} - 0.10\text{mol/ L} \times 10\text{ml}}{30\text{ml}} = \frac{1}{30}\text{mol/ L}$$

The concentration of generated NaAc is:

$$c(\text{Ac}^-) = \frac{0.10\text{mol/ L} \times 10\text{ml}}{30\text{ml}} = \frac{1}{30}\text{mol/ L}$$

Substituting into (4-12):

$$\text{pH} = \text{p}K_a^{\ominus} + \lg\frac{c(\text{conjugate base})}{c(\text{conjugate acid})} = 4.76 + \lg\frac{c(\text{Ac}^-)}{c(\text{HAc})} = 4.76 + \lg\frac{\frac{1}{30}}{\frac{1}{30}} = 4.76$$

In the derivation of formula (4-12): ① the H^+ and OH^- provided by the proton transfer equilibra of weak acid and weak base are ignored. ② Interactions between ions are ignored. If the ratio of conjugate acid and base is two large or the solution concentration is too dilute, the H^+ and OH^- provided by the proton transfer equilibrium of the weak acid and weak base cannot be ignored, and the initial concentration

cannot be used to calculate the equilibrium concentration. When the conjugate acid and base are ions and the concentration is large, the influence of ionic strength must be considered, and the concentration cannot be used instead of the activity calculation. In both cases, an exact formula must be used to solve the problem, which is not required in this book.

3.2.2 The Buffer Range

There is a limit to the buffer capacity (缓冲能力) of any buffer solution. When the ratio of the conjugate acid/base in buffer solution is too large or the amount of strong acid and base added is too large, the pH of the solution will change greatly and the buffer solution will lose its buffering effect. In general, the maximum buffer capacity exists when the concentrations of a weak acid and its conjugate base are kept large and approximately equal to each other.

The approximate calculation formula of the buffer solution shows that pH change of the buffer solution is caused by the change of the concentration ratio of the buffer pair (缓冲对). Therefore, after adding a small amount of acid or base, the smaller the change in the concentration ratio of the buffer pair, the smaller the pH change and the stronger the buffering capacity are. Conversely, the buffering capacity is weaker. After adding a small amount of acid or base, the concentration ratio change of the buffer pair depends on the following two factors.

(1) When the concentration ratio of buffer pair is fixed, the larger the total concentration of the buffer solution, the greater the buffer capacity.

(2) When the total concentration of the buffer solution is fixed, the closer the component concentrations are to each other, the greater the capacity. When the concentration ratio of the buffer pair is equal to 1, the buffering ability is the strongest.

Therefore, in order to prepare the buffer solution with a large buffer capacity, both a larger total concentration and the concentration ratio of the buffer pair should be considered. Generally, if one component concentration is more than 10 times the other—buffering action is poor. So the ratio should be controlled between 0.1~10. So buffers have a usable range within ±1 pH unit of the $pK_a^\ominus$ of the acid component:

$$pH = pK_a^\ominus \pm 1 \tag{4-13}$$

Equation (4-13) refers to the effective pH range of the buffer solution. The buffer solution has strong buffer ability in this pH range. Table 4-3 lists the buffer ranges of common buffer solutions.

Table 4-3 Common buffer solutions and their buffer ranges

Buffer Solutions	Buffer Pairs	$pK_a^\ominus$	Buffer Ranges
HCOOH-HCOONa	$HCOOH-HCOO^-$	3.75	2.75~4.75
HAc-NaAc	$HAc-Ac^-$	4.76	3.76~5.76
Hexamethylenetetramine (六次甲基四胺)-HCl	$(CH_2)_6N_4H^+$- $(CH_2)_6N_4$	5.15	4.15~6.15
NaH_2PO_4-Na_2HPO_4	$H_2PO_4^-$-HPO_4^{2-}	7.21	6.21~8.21
$Na_2B_4O_7$-HCl	H_3BO_3-$H_2BO_3^-$	9.27	8.27~10.27
$NH_3\cdot H_2O$-NH_4Cl	NH_4^+-NH_3	9.24	8.24~10.24
$NaHCO_3$-Na_2CO_3	HCO_3^--CO_3^{2-}	10.33	9.33~11.33
Na_2HPO_4-Na_3PO_4	HPO_4^{2-}-PO_4^{3-}	12.32	11.32~13.32

3.3 Preparation of Buffer Solution

Henderson-Hasselbalch equation is useful in both calculating pH and preparing a buffer solution. In practice, the following steps can be taken.

(1) The selected buffer pair can't interact with the reactants and products. For medicinal buffer pair, it is also necessary to consider that buffer pair can't be incompatible with remedium cardinale. And the buffer pair should be stable and not toxic during the heating sterilization and storage period.

(2) The appropriate buffer pair is selected so that the $pK_a^{\ominus}$ of conjugated acid is equal to or similar to the desired pH, and the concentration ratio of buffer pair is between 0.1~10 to ensure stronger buffer capacity.

(3) Controlling the appropriate total concentration of the buffer: For most laboratory applications, concentrations from 0.05mol/L to 0.5mol/L are suitable.

(4) Based on the desired pH and total concentration, the amount of conjugated acid and base is calculated using the equation (4-12).

(5) Finally, the pH of the buffer is measured and calibrated with pH meter.

Example 4-13 To prepare a 500ml buffer solution with pH = 4.70, how much volume of 1.0mol/L NaOH mixed with 100mL 1.0mol/L HAc? How to prepare it? (1.0mol/L HAc, $pK_a^{\ominus} = 4.76$)

Solution We substitute the $pK_a^{\ominus} = 4.76$ of HAc in formula (4-12):

$$4.70 = 4.76 + \lg\frac{c(Ac^-)}{c(HAc)} \qquad \frac{c(Ac^-)}{c(HAc)} = 0.87$$

Since HAc is neutralized by NaOH, the desired $c(Ac^-)$ is the concentration of NaOH in the buffer solution.

$$(Ac^{\ }) \qquad \frac{(NaOH)\times 1.0mol/L}{0.50L}$$

$$c(HAc) = \frac{1.0mol/L \times [0.10L - V(NaOH)]}{0.50L}$$

Substituting: $$\frac{V(NaOH)\times 1.0mol/L}{0.50L} = 0.87 \times \frac{1.0mol/L \times [0.10L - V(NaOH)]}{0.50L}$$

$$V(NaOH) = 0.047L = 47ml$$

After mixing 100ml of 1.0mol/L HAc and 47ml of 1.0mol/L NaOH, the buffer solution of pH = 4.70 is obtained by water dilution to 500ml and calibrated with pH meter finally.

Example 3-14 How to prepare 1L buffer solution with pH = 9.50 and $c(NH_3 \cdot H_2O) = 1.0mol/L$? The chemicals: ammonia aqueous solution (15mol/L $NH_3 \cdot H_2O$) and solid NH_4Cl. $pK_a^{\ominus} = 4.76$

Solution Owing to $pK_a^{\ominus} = 4.76$ of $NH_3 \cdot H_2O$, the $pK_a^{\ominus} = 9.24$ of NH_4^+ can be obtained. Substituting into (4-12):

$$9.50 = 9.24 + \lg\frac{1.0}{c(NH_4^+)} \qquad c(NH_4^+) = 0.55$$

$$m(NH_4Cl) = 1.0L \times 0.55mol/L \times 53.5g/mol = 29g$$

$$V(NH_3 \cdot H_2O) = \frac{1.0mol/L \times 1000ml}{15mol/L} = 67ml$$

29g solid NH_4Cl is dissolved in a small amount of water, and 67ml concentrated ammonia is added in the solution. Then the buffer solution with pH = 9.50 is obtained after diluted to 1.0 L using water and calibrated by pH meter.

3.4 Application of Buffer Solution in Medicine

The most complex metabolism of substances in the body is controlled by various enzymes, and each enzyme is active only in body fluids with a certain pH range. Therefore, all body fluids must be constant in a certain pH range before the metabolic reaction can proceed normally. The important function of the buffer solution is to control the pH of the solution. How to keep pH of various body fluids in a certain range? It depends on the buffer solutions. Taking the buffer system in human blood as example, the application of buffer solution in medicine is discussed.

Blood is a buffer system which is consisted of many buffer solutions. The followings are some mainly buffer systems:

In plasma: H_2CO_3-HCO_3^-, $H_2PO_4^-$-HPO_4^{2-}, H_nP-$H_{n-1}P^-$ (H_nP means protein)

In red blood cell: H_2b-Hb^- (H_2b血红蛋白), H_2bO_2-HbO_2^- (H_2bO_2氧合血红蛋白), H_2CO_3- HCO_3^-, $H_2PO_4^-$- HPO_4^{2-}

H_2CO_3- HCO_3^- buffer pair is the most important buffer system in the blood. Carbonic acid (碳酸) mainly exists in the form of dissolved CO_2 in the solution, and there is a equilibrium in the H_2CO_3- HCO_3^- buffer system:

$$CO_2(\text{dissolved}) + H_2O \rightleftharpoons H_2CO_3 \rightleftharpoons H^+ + HCO_3^-$$

When $[H^+]$ increases, HCO_3^- combines with H^+ and the equilibrium shifts to the left. So pH has no significant change. When $[H^+]$ decreases, the equilibrium shifts to the right. So pH has no significant change.

The normal pH range of blood is 7.35 ~7.45. If the pH of blood is less than 7.35, it will cause acidosis (酸中毒); if the pH of blood is greater than 7.45, alkalosis (碱中毒) will occur.

The concentration ratio of HCO_3^--CO_2 is 20:1 in normal plasma, which is beyond the range of effective concentration ratio of buffer solution (10:1~1~10), so the buffer capacity of this buffer system seems weak. Actually, the buffer capacity of this buffer system is strong. Because changes in $[HCO_3^-]$ or $[CO_2]$(dissolved) can be regulated by respiration and kidney after buffering in body (not in vitro). So the concentration of HCO_3^- and CO_2(dissolved) can remain relatively stable. Therefore, H_2CO_3- HCO_3^- always has strong buffer capacity.

Here is the relationship between the buffer effect of H_2CO_3- HCO_3^- buffer system in plasma and the regulation of lung and kidney:

$$H_2CO_3 \underset{+H^+}{\overset{+OH^-}{\rightleftharpoons}} HCO_3^-$$

$$\Updownarrow \qquad\qquad \Updownarrow$$

$$\text{肺} \rightleftharpoons CO_2 + H_2O \qquad \text{肾}$$

H_2b-Hb^- and H_2bO_2-HbO_2^- are also important buffer systems in the red blood cell. The buffer effect of blood on the large amount of CO_2 produced in the metabolic process of human body is mainly achieved by them. CO_2 interacts with Hb^- firstly:

$$CO_2 + H_2O + Hb^- \rightleftharpoons HHb + HCO_3^-$$

Then the HCO_3^- is transported by blood to the lung and interacts with oxyhemoglobin:

$$HCO_3^- + HHbO_2 \rightleftharpoons HbO_2^- + H_2O + CO_2$$

CO_2 exhales from the lungs. This shows that due to the buffer effects of hemoglobin and oxyhemoglobin, the pH of the blood will not be greatly affected during the transport of a large amount of CO_2 from tissue cell to the lung.

Due to the buffer effects of various buffer systems in the blood and the regulating effects of the lung and kidney, the pH of normal human blood is maintained in a narrow range of 7.35 ~ 7.45.

Many diseases can cause a temporary increase in blood acidity or basicity. Acidosis or alkalosis can occur if the blood pH is not prevented by a buffer system and compensation mechanism. For example, insufficient ventilation in the lungs caused by emphysema (肺气肿), congestive heart failure and bronchitis (支气管炎), diabetes (糖尿病) and consumption of low-carbohydrate and high-fat foods cause an increase in metabolic acid, and excessive intake of acid will cause an increase in $[H^+]$ in the blood. The body first eliminates excess CO_2 by accelerating the rate of breathing, followed by accelerating the excretion of H^+ and prolonging the residence time of HCO_3^- in the kidney (例如，肺气肿引起的肺部换气不足，充血性心力衰竭和支气管炎，糖尿病和食用低碳水化合物、高脂肪食物引起代谢酸的增加，摄食过多的酸等都会引起血液中 H^+ 的增加，然而身体首先通过加快呼吸的速度来排除多余的 CO_2，其次是加速 H^+ 的排泄和延长肾里的 HCO_3^- 的停留时间). Therefore, pH of blood can return to normal level. If the loss of HCO_3^- caused by severe diarrhea (腹泻) is excessive, or H^+ excretion is reduced due to renal failure (肾衰竭), neither the buffer system nor the compensatory function of the body can effectively prevent the pH of the blood from falling, then acidosis is caused. High fever, asthma (哮喘), hyperventilation (换气过度), excessive intake of basic substances, and severe vomiting (呕吐) can make blood basicity increase. The body's compensation mechanism cooperates with the buffer system by reducing the emission of CO_2 from the lung and increasing the excretion of HCO_3^- in the kidney to restore the pH to normal level. At this time, alkaline urine is caused by the increase in HCO_3^- concentration in the urine. If the increase in pH of the blood cannot be prevented by the buffer system and the compensation mechanism, alkalosis is caused.

重点小节

1. 酸碱理论 酸碱理论包括：酸碱离子论（阿累尼乌斯理论），酸碱质子论（布朗斯特－劳莱理论）和酸碱电子论（路易斯理论）

2. 弱电解质的质子传递平衡 纯水具有微弱的导电性，水分子间存在质子自递平衡。298K时，水溶液中，$pK_W^\ominus = pH + pOH = 14$ 弱酸和弱碱在水中的电离平衡常数分别用 $K_a^\ominus$ 和 $K_b^\ominus$ 表示。$K_a^\ominus$、$K_b^\ominus$ 只与温度有关，而与浓度无关。水溶液中，共轭酸碱对的电离平衡常数的乘积等于水的离子积，即 $K_a^\ominus \times K_b^\ominus = K_W^\ominus$。弱酸、弱碱在溶液中的解离程度可用电离度（解离度）表示，符号 α。对于同一弱电解质溶液，浓度越稀，解离度越大，这个规律称为稀释定律：$\alpha = \sqrt{\dfrac{K^\ominus}{c}}$。

对于一元弱酸，当 $c \cdot K_a^\ominus \geqslant 20K_W^\ominus$ 忽略水的电离；当 $\dfrac{c}{K_a^\ominus} \geqslant 400$ 或 $\alpha < 5\%$ 时，$[H^+] = \sqrt{c \cdot K_a^\ominus}$。

对于一元弱碱，当 $c \cdot K_b^\ominus \geqslant 20K_W^\ominus$ 忽略水的电离；当 $\frac{c}{K_b^\ominus} \geqslant 400$ 或 $\alpha < 5\%$ 时，$[OH^-] = \sqrt{c \cdot K_b^\ominus}$。

当 $\frac{c}{K_a^\ominus} < 400$（$\frac{c}{K_b^\ominus} < 400$）或 $\alpha > 5\%$，需要解一元二次方程。

多元弱酸（碱）在溶液中的质子传递是分步进行的，溶液中存在多步质子传递平衡。多元弱酸（碱）的质子传递均是逐级减弱，酸（碱）常数逐级减小，一般都彼此相差 10^4~10^5。当多元弱酸（碱）的 $K_{a1}^\ominus >> K_{a2}^\ominus$，$K_{b1}^\ominus >> K_{b2}^\ominus$ 时，求 $[H^+]$、$[OH^-]$ 可当作一元弱酸（碱）处理，衡量多元弱酸碱的强度可用 $K_{a1}^\ominus$、$K_{b1}^\ominus$。

酸碱质子论中，把既能给出质子又能接受质子的物质称为两性物质，其氢离子浓度计算如下：

$$[H^+] = \sqrt{K_a^\ominus \cdot K_a^{\ominus *}}$$

$$pH = \frac{1}{2}pK_a^\ominus + \frac{1}{2}pK_a^{\ominus *}$$

在弱电解质溶液中，加入与该弱电解质含有相同离子的强电解质，使弱电解质解离度降低的现象称为同离子效应。在弱电解质溶液中加入与该弱电解质不含相同离子的强电解质，使弱电解质解离度略微增大的作用称为盐效应。

3. 缓冲溶液 能抵抗外来少量强酸、强碱或水的稀释而保持本身 pH 基本不变的溶液称为缓冲溶液。缓冲溶液由一对共轭酸碱组成。组成缓冲溶液的一对共轭酸碱称为缓冲对或缓冲系。常见的缓冲对有以下几类：弱酸及其共轭碱；弱碱及其共轭酸；两性物质及其共轭酸、碱。缓冲溶液中 pH 的近似计算等同于同离子效应中 pH 的近似计算，$pH = pK_a^\ominus + \lg\frac{c(共轭碱)}{c(共轭酸)}$。缓冲溶液的有效 pH 范围：$pH = pK_a^\ominus \pm 1$。选择缓冲溶液时，所选用的缓冲对不能与反应物、生成物发生作用。对于药用缓冲对，还要考虑缓冲对物质不能与主药发生配伍禁忌，在加温灭菌和贮存期内要稳定，不能有毒性等。

Exercises

1. Give the conjugate acid of the following base: H_2O, HPO_4^{2-}, $H_2PO_4^-$, CO_3^{2-}, HS^-.

2. Give the conjugate base of the following acid: H_2O, H_2CO_3, HCN, NH_4^+, CH_3COOH.

3. Calculate the pH of the following mixed solutions at 298K:

(1) Equal volume pH=1.00 HCl and pH=3.00 HCl

(2) Equal volume pH=1.00 HCl and pH=12.00 NaOH

4. The pH of 0.010mol/L weak acid HA is 4.00, calculate the $K_a^\ominus$ and α of HA.

5. Calculate the pH and degree of dissociation of 0.05mol/L $NH_3 \cdot H_2O$ at 20°C. $K_b^\ominus = 1.74 \times 10^{-5}$.

6. At room temperature, degree of dissociation of HAc is 2.0%, $K_a^\ominus$=1.75×10^{-5}, calculate the concentration of HAc.

7. Calculate the $[H^+]$、$[HCO_3]$、$[CO_3^{2-}]$ and pH of the 0.034mol/L H_2CO_3. $K_{a1}^\ominus = 4.47 \times 10^{-7}$, $K_{a2}^\ominus = 4.68 \times 10^{-11}$

8. Add some solid NaAc into 0.10mol/L HAc solution to make the concentration of NaAc reaches 0.20mol/L. Calculate the $[H^+]$ and a of the HAc solution. (For HAe, $K_a^\ominus$=1.75×10^{-5})

9. To prepare a buffer solution with pH = 4.90, how many grams of solids $NaAc\cdot 3H_2O$ should be dissolved into 500ml 0.100mol/L HAc? ($pK_a^\ominus = 4.76$)

10. To prepare 200ml buffer solution with pH = 9.00, how many milliliters of 3.0mol/L HCl must be added to 100ml 0.1mol/L $NH_3\cdot H_2O$? (For $NH_3\cdot H_2O$, $K_b^\ominus = 1.74\times10^{-5}$)

11. 20ml 0.500mol/L weak monoprotic acid was mixed with 20ml 0.200mol/L NaOH solution. Then when the mixture was diluted to 1L, pH = 4.93. Calculate the $K_a^\ominus$ of the weak monoprotic acid.

12. 30ml 0.50mol/L weak acid HB solution mixed with 10ml 0.50mol/L NaOH solution, after diluting to 100ml by water, what is the pH? (For HB, $K_a^\ominus = 5.0\times10^{-7}$)

13. Calculate the pH of following solution. (For $NH_3\cdot H_2O$, $K_b^\ominus = 1.74\times10^{-5}$)

(1) 10ml 0.20mol/L HCl is added to 20ml 0.20mol/L$NH_3\cdot H_2O$.

(2) 20ml 0.20mol/L HCl is added to 20ml 0.20mol/L$NH_3\cdot H_2O$.

(3) 30ml 0.20mol/L HCl is added to 20ml 0.20mol/L$NH_3\cdot H_2O$.

目标检测

1. 给出下列碱的共轭酸：H_2O, HPO_4^{2-}, $H_2PO_4^-$, CO_3^{2-}, HS^-.

2. 给出下列酸的共轭碱：H_2O, H_2CO_3, HCN, NH_4^+, CH_3COOH.

3. 计算 298K 时，下列混合溶液的 pH。

（1）pH=1.00 和 pH=3.00 的 HCl 等体积混合；

（2）pH=1.00 的 HCl 和 pH=12.00 的 NaOH 等体积混合。

4. 已知 0.010mol/L 弱酸 HA 溶液的 pH 值为 4.00，计算 HA 的电离常数 $K_a^\ominus$ 和电离度 α。

5. 计算 20℃时 0.05mol/L $NH_3\cdot H_2O$ 溶液的 pH 值及电离度。（已知 20℃时 $K_b^\ominus = 1.74\times10^{-5}$）

6. 已知室温时醋酸的电离度为 2.0%，其电离平衡常数 $K_a^\ominus = 1.75\times10^{-5}$，试计算 HAc 的浓度。

7. 计算 0.034mol/L H_2CO_3 溶液的 $[H^+]$、$[HCO_3^-]$、$[CO_3^{2-}]$ 及溶液的 pH 值。（已知 H_2CO_3 的 $K_{a1}^\ominus = 4.47\times10^{-7}$，$K_{a2}^\ominus = 4.68\times10^{-11}$）

8. 如果在 0.10mol/L 的 HAc 溶液中加入固体 NaAc，使 NaAc 的浓度达到 0.20mol/L，求该 HAc 溶液中的 $[H^+]$ 和电离度 α。（已知 HAc 的 $K_a^\ominus=1.75\times10^{-5}$）

9. 要配制 pH 为 4.90 的缓冲溶液，需称取多少克 $NaAc\cdot 3H_2O$ 固体溶解于 500ml 0.100mol/L 的 HAc 溶液中？（已知 HAc 的 $pK_a^\ominus = 4.76$）

10. 用 100ml 0.1mol/L $NH_3\cdot H_2O$ 溶液和 3.0mol/L HCl 溶液配制 200ml pH=9.00 的缓冲溶液，计算需此盐酸溶液多少毫升？（已知 $NH_3\cdot H_2O$ 的 $K_b^\ominus = 1.74\times10^{-5}$）

11. 0.500mol/L 的一元弱酸溶液 20ml 与 0.200mol/L NaOH 溶液 20ml 混合，稀释到 1L 后，测得 pH 值为 4.93，求此一元弱酸的电离平衡常数 $K_a^\ominus$。

12. 取 0.50mol/L 某弱酸 HB 溶液 30ml 与 0.50mol/L NaOH 溶液 10ml 混合，并用水稀释至 100ml。测得该溶液的 $K_a^\ominus$ 值为 5.0×10^{-7}，求 pH 值。

13. 计算下列混合溶液的 pH 值。（已知氨水的 $K_b^\ominus = 1.74\times10^{-5}$）

（1）20ml 0.20mol/l NH_3 水溶液加入 10ml 0.20mol/LHCl 溶液。

（2）20ml 0.20mol/l NH_3 水溶液加入 20ml 0.20mol/LHCl 溶液。

（3）20ml 0.20mol/l NH_3 水溶液加入 30ml 0.20mol/LHCl 溶液。

Chapter 5 The Equilibrium of Precipitation and Dissolution
第五章 沉淀－溶解平衡

学习目标

知识要求

1．**掌握** 溶度积和溶解度的基本概念以及二者之间的换算；溶度积规则以及应用；溶度积规则判断沉淀的生成和溶解。

2．**熟悉** 难溶强电解质的同离子效应和盐效应。

3．**了解** 沉淀－溶解平衡在药学研究中的应用。

能力要求

利用难溶强电解质的沉淀－溶解平衡理论解释医药学领域的沉淀溶解现象，应用沉淀－溶解平衡原理进行离子鉴别、分离和含量测定等相关研究。

As a rule of thumb, compounds which dissolve to the extent of 0.020mol/L or more in water are classified as soluble (可溶的). Refer to the solubility guidelines (Table 5-1) as necessary. So far we have discussed mainly compounds that are quite soluble in water. Although most compounds dissolve in water to some extent, many are so slightly soluble that they are called "insoluble compounds" (不溶性化合物). Strictly speaking, the solubility of a substance is relative. There is no substance that is absolutely insoluble in water (没有绝对不溶于水的物质). Generally, substances with solubility less than 0.01g/100g (H_2O) are called "slightly soluble compounds" (难溶、几乎不溶或不溶解的化合物).

Table 5-1 Solubility guidelines for common ionic compounds in water

Generally soluble	Exceptions
Na^+, K^+, NH_4^+ compounds	No common exceptions
fluorides (F^-)	Insoluble: MgF_2, CaF_2, SrF_2, BaF_2, PbF_2
chlorides (Cl^-)	Insoluble: AgCl, Hg_2Cl_2
	Soluble in hot water: $PbCl_2$
bromides (Br^-)	Insoluble: AgBr, Hg_2Br_2, $PbBr_2$
	Moderately soluble: $HgBr_2$
iodides (I^-)	Insoluble: many heavy-metal iodides

(continued)

Generally soluble	Exceptions
sulfates (SO_4^{2-})	Insoluble: $BaSO_4$, $PbSO_4$, $HgSO_4$
	Moderately soluble: $CaSO_4$, $SrSO_4$, Ag_2SO_4
nitrates (NO_3^-), nitrites (NO_2^-)	Moderately soluble: $AgNO_2$
chlorates (ClO_3^-), perchlorates (ClO_4^-)	No common exceptions
acetates (CH_3COO^-)	Moderately soluble: $AgCH_3COO$
sulfides (S^{2-})**	Soluble: those of NH_4^+, Na^+, K^+, Mg^{2+}, Ca^{2+}
oxides (O^{2-}), hydroxides (OH^-)	Soluble: Li_2O*, $LiOH$, Na_2O*, $NaOH$, K_2O*, KOH, BaO*, $Ba(OH)_2$
	Moderately soluble: CaO*, $Ca(OH)_2$, SrO*, $Sr(OH)_2$
carbonates (CO_3^{2-}), phosphates (PO_4^{3-}), arsenates (AsO_4^{3-})	Soluble: those of NH_4^+, Na^+, K^+

* Dissolves with evolution of heat and formation of hydroxides.

** Dissolves with formation of HS^- and H_2S.

Slightly soluble compounds are important in many natural phenomena. Our bones and teeth are mostly calcium phosphate, $Ca_3(PO_4)_2$, a slightly soluble compound. Also, many natural deposits of $Ca_3(PO_4)_2$ rock are mined and converted into agricultural fertilizer Limestone caves have been formed by acidic water slowly dissolving away calcium carbonate, $CaCO_3$. Sinkholes are created when acidic water dissolves away most of the underlying $CaCO_3$. The remaining limestone can no longer support the weight above it, so it collapses, and a sinkhole is formed.

The complete ionization of the slightly soluble compounds will occur when dissolved in water, which is called the slightly soluble strong electrolyte (难溶强电解质). In this chapter, we shall consider those that are only very slightly soluble.

The equilibrium of precipitation and dissolution (沉淀溶解平衡) of the slightly soluble strong electrolyte is the balance between the slightly soluble strong electrolyte and their dissolved ions. For example, AgCl is a kind of slightly soluble strong electrolyte. After a small amount of AgCl dissolves in water, Ag^+ and Cl^- are obtained. There is a multiple phase dynamic equilibrium between them and AgCl.

1. Solubility Product Principle

PPT

微课

1.1 Solubility Product

At a certain temperature, mixing the insoluble AgCl solid with water, some Ag^+ and Cl^- on the surface of AgCl solid enter the water in the form of hydrated ions under the action of polar water molecules. This process is called dissolution (溶解). At the same time, Ag^+ (aq) and Cl^- (aq) are in a continuous disordered motion and Ag^+ (aq) and Cl^- (aq) will be redeposited on the surface of AgCl under the action of positive and negative ions on the surface of the solid. This process is called precipitation (沉淀). When the rate of dissolution and precipitation is equal, the system reaches dynamic equilibrium,

which is called the Equilibrium of Precipitation and Dissolution (沉淀-溶解平衡). This balance can be expressed as:

$$AgCl(s) \underset{\text{precipitation}}{\overset{\text{dissolution}}{\rightleftharpoons}} Ag^{+}(aq) + Cl^{-}(aq)$$

We usually omit H_2O over the arrows in such equations; the (aq) phase of the ions means that we are dealing with aqueous solution. According to the law of chemical equilibrium, the expression of the above equilibrium constant is:

$$K_{ap}^{\ominus}(AgCl) = a(Ag^{+}) \times a(Cl^{-})$$

When the equilibrium of precipitation and dissolution is reached, the activity of each ion in the solution does not change with time, and the solution is the saturated solution of AgCl at this temperature. The equilibrium constant $K_{ap}^{\ominus}$ is the product of the activity of hydrated ions in saturated solution, which is called activity product constant (活度积). slightly soluble strong electrolyte is discussed whose solubility is very small, means the concentration of ions in the solution is less, and the interaction between ions can be ignored, so concentration can be substituted for activity, and solubility product can take the place of activity product.

Therefore, the equation of above mentioned equilibrium constant can be expressed as:

$$K_{sp}^{\ominus} = [Ag^{+}]/c^{\ominus} \cdot [Cl^{-}]/c^{\ominus}$$

Abbreviated as: $K_{sp}^{\ominus} = [Ag^{+}][Cl^{-}]$

$K_{sp}^{\ominus}$ is the equilibrium constant of equilibrium of precipitation and dissolution of slightly soluble compounds, which reflects the dissolubility of substances, so it is called solubility product constant (溶度积常数) The constant uses "sp" as the abbreviation of solution product. The solubility product constant for AgCl is the product of the concentrations of its constituent ions in a saturated solution. In the expression, $[Ag^{+}]$ and $[Cl^{-}]$ are the equilibrium concentration, whose unit is mol/L, $c^{\ominus}$=1mol/L。Each kind of slightly soluble strong electrolyte has its own solubility product at a certain temperature, and the slightly soluble strong electrolytes of different types have different solubility product expressions.

For example: $ZnS(s) \rightleftharpoons Zn^{2+} + S^{2-}$

(for simplicity, the hydration symbol may be omitted)

When reaching equilibrium, $K_{sp}^{\ominus} = [Zn^{2+}] \times [S^{2-}]$

If reaction: $PbCl_2(s) \rightleftharpoons Pb^{2+} + 2Cl^{-}$

Achieve equilibrium, $K_{sp}^{\ominus} = [Pb^{2+}][Cl^{-}]^2$

In summary, it can be expressed by general formula:

$$A_mB_n(s) \rightleftharpoons mA^{n+} + nB^{m-}$$

When the balance is reached, $K_{sp}^{\ominus}(A_mB_n) = [A^{n+}]^m[B^{m-}]^n$

The upper formula shows that it is a constant in the saturated solution of the insoluble strong electrolyte, at the certain temperature.

Like other chemical equilibrium constants, $K_{sp}^{\ominus}$ is only related to the nature of the slightly soluble strong electrolyte and temperature, but not to the ion concentrations. Generally, $K_{sp}^{\ominus}$ does not change much with the temperature. In general, $K_{sp}^{\ominus}$ with 298.15K is used, and some common slightly soluble strong electrolytes (298.15K) are listed in Appendix 4.

1.2 Solubility Product and Solubility

The solubility product $K_{sp}^{\ominus}$ indicates the dissolution trend of the slightly strong electrolyte. And the solubility (s) is the saturated concentration of insoluble substances, which can also indicate the dissolution degree of them. There is an inevitable relationship between the two. The quantitative relationship between solubility product and solubility is different for different types of slightly insoluble strong electrolytes.

1.2.1 The AB Type of Slightly Soluble Strong Electrolyte

This type includes: AgBr, $BaSO_4$, $PbCrO_4$, $CaCO_3$, etc. When the precipitation-dissolution equilibrium is reached, the amount of positive and negative ions is equal. Therefore, the relative concentration of positive and negative ions in the solution is numerically equal to the solubility of the substance.

If the solubility of the AB type of slightly soluble strong electrolyte is *s* mol/L, then:

$$AB(s) \rightleftharpoons A^{+} + B^{-}$$

Equilibrium concentration (mol/L) $\quad s \quad s$

$$K_{sp}^{\ominus} = [A^{+}][B^{-}] = s^2$$

$$s = \sqrt{K_{sp}^{\ominus}}$$

Example 5-1 It is known that $K_{sp}^{\ominus}$ for AgCl is 1.77×10^{-10}, and the solubility of AgCl is calculated.

Solution the following equilibrium exists in the saturated solution of AgCl:

$$AgCl \rightleftharpoons Ag^{+} + Cl^{-}$$

Equilibrium concentration (mol/L) $\quad s \quad s$

$$K_{sp}^{\ominus}(AgCl) = [Ag^{+}][Cl^{-}] = s^2$$

$$s = \sqrt{K_{sp}^{\ominus}(AgCl)} = \sqrt{1.77\times10^{-10}} = 1.33\times10^{-5} mol/L$$

1.2.2 The AB_2 or A_2B Type of Slightly Soluble Strong Electrolyte

This type includes: $Mn(OH)_2$, $PbCl_2$, Ag_2CrO, etc. The relationship between its solubility product and solubility is as follows:

Taking the type of AB_2 as an example, if its solubility is *s*, then:

$$AB_2(s) \rightleftharpoons A^{2+} + 2B^{-}$$

Equilibrium concentration (mol/L) $\quad s \quad 2s$

$$K_{sp}^{\ominus} = [A^{2+}][B^{-}]^2 = s(2s)^2 = 4s^3$$

$$s = \sqrt[3]{\frac{K_{sp}^{\ominus}}{4}}$$

Example 5-2 At 298K, the solubility product of $Mg(OH)_2$ is 5.61×10^{-12}. If $Mg(OH)_2$ dissociates completely in saturated solution, try to calculate the solubility of $Mg(OH)_2$ in water and the concentrations of Mg^{2+} and OH^{-}.

Solution Let the solubility of $Mg(OH)_2$ in water be s

$$Mg(OH)_2(s) \rightleftharpoons Mg^{2+}(aq) + 2OH^-(aq)$$

Equilibrium concentration (mol/L) s $2s$

$$K_{sp}^{\ominus} = s(2s)^2 = 5.61\times10^{-12}$$

$$s = \sqrt[3]{\frac{K_{sp}^{\ominus}}{4}} = \sqrt[3]{\frac{5.61\times10^{-12}}{4}} = 1.12\times10^{-4}$$

Concentration of each ion in the solution:

$$[Mg^{2+}] = s = 1.12\times10^{-4} mol/L$$

$$[OH^-] = 2s = 2.24\times10^{-4} mol/L$$

1.2.3 The AB_3 or A_3B Type of Slightly Soluble Strong Electrolyte

The relationship between its solubility and solubility product is as follows: When the solution of slightly soluble strong electrolyte reaches equilibrium with precipitation, the solubility is set as s mol/L

$$AB_3 \rightleftharpoons A^{3+} + 3B^-$$

Equilibrium concentration (mol/L) s $3s$

$$K_{sp}^{\ominus} = s(3s)^3 = 27s^4$$

$$s = \sqrt[4]{\frac{K_{sp}^{\ominus}}{27}}$$

For the general precipitation reaction of slightly soluble strong electrolyte, it can be expressed by A_mB_n general formula. If its solubility is s at a certain temperature, the following equilibrium exists in the saturated solution:

$$A_mB_n(s) \rightleftharpoons mA^{n+} + nB^{m-}$$

Equilibrium concentration (mol/L) ms ns

$$K_{sp}^{\ominus} = [A^{n+}]^m[B^{m-}]^n$$

$$=(ms)m \cdot (ns)n$$

$$=m^m \cdot n^n \cdot s^{m+n}$$

$$s = \sqrt[m+n]{\frac{K_{sp}^{\ominus}}{m^m n^n}}$$

The conversion relationship between solubility and solubility product shall meet the following conditions.

① It is only suitable for the slightly soluble strong electrolyte with very small solubility. The solubility of slightly soluble strong electrolyte is small, the concentration of ions in saturated solution is small, and the interaction between ions is weak, so the solubility can be calculated by using concentration instead of activity.

② It is only suitable for the slightly soluble strong electrolyte which is ionized after dissolution and ionized ions do not have any chemical reaction in the aqueous solution, not suitable for the slightly soluble electrolyte which is easy to hydrolyze.

For example, for some slightly soluble sulfides, carbonate and phosphoric acid solutions, the

hydrolysis of anions cannot be ignored.

③ It is only suitable for the slightly soluble strong electrolyte which is completely ionized in one step after dissolution, not suitable for weak electrolyte.

For example, $Fe(OH)_3$ is ionized in three steps in aqueous solution:

Table 5-2 $K_{sp}^{\ominus}$ and solubility of different types of slightly soluble electrolytes (298K)

Type	Slightly soluble electrolyte	$K_{sp}^{\ominus}$	s/(mol/L)
AB	AgCl	1.77×10^{-10}	1.33×10^{-5}
	AgBr	5.35×10^{-13}	7.33×10^{-7}
	AgI	8.52×10^{-17}	9.25×10^{-9}
AB_2	MgF_2	5.16×10^{-12}	3.72×10^{-4}
A_2B	Ag_2CrO_4	1.12×10^{-12}	6.54×10^{-5}

$$Fe(OH)_3(s) \rightleftharpoons Fe(OH)_2^+ + OH^- \qquad K_1^{\ominus}$$
$$Fe(OH)_2^+ \rightleftharpoons Fe(OH)^{2+} + OH^- \qquad K_2^{\ominus}$$
$$Fe(OH)^{2+} \rightleftharpoons Fe^{3+} + 3OH^- \qquad K_3^{\ominus}$$

Relative to total ionization equilibrium, although the relationship of $[Fe^{3+}]\times[OH^-]^3 = K_{sp}^{\ominus}$ exists, the ratio of $[Fe^{3+}]$ to $[OH^-]$ in the solution is not equal to 1:3; and the covalent slightly soluble electrolyte such as $HgCl_2$ whose dissolved parts are not all simple ions.

To sum up, the value of $K_{sp}^{\ominus}$ can be used to compare the solubility of slightly soluble strong electrolytes. At a certain temperature, for the same type of slightly soluble electrolytes, the higher $K_{sp}^{\ominus}$, the higher its solubility. For different types of slightly soluble electrolytes, $K_{sp}^{\ominus}$ cannot be used to compare the solubility. Only after conversion can we draw a conclusion. The solubility is related to the presence of ions in the solution, but $K_{sp}^{\ominus}$ does not change.

Only the same type of slightly soluble strong electrolyte, which is basically non-hydrolyzed, can be compared with the solubility directly according to the solubility product as shown in Table 5-2.

1.3 Solubility Product Principle

According to solubility product constant, the direction of precipitation and dissolution reaction can be determined. In some slightly soluble electrolyte solution, the product of the power of ion concentrations is called ion product (离子积), which is expressed in Q.

For example, the ion product of $BaSO_4$ is:

$$Q = [Ba^{2+}][SO_4^{2-}]$$

$K_{sp}^{\ominus}$ refers in particular to the product of the power of the relative concentration of each ion of the slightly soluble strong electrolyte in the saturated solution, which is a constant at a certain temperature. And Q is the product of the power of the relative concentration of each ion in any case. At a certain temperature, its value is uncertain. $K_{sp}^{\ominus}$ is just a special case of Q.

In any given solution:

When $Q = K_{sp}^{\ominus}$, the solution is saturated, that is to say, it reaches the state of equilibrium. of precipitation-dissolution.

When $Q < K_{sp}^{\ominus}$, the solution is unsaturated without precipitation; If there is a solid in the system, the precipitate will dissolve until it reaches a new equilibrium (saturation).

When $Q > K_{sp}^{\ominus}$, the solution is supersaturated and precipitation precipitates from the solution until it is saturated.

As mentioned above, the relation between Q and $K_{sp}^{\ominus}$ and its conclusion are called solubility product principle (溶度积原理). It is a summary of the moving law of precipitation-dissolution equilibrium, which can be used to judge the formation and dissolution of precipitation or whether the precipitation and solution are in equilibrium. It is the basic rule of precipitation reaction. The solubility product expression for a compound is the product of the concentrations of its constituent ions, each raised to the power that corresponds to the number of ions in one formula unit of the compound. The quantity is constant at constant temperature for a saturated solution of the compound. This statement is the solubility product principle (溶度积原理).

2. Equilibrium Shift between Precipitation and Dissolution

PPT

Precipitation dissolution equilibrium is a dynamic equilibrium, when the equilibrium conditions change, the equilibrium will shift, formation of precipitation or dissolution of precipitates will occur.

2.1 Dissolution of Precipitates

According to the solubility product principle, the necessary condition for precipitation dissolution is $Q < K_{sp}^{\ominus}$ .Therefore, decrease the concentration of the relevant ions in the solution so that $Q < K_{sp}^{\ominus}$, the precipitate will dissolve. Several methods are commonly used as follows.

2.1.1 Dissolution of Precipitates due to the Formation of Weak Electrolytes

When acid, base and salt solutions are added to the slightly soluble and strong electrolytes, they dissolve due to the reaction to form weak electrolytes such as H_2O, weak acid, weak base and salt that are hard to ionize.

(1) Dissolution of Precipitates due to the Formation of Water

Some hydroxides that is difficult to dissolve, such as $Mg(OH)_2$, $Cu(OH)_2$, $Fe(OH)_3$, $Al(OH)_3$, react with the acid to form weak electrolyte water, which reduces the solubility of OH^- in the solution and makes the hydroxides $Q < K_{sp}^{\ominus}$, then the equilibrium will shift towards the dissolution of precipitation. Such as:

$$\begin{array}{c} Al(OH)_3 \rightleftharpoons Al^{3+} + 3\,OH^- \\ + \\ 3HCl \rightarrow 3Cl^- + 3H^+ \\ \upharpoonleft\downharpoonright \\ 3H_2O \end{array}$$

the reaction equation of dissolution: $Al(OH)_3 + 3H^+ \rightleftharpoons Al^{3+} + 3H_2O$

the equilibrium constant:

$$K^{\ominus}=\frac{[Al^{3+}]}{[H^+]^3}=\frac{[Al^{3+}][OH^-]^3}{[H^+]^3[OH^-]^3}=\frac{K_{sp}^{\ominus}(Al(OH)_3)}{(K_W^{\ominus})^3}$$

$$=\frac{1.3\times10^{-33}}{(1.0\times10^{-14})^3}=1.3\times10^9$$

The equilibrium constant of the reaction $K^{\ominus}>10^6$, so the reaction is complete. The equilibrium constant $K^{\ominus}$ of the dissolution reaction of the slightly soluble hydroxide is related to the solubility of the slightly soluble electrolyte, the strength of the acid and the ionic product of water.

(2) Dissolution of Precipitates due to the Formation of Weak Acid

Many slightly soluble strong electrolytes, such as $CaCO_3$, $BaCO_3$, FeS, MnS, can be dissolved in strong acid solutions. For example:

$$MnS(s)\rightleftharpoons Mn^{2+}+S^{2-}$$
$$+$$
$$2HCl\rightleftharpoons 2Cl^-+2H^+$$
$$\upharpoonleft\downharpoonright$$
$$H_2S$$

The S^{2-} and H^+ ionized by HCl form weak acid H_2S, which reduces the solubility of S^{2-} in the solution and makes the solution $Q<K_{sp}^{\ominus}$, then the precipitate MnS will dissolve. The reaction equation of dissolution: $MnS(s)+2H^+\rightleftharpoons Mn^{2+}+H_2S$

the equilibrium constant:

$$K^{\ominus}=\frac{[Mn^{2+}][H_2S]}{[H^+]^2}=\frac{[Mn^{2+}][H_2S][S^{2-}][HS^-]}{[H^+]^2[S^{2-}][HS^-]}$$

$$=\frac{K_{sp}^{\ominus}(MnS)}{K_{a1}^{\ominus}(H_2S)\cdot K_{a2}^{\ominus}(H_2S)}=\frac{2.5\times10^{-13}}{8.91\times10^{-27}}=2.8\times10^{13}$$

The equilibrium constant of the reaction $K^{\ominus}>10^6$, so the reaction is complete. If the above-mentioned sulfide is replaced by CuS, the reaction equation of dissolution.

$$CuS(s)+2H^+\rightleftharpoons Cu^{2+}+H_2S$$

the equilibrium constant:

$$K^{\ominus}=\frac{[Cu^{2+}][H_2S]}{[H^+]^2}$$

$$=\frac{K_{sp}^{\ominus}(CuS)}{K_{a1}^{\ominus}(H_2S)\cdot K_{a2}^{\ominus}(H_2S)}=\frac{6.3\times10^{-36}}{8.91\times10^{-27}}=7.1\times10^{10}$$

The equilibrium constant of the reaction $K^{\ominus}>10^6$, so the reaction is almost impossible.

Obviously, the equilibrium constant is determined by $K_{sp}^{\ominus}$ and $K_a^{\ominus}$. The larger the $K_{sp}^{\ominus}$ of precipitation, or the smaller the $K_a^{\ominus}$ of the weak acid, the more complete the dissolution reaction.

(3) Dissolution of Precipitates due to the Formation of Weak Base

Some hydroxides that is difficult to dissolve, such as $Mg(OH)_2$, $Mn(OH)_2$, can be dissolved in the solution of NH_4Cl. The OH^- and NH_4^+ ionized by NH_4Cl form weak base $NH_3\cdot H_2O$, which reduces the solubility of OH^- in the solution and makes the hydroxides $Q<K_{sp}^{\ominus}$, then the equilibrium will shift

towards the dissolution of precipitation. Such as:

$$Mg(OH)_2(s) \rightleftharpoons Mg^{2+} + 2OH^-$$
$$+$$
$$2NH_4Cl \rightleftharpoons 2Cl^- + 2NH_4^+$$
$$\upharpoonleft\downharpoonright$$
$$2NH_3 \cdot H_2O$$

the reaction equation of dissolution:

$$Mg(OH)_2(s) + 2NH_4^+ \rightleftharpoons Mg^{2+} + 2NH_3 \cdot H_2O$$

the equilibrium constant:

$$K^\ominus = \frac{[Mg^{2+}][NH_3 \cdot H_2O]^2}{[NH_4^+]^2} = \frac{[Mg^{2+}][OH^-]^2[NH_3 \cdot H_2O]^2}{[NH_4^+]^2[OH^-]^2} = \frac{K_{sp}^\ominus(Mg(OH)_2)}{[K_b^\ominus(NH_3 \cdot H_2O)]^2}$$

$$= \frac{5.61 \times 10^{-12}}{(1.74 \times 10^{-5})^2} = 1.85 \times 10^{-2}$$

The equilibrium constant of the reaction is not very large, but the precipitation will dissolve with enough NH_4Cl. Whether the slightly soluble hydroxide can be dissolved by the formation of a weak base depends on the $K_{sp}^\ominus$ of the slightly soluble hydroxide and the $K_b^\ominus$ of the weak base. The larger the $K_{sp}^\ominus$ of precipitation, or the smaller the $K_b^\ominus$ of the weak base, the easier the dissolution reaction. Some hydroxides that difficult to dissolve, such as $Al(OH)_3$ and $Fe(OH)_3$, are soluble in strong acid solutions, but slightly soluble in ammonium solution.

(4) Dissolution of Precipitates due to the Formation of Salt that is Difficult to Ionization

The precipitate $PbSO_4$ can be dissolved in saturated NaAc solution, and the reaction is as follows:

$$PbSO_4(s) \rightleftharpoons Pb^{2+} + SO_4^{2-}$$
$$+$$
$$2NaAc \rightleftharpoons 2Ac^- + 2Na^+$$
$$\upharpoonleft\downharpoonright$$
$$Pb(Ac)_2$$

The Pb^{2+} and Ac^- ionized by NaAc form weak electrolyte $Pb(Ac)_2$, which reduces the solubility of Pb^{2+} in the solution and makes the solution $Q < K_{sp}^\ominus$, then the precipitate $PbSO_4$ will dissolve.

2.1.2 Dissolution of Precipitates due to Redox Reaction

The precipitates with very small $K_{sp}^\ominus$, such as CuS and Ag_2S, are slightly soluble in non-oxidizing strong acids, because the addition of high concentration of non-oxidizing strong acid cannot effectively reduce the concentration of S^{2-} ion. However, if strong oxidizing acids, such as hot and dilute HNO_3, are added, S^{2-} ion can be oxidized to elemental S, thus greatly reducing the concentration of S^{2-} ions and dissolving the precipitation , so that $Q < K_{sp}^\ominus$, precipitation will dissolve. Reactions can be expressed as follows:

$$3CuS + 8HNO_3(dilute) = 3Cu(NO_3)_2 + 2NO\uparrow + 3S\downarrow + 4H_2O$$
$$3Ag_2S + 8HNO_3(dilute) = 6AgNO_3 + 2NO\uparrow + 3S\downarrow + 4H_2O$$

2.1.3 Dissolution of Precipitates due to the Formation of Complex Ion

Some slightly soluble strong electrolytes, such as AgX, are slightly soluble in either non-oxidizing or oxidizing acids. However, they can form complexes with some complexants, and the concentration of ions in the solution is greatly reduced, so that $Q < K_{sp}^\ominus$, precipitation will dissolve. For example, AgCl is

soluble in $NH_3 \cdot H_2O$, AgBr is soluble in solution of $Na_2S_2O_3$, and AgI is soluble in solution of KCN.

$$AgCl(s) + 2NH_3 \rightleftharpoons [Ag(NH_3)_2]^+ + Cl^-$$

$$AgBr(s) + 2S_2O_3^{2-} \rightleftharpoons [Ag(S_2O_3)_2]^{3-} + Br^-$$

$$AgI(s) + 2CN^- \rightleftharpoons [Ag(CN)_2]^- + I^-$$

Whether the precipitation dissolve or not, and how much precipitation is dissolved, depend not only on the $K_{sp}^{\ominus}$ of the precipitate, but also on the stability of the formation of the complex ion.

In some cases, it is sometimes necessary to use two or more methods at the same time in order to dissolve the extremely slightly soluble strong electrolyte. For example, HgS ($K_{sp}^{\ominus} = 4.0 \times 10^{-53}$) must be dissolved by adding aqua regia, using both redox reaction and formation of complex ion. Specific reactions are as follows:

$$3HgS + 2NO_3^- + 12Cl^- + 8H^+ = 3[HgCl_4]^{2-} + 3S\downarrow + 2NO\uparrow + 4H_2O$$

During the reaction, S^{2-} is oxidized by HNO_3 to S, Hg^{2+} and Cl^- form complex ion $[HgCl_4]^{2-}$, the solubility of S^{2-} and Hg^{2+} in the solution is greatly reduced, $Q < K_{sp}^{\ominus}$, so precipitation will dissolve.

2.2 Formation of Precipitation

According to the solubility product principle, the necessary condition for formation of precipitation is $Q > K_{sp}^{\ominus}$. By adding a precipitant, the concentration of ions can be increased so that the equilibrium will shift to precipitate.

Example 5-3 The concentration of calcium ion in the blood plasma is 0.0025mol/L. If the concentration of oxalate ion is 0.001mol/L, do you expect calcium oxalate to precipitate? $K_{sp}^{\ominus}$ for calcium oxalate is 2.32×10^{-9}.

Solution

$$Q = [Ca^{2+}][C_2O_4^{2-}] = (0.0025) \times (0.001) = 2.5 \times 10^{-6} > K_{sp}^{\ominus}(CaC_2O_4 \cdot H_2O)$$

According to the solubility product principle, $Q > K_{sp}^{\ominus}$, so we expect precipitation to occur.

2.3 Fractional Precipitation

微课

When there are multiple ions in the solution, and they all react with the same precipitant to form a precipitate, what is the precipitate order? When the second ion precipitates, to what extent does the first one precipitate? For example, suppose a solution is 0.1mol/L Ba^{2+} and 0.1mol/L Sr^{2+}. As we will show in the following paragraphs, when we slowly add a concentrated solution of potassium chromate, K_2CrO_4, to the solution of Ba^{2+} and Sr^{2+} ions, barium chromate precipitates first. After most of the Ba^{2+} ion have precipitated, strontium chromate begins to come out of solution. It is therefore possible to separate Ba^{2+} and Sr^{2+} ions from a solution by fractional precipitation using K_2CrO_4.

To understand why Ba^{2+} and Sr^{2+} ions can be separated this way, let us calculate (1) the concentration of CrO_4^{2-} necessary to just begin the precipitation of $BaCrO_4$ and (2) that necessary to just begin the precipitation of $SrCrO_4$. We will ignore any volume change in the solution of Ba^{2+} and Sr^{2+} ions resulting from the addition of the concentrated K_2CrO_4 solution. To calculate the CrO_4^{2-} concentration when $BaCrO_4$ begins to precipitate, we substitute the initial Ba^{2+} concentration into the solubility product equation. $K_{sp}^{\ominus}$ for $BaCrO_4$ is 1.17×10^{-10}.

$$[Ba^{2+}][CrO_4^{2-}] = K_{sp}^{\ominus}(BaCrO_4)$$

$$(0.10)[CrO_4^{2-}] = 1.17 \times 10^{-10}$$

$$[CrO_4^{2-}] = \frac{1.17 \times 10^{-10}}{0.10} = 1.17 \times 10^{-9}\,\text{mol/L}$$

In the same way, we can calculate the CrO_4^{2-} concentration when $SrCrO_4$ begins to precipitate. $K_{sp}^{\ominus}$ for $SrCrO_4$ is 2.2×10^{-5}.

$$[Sr^{2+}][CrO_4^{2-}] = K_{sp}^{\ominus}(SrCrO_4)$$

$$(0.10)[CrO_4^{2-}] = 2.2 \times 10^{-5}$$

$$[CrO_4^{2-}] = \frac{2.2 \times 10^{-10}}{0.10} = 2.2 \times 10^{-4}\,\text{mol/L}$$

Note the $BaCrO_4$ precipitates first because the CrO_4^{2-} concentration necessary to form the $BaCrO_4$ precipitate is smaller.

These results reveal that as the solution of K_2CrO_4 is slowly added to the solution of Ba^{2+} and Sr^{2+}, $BaCrO_4$ begins to precipitate when the CrO_4^{2-} concentration reaches to 1.17×10^{-9}mol/L. $BaCrO_4$ continues to precipitate as K_2CrO_4 is added. When the concentration of CrO_4^{2-} reaches to 2.2×10^{-4}mol/L, $SrCrO_4$ begins to precipitate.

What is the concentration of Ba^{2+} ion remaining just as $SrCrO_4$ begins to precipitate? Let us first calculate the concentration of Ba^{2+} at this point. We write the solubility product equation and substitute $[CrO_4^{2-}] = 2.2 \times 10^{-4}$mol/L, which is the concentration of CrO_4^{2-} when $SrCrO_4$ begins to precipitate.

$$[Ba^{2+}][CrO_4^{2-}] = K_{sp}^{\ominus}(BaCrO_4)$$

$$[Ba^{2+}](2.2 \times 10^{-4}) = 1.17 \times 10^{-10}$$

$$[Ba^{2+}] = 5.3 \times 10^{-7}\text{mol/L} < 1.0 \times 10^{-5}\text{mol/L}$$

The concentration of Ba^{2+} ion remaining is quite low, less than 1.0×10^{-5}mol/L, so the Ba^{2+} ion has precipitated completely by the time $SrCrO_4$ begins to precipitate.

This phenomenon of adding a precipitant to precipitate a variety of ions in solution according to the order of the solubility product is called fractional precipitation (分步沉淀). Fractional precipitation is the technique of separating two or more ions from a solution by adding a reactant that precipitates first one ion, then another, and so forth. In general analysis, when the ion concentration is less than or equal to 1.0×10^{-5}mol/L, the ion can be considered to have precipitated completely.

Remarkably, the order of ion precipitation depends on the $K_{sp}^{\ominus}$ of the precipitate and the solubility of the precipitated ion. For precipitation of the same type, if the difference of $K_{sp}^{\ominus}$ in precipitation is relatively large, the precipitation of $K_{sp}^{\ominus}$, smaller will precipitate first, and the precipitation of $K_{sp}^{\ominus}$, larger will precipitate later. For precipitation of the different types, the sequence of precipitation cannot be directly determined according to $K_{sp}^{\ominus}$, but must be determined according to the calculation results. For example, when we slowly add a concentrated solution of $AgNO_3$, to the solution of 0.01mol/L Cl^- and 0.01mol/L CrO_4^{2-} ions, the concentration of Ag^+ when AgCl and Ag_2CrO_4 begin to precipitate respectively are:

$$c_1(Ag^+) = \frac{K_{sp}(AgCl)}{c(Cl^-)} = \frac{1.77 \times 10^{-10}}{0.01} = 1.77 \times 10^{-8}\,\text{mol/L}$$

$$c_2(Ag^+) = \sqrt{\frac{K_{sp}(Ag_2CrO_4)}{c(CrO_4^{2-})}} = \sqrt{\frac{1.12 \times 10^{-12}}{0.01}} = 1.06 \times 10^{-5}\,\text{mol/L}$$

Although $K_{sp}^{\ominus}(AgCl) > K_{sp}^{\ominus}(Ag_2CrO_4)$, the concentration of Ag^+ to form the AgCl precipitate is smaller, so AgCl will precipitate first.

2.4 Transformation of Precipitation

Transformation of precipitation (沉淀的转化) is called a process of converting one precipitation into another by chemical reaction. The transformation generally has two conditions as following:

$$MX(s) + Y \rightleftharpoons MY(s) + X$$

2.4.1 Precipitate of Greater Solubility Can be Transformed to That of Less Solubility

The transformation of precipitation is significant in practice. For example, a light yellow solution of K_2CrO_4 is added into the tube which contains a white $BaCO_3$ precipitate. $BaCO_3$ precipitate can be transformed to $BaCrO_4$ precipitate.

$$BaCO_3(s) + CrO_4^{2-} \rightleftharpoons BaCrO_4(s) + CO_3^{2-}$$

$$K^{\ominus} = \frac{[CO_3^{2-}]}{[CrO_4^{2-}]} = \frac{[CO_3^{2-}][Ba^{2+}]}{[CrO_4^{2-}][Ba^{2+}]} = \frac{K_{sp}(BaCO_3)}{K_{sp}(BaCrO_4)} = \frac{2.58\times10^{-9}}{1.17\times10^{-10}} = 22$$

The likelihood of conversion depends on the equilibrium constant of the reaction $K^{\ominus}$. When $K^{\ominus}$ is very large ($K^{\ominus} \geqslant 10^6$), the reaction can completely occur spontaneously. When $K^{\ominus} > 1$, the conversion can be occurred.

$CaSO_4$ in pot scale can easily be transformed to $CaCO_3$ with Na_2CO_3 solution. $PbCl_2$ in industrial residues can be transformed to $PbSO_4$ with treatment of Na_2SO_4 solution, and so on.

$$Na_2CO_3 + CaSO_4\ (s) \rightleftharpoons CaCO_3\downarrow + Na_2SO_4$$

$$PbCl_2(s) + Na_2SO_4 \rightleftharpoons PbSO_4\downarrow + 2NaCl$$

2.4.2 Slightly soluble Strong Electrolyte Can be Transformed to Slightly Soluble, Strong Electrolyte.

In general, when $K^{\ominus} > 1$, the transformation of precipitation can happen. When $K^{\ominus} < 1$, the transformation is difficult to happen. In other words, the precipitation with low solubility is difficult to be transformed to those with greater solubility. However, the transformation can occur between two precipitations with similar solubility only when we control certain conditions.

For example, $BaSO_4$, one of the important mineral resources of radon, is slightly soluble in various acids (e.g. hydrochloric acid, nitric acid, acetic acid, etc.). We need to control certain conditions to convert $BaSO_4$ into $BaCO_3$ which can be dissolved in hydrochloric acid solution.

$$BaSO_{4(s)} + CO_3^{2-} \rightleftharpoons BaCO_3(s) + SO_4^{2-}$$

$$K = \frac{[SO_4^{2-}]}{[CO_3^{2-}]} = \frac{K_{sp}(BaSO_4)}{K_{sp}(BaCO_3)} = \frac{1.08\times10^{-10}}{2.58\times10^{-9}} = \frac{1}{24}$$

$K^{\ominus} < 1$, the transformation is difficult. We need to treat $BaSO_4$ precipitation for 3-4 times with saturated Na_2CO_3 solution and the reaction is relatively complete.

2.5 Common Eon Effect and Salt Effect

2.5.1 Common Ion Effect

The common ion effect (同离子效应) is a term used to describe the effect on the decreasing solubility of the slightly soluble strong electrolyte in a solution of soluble electrolyte that contains the same ion or ions. A common ion is an ion that enters the solution from two different electrolyte sources with insolubility and solubility. For example, $AgNO_3$ is added to a saturated solution of AgCl, it is often

called as a source of a common ion of the Ag^+, which decreases the solubility of AgCl.

$$AgNO_3 \rightarrow Ag^+ + NO_3^-$$

$$AgCl(s) \rightleftharpoons Ag^+ + Cl^-$$

The Ag^+ concentration in the solution increases and the above reaction will shift left. More AgCl precipitation will generate until a new balance is established. The solubility of AgCl ultimately reduced.

Example 5-3 At 298.15*K*, the solubility of AgCl in pure water is 1.33×10^{-5}mol/L, calculate its solubility in 0.10mol/L HCl and 0.20mol/L $AgNO_3$, respectively. ($K_{sp}^{\ominus}$AgCl) = 1.77×10^{-10}

Solution Let s_1 be the molar solubility of AgCl in 0.10mol/L HCl solution, then

$$AgCl(s) \rightleftharpoons Ag^+ + Cl^-$$

Concentration in balance (mol/L) $\quad s_1 \quad s_1+0.10$

According to rules of the solubility product: $K_{sp}^{\ominus}(AgCl) = s_1(s_1 + 0.10) \approx 0.10s_1$

$$s_1 = 1.77 \times 10^{-9} \text{mol/L}$$

Let s_2 be the molar solubility of AgCl in 0.20mol/L $AgNO_3$ solution, then

$$[Ag+] = 0.20 + s_2 \approx 0.20 \qquad [Cl^-]=s_2$$

According to rules of the solubility product: $K_{sp}^{\ominus}(AgCl) = [Ag^+][Cl^-] = 0.20s_2$

$$s_2 = 8.8 \times 10^{-10} \text{mol/L}$$

2.5.2 Salt Effect

Salt effect (盐效应) is a term used to describe a phenomenon that the solubility of increasing slightly soluble strong electrolytes increases slightly by adding soluble strong electrolytes which does not contain the same ion.

Examples: KNO_3 is added in a $BaSO_4$ saturated solution, then:

$$Ba(SO_4)(s) \rightleftharpoons Ba^{2+} + SO_4^{2-}$$

$$KNO_3 \rightarrow K^+ + NO_3^-$$

The ion concentration of the solution increases, the interaction between ions increases, the effective ion concentration (i.e. activity) decreases and eventually leads to the right shift of the balance, resulting in a slight increase in $BaSO_4$ solubility. A common ion effect is also present in such solution. The presence of the common ion effect suppresses the salt effect which can be ignored.

As mentioned above, we need to add excess precipitation agent about 20 to 50 % in order to make the complete precipitation. However, if the concentration of the precipitation agent is too large, it may sometimes cause salt effects which increase the solubility of precipitation.

2.6 Applications of Precipitation Reaction

The precipitation reaction can be applied in drug production, quality control of drug and separation of precipitate, etc.

2.6.1 Application in Drug Production

Slightly soluble electrolytes are often prepared from two soluble electrolytes under the appropriate conditions (e.g. concentration, temperature, pH, mixing speed and time, etc.). For example, drugs of $BaSO_4$, $Al(OH)_3$ and NaCl can be prepared using precipitation reaction in Chinese Pharmacopeia.

(1) $BaSO_4$ is often used as an X-ray agent to diagnose gastrointestinal diseases. Since Ba^{2+} is toxic to humans, $BaCl_2$ and $Ba(NO_3)_2$ cannot be used. $BaCO_3$ is also not available for medicinal use because it can dissolve in stomach acid. $BaSO_4$ is generally prepared by mixing $BaCl_2$ solution with Na_2SO_4 solution or

sulfuric acid under the appropriate conditions. The equation is as follows:

$$BaCl_2 + Na_2SO_4 = BaSO_4{\downarrow} + 2NaCl$$

(2) $Al(OH)_3$ is often available for medicinal use to treat gastric acid, stomach and duodenal ulcers and other diseases. As a common anti-acid drug, it has no risk of alkali poisoning. $AlCl_3$ can be generated after the reaction of $Al(OH)_3$ with gastric acid and has local hemostatic effect. $Al(OH)_3$ is generally prepared from bauxite (main component: Al_2O_3) which is dissolved in sulfuric acid and then acted with Na_2CO_3 solution under pH of 8~8.5. The equation is as follows:

$$Al_2O_3 + 3H_2SO_4 = Al_2(SO_4)_3 + 3H_2O$$

$$Al_2(SO_4)_3 + 2Na_2CO_3 + 3H_2O = 2Al(OH)_3{\downarrow} + 3Na_2SO_4 + 3CO_2$$

2.6.2 Quality Control of Drug

Precipitation reaction is often used in quality control of drug such as impurity inspection and content determination. For example, the impurity of Cl^- and SO_4^{2-} are often inspected using precipitation reaction. The inspection protocol of Cl^- is as follows: In 50mL of water sample, add 5 drops of dilute HNO_3 (2mol/L), 1 ml of $AgNO_3$ solution (0.1mol/L), place for half a minute, it should not be turbid. This reaction is based on the principle that slightly soluble AgCl precipitate can be formed between Ag^+ and Cl^- ions. The role of nitric acid is to prevent interference from CO_3^{2-} and OH^- ions. The reaction equation is as follows:

$$Ag^+ + Cl^- \rightleftharpoons AgCl{\downarrow}$$

$$2Ag^+ + CO_3^{2-} \rightleftharpoons Ag_2CO_3{\downarrow}$$

$$2Ag^+ + 2OH^- = 2AgOH{\downarrow} \rightarrow Ag_2O{\downarrow} + H_2O$$

$$[Ag^+] = \frac{1}{50+1} \times 0.1 = 2 \times 10^{-3} mol/L$$

$$K_{sp}^{\ominus} = [Ag^+] \times [Cl^-] = 1.8 \times 10^{-10}$$

$$[Cl^-] = K_{sp}^{\ominus}/[Ag^+] = \frac{1.8\times10^{-10}}{2\times10^{-3}} = 9 \times 10^{-8} mol/L$$

When the concentration of Cl^- ion is more than 9×10^{-8}mol/L, slightly soluble precipitate of AgCl will be formed and the turbid solution happens.

2.6.3 Separation of Precipitate

Many metal sulfides can be separated by controlling the acidity of the solution with H_2S gas. When $Q > K_{sp}^{\ominus}$, metal sulfide precipitate can be formed according to the rule of solubility product. H_2S is a binary weak acid. The concentration of S^{2-} ion can be calculated by its equilibrium constant as following.

$$\frac{([H^+]^2[S^{2-}])}{([H_2S])} = K_{a1}^{\ominus} \cdot K_{a2}^{\ominus}$$

$$[S^{2-}] = K_{a1}^{\ominus} \cdot K_{a2}^{\ominus} \frac{[H_2S]}{[H^+]^2}$$

For common bivalent metal sulfide,

$$MS(s) \rightleftharpoons M^{2+} + S^{2-}$$

$$K_{sp}^{\ominus} = [M^{2+}][S^{2-}], \text{ then, } [S^{2-}] = K_{sp}^{\ominus}/[M^{2+}]$$

$$[H^+]^2 = K_{a1}^{\ominus} \cdot K_{a2}^{\ominus}[H_2S][M^{2+}]/K_{sp}^{\ominus}(MS) \text{ 或 } [H^+] = \sqrt{K_{a1}^{\ominus}K_{a2}^{\ominus}[H_2S][M^{2+}]/K_{sp}^{\ominus}(MS)}$$

($K_{a1}^{\ominus} = 1.32 \times 10^{-7}$ $K_{a2}^{\ominus} = 7.08 \times 10^{-15}$ the concentration of H_2S saturated solution is 0.10mol/L)

Example 5-4 In a certain solution that the concentrations of Zn^{2+} and Pb^{2+} ions are 0.20mol/L, it is saturated by passing through H_2S gas and then hydrochloric acid is added. Calculate the pH of the solution that PbS precipitate can be formed and Zn^{2+} cannot formed to ZnS precipitate? $K_{sp}^{\ominus}$ (PbS)= 8.0×10^{-28}, $K_{sp}^{\ominus}$ (ZnS)= 2.5×10^{-22}

Solution $[Zn^{2+}] = [Pb^{2+}] = 0.20\text{mol/L}$

PbS precipitate can be formed, then $[S^{2-}] \geqslant K_{sp}^{\ominus}(\text{PbS})/[Pb^{2+}]$

$$[S^{2-}] \geqslant 4.0\times10^{-27}\text{mol/L}$$

ZnS precipitate cannot be formed, then $[S^{2-}] \leqslant K_{sp}^{\ominus}(\text{ZnS})/[Zn^{2+}]$

$$[S^{2-}] \leqslant 1.2\times10^{-21}\text{mol/L}$$

In case of $[S^{2-}] = 1.2\times10^{-21}\text{mol/L}$, $K_{a1}^{\ominus}\cdot K_{a2}^{\ominus} = \dfrac{[H^+]^2[S^{2-}]}{[H_2S]}$

$$[H^+] \geqslant \sqrt{\frac{1.32\times10^{-7}\times7.08\times10^{-15}\times0.1}{1.2\times10^{-21}}} = 0.279\text{mol/L}$$

Then, when pH $\leqslant$ 0.55, Zn^{2+} cannot formed to ZnS precipitate.

In case of $[S^{2-}] = 5.0\times10^{-28}\text{mol/L}$, $[H^+] \leqslant \sqrt{\dfrac{1.1\times10^{-22}}{5.0\times10^{-28}}} = 432\ \text{mol/L}$

The above condition is impossible. When pH $\leqslant$ 0.55, only PbS precipitate can be formed without ZnS precipitate.

重点小结

1. 溶度积常数 一定温度下，在水溶液中，固体难溶强电解质达到沉淀与溶解平衡时，难溶强电解质由于溶解而电离产生的离子相对平衡浓度幂的乘积为一常数，该常数称为溶度积常数，简称溶度积，用 $K_{sp}^{\ominus}$ 表示，它代表难溶强电解质溶解趋势的大小或生成沉淀的难易。

2. 溶度积规则

$Q=K_{sp}^{\ominus}$ 时，饱和溶液，处于平衡状态；

$Q<K_{sp}^{\ominus}$ 时，不饱和溶液，无沉淀析出或有沉淀溶解；

$Q>K_{sp}^{\ominus}$ 时，过饱和溶液，沉淀析出，直至饱和为止。

当溶液中 $Q>K_{sp}^{\ominus}$ 时，将会有沉淀析出。可加入沉淀剂，增大离子浓度，使平衡向生成沉淀的方向移动；当溶液中 $Q<K_{sp}^{\ominus}$ 时，沉淀将会溶解。降低溶液中相关离子浓度，可促进沉淀溶解，例如，生成弱电解质(水、弱酸、弱碱、难电离的盐等等、发生氧化还原反应，生成配位化合物的方法均可促进沉淀溶解，沉淀溶解的程度可以通过计算溶解反应的 $K^{\ominus}$ 值来预测。

3. 分步沉淀 加入一种沉淀剂，使得溶液中多种离子按照到达溶度积的先后顺序沉淀出来的现象。利用分步沉淀原理，适当地控制条件，可以实现多种离子的分离。

离子沉淀的顺序取决于沉淀的 $K_{sp}^{\ominus}$ 值与被沉淀离子的浓度。对于同类型的沉淀，一般来说，$K_{sp}^{\ominus}$ 小的先沉淀，$K_{sp}^{\ominus}$ 大的后沉淀。对于不同类型的沉淀，必须根据计算结果确定，不能根据 $K_{sp}^{\ominus}$ 值判断。

4. 沉淀的转化 溶解度较大的沉淀转化为溶解度较小的沉淀。根据转化平衡常数的大小可以判断沉淀转化的可能性。沉淀转化的平衡常数 $K^{\ominus} \geqslant 10^6$，转化反应不仅能自发进行，而且进行得很完全。$K^{\ominus}>1$，转化可以进行；$K^{\ominus}$ 越大，转化进行的程度越大。$K^{\ominus}<1$，转化比较困难，若两种沉淀溶解度相差不大，控制一定的条件，转化仍然可以进行。

5. 同离子效应与盐效应 在难溶强电解质溶液中，加入与难溶强电解质具有相同离子的易溶强电解质，使难溶强电解质的溶解度减小的效应，称沉淀 - 溶解平衡中的同离子效应。盐效应是指在难溶强电解质的饱和溶液中加入与其不含有相同离子的易溶强电解质时，难溶强电解质的溶解度略有增大的现象。一般同离子效应和盐效应共存时，以同离子效应的影响为主，而忽略盐效应的影响。

6. 沉淀溶解平衡的应用 沉淀反应可用于药品生产、药品检验以及难溶硫化物、难溶氢氧化物的分离等方面。例如，当 $K_{sp}^{\ominus} > Q$ 时，就可生成金属硫化物沉淀。溶液中 S^{2-} 离子浓度与溶液中 $[H^+]$ 离子浓度直接有关，因此控制溶液的酸度，通入 H_2S 气体就可达到硫化物沉淀的分离。

题库

Exercises

1. True or False

(1) At the certain temperature, the product of Ag^+ and Br^- concentration is a constant in AgBr aqueous solution.

(2) When a precipitant is added to the mixed ionic solution, the slightly soluble electrolyte with the less $K_{sp}^{\ominus}$ will first precipitate.

(3) At the certain temperature, the solubility of the slightly soluble electrolyte is higher which $K_{sp}^{\ominus}$ is less.

(4) The size of the solubility product depends on the nature and temperature of the substance, independent of the concentration.

2. One-Choice Questions

(1) Which is the Solubility product expression of Fe_2S_3?

A. $K_{sp}^{\ominus} = [Fe^{3+}][S^{2-}]$　　B. $K_{sp}^{\ominus} = 3[Fe_3^{3+}] \times 2[S_3^{2-}]$　　C. $K_{sp}^{\ominus} = 2[Fe^{3+}] \times 3[S^{2-}]$

D. $K_{sp}^{\ominus} = [Fe^{3+}]^2[S^{2-}]^3$　　E. $K_{sp}^{\ominus} = [Fe^{3+}]^3[S^{2-}]^2$

(2) The slightly soluble strong electrolyte A_2B (molar mass is 80g/mol), the solubility in water at room temperature is 2.4×10^{-3}g/L, the solubility product of A_2B is

A. 1.1×10^{-13}　　B. 2.7×10^{-14}　　C. 1.8×10^{-9}

D. 9.0×10^{-10}　　E. 8.4×10^{-8}

(3) It is known that the solubility product constant of AB_2 is 1.08×10^{-13} and the solubility is

A. 4.76×10^{-5}mol/L　　B. 3.00×10^{-5}mol/L　　C. 3.29×10^{-7}mol/L

D. 1.64×10^{-7}mol/L　　E. 2.57×10^{-6}mol/L

(4) In the following solutions, the most soluble of $BaSO_4$ is

A. 1mol/L NaCl　　B. 1mol/L H_2SO_4　　C. 2mol/L $BaCl_2$

D. Water　　E. Ethanol

(5) The slightly soluble strong electrolyte A_2B, when the solution equilibrium is reached in aqueous solution, $[A^+]=x$ mol/L, $[B^{2-}]=y$ mol/L, the $K_{sp}^{\ominus}$ is

A. $K_{sp}^{\ominus} = xy$　　B. $K_{sp}^{\ominus} = (2x)^2y$　　C. $K_{sp}^{\ominus} = x^2y$

D. $K_{sp}^{\ominus} = x^2y/2$　　E. $K_{sp}^{\ominus} = 4x^3y$

(6) Add enough S^{2-} ions to the mixed solution with the same concentration of ions, the most ions left in the solution is

A. Cd^{2+}　　B. Zn^{2+}

C. Fe^{2+}　　D. Cu^{2+}

(7) When the saturated solution of $BaSO_4$ and $CaSO_4$ is mixed in equal volume and sufficient solid Na_2SO_4 is added, the phenomenon is ($K_{sp}^{\ominus}(BaSO_4) = 1.08 \times 10^{-10}$, $K_{sp}^{\ominus}(CaSO_4) = 4.93 \times 10^{-6}$)

A. the two kinds of precipitation are precipitated in equal amount.

B. both are precipitated, but $BaSO_4$ is the main one.

C. both are precipitated, but $CaSO_4$ is the main one.

D. only one precipitation is produced.

(8) In the following solutions, the most soluble of $BaCO_3$ is

A. HAc　　B. H_2O

C. $BaCl_2$　　D. Na_2CO_3

(9) The solubility product expression of Ag_3PO_4 is

A. $K_{sp}^{\ominus} = [Ag^+][PO_4^{3-}]$　　B. $K_{sp}^{\ominus} = 3[Ag^+][PO_4^{3-}]$

C. $K_{sp}^{\ominus} = [Ag^+]^3[PO_4^{3-}]$　　D. $K_{sp}^{\ominus} = [Ag^+][PO_4^{3-}]^3$

(10) $NH_3 \cdot H_2O$ is added to the mixture of Cu^{2+} and Mg^{2+} with concentrations of 0.1mol/L. To separate the two ions, the pH should be

(The initial and complete precipitation pH values of Cu^{2+} are 5.21 and 7.21, respectively. The initial and complete precipitation pH values of Mg^{2+} are 8.87 and 10.87, respectively.)

A. 5.21~7.21　　B. 7.21~8.87

C. 8.87~10.87　　D. unknown

3. Calculation Question

(1) Calculate the solubility of Ag_2CrO_4, when it is in the 0.10mol/L $AgNO_3$ solution and in the 0.10mol/L Na_2CrO_4 solution. It is known that $K_{sp}^{\ominus}(Ag_2CrO_4)= 1.12\times10^{-12}$。

(2) It is known that the solubility product of $Zn(OH)_2$ is 3×10^{-17} at 298.15K, calculate:

① the solubility of $Zn(OH)_2$ in water

② the concentrations of Zn^{2+} and OH^- in the $Zn(OH)_2$ saturated solution, and its pH.

③ the solubility of $Zn(OH)_2$ in 0.10mol/L NaOH.

④ the solubility of $Zn(OH)_2$ in 0.10mol/L $ZnSO_4$.

目 标 检 测

1. 判断题

（1）一定温度下 AgBr 的水溶液中，其 Ag^+ 与 Br^- 浓度的乘积是一常数。

（2）在混合离子溶液中加入沉淀剂，$K_{sp}^{\ominus}$ 值小的难溶电解质首先生成沉淀。

（3）温度一定时，难溶电解质的 $K_{sp}^{\ominus}$ 越小，其溶解度越大。

（4）溶度积的大小取决于物质的本性和温度，与浓度无关。

2. 单项选择题

（1）Fe_2S_3 的溶度积表达式是

A. $K_{sp}^{\ominus} = [Fe^{3+}][S^{2-}]$　　B. $K_{sp}^{\ominus} = 3[Fe_3^{3+}] \times 2[S_3^{2-}]$　　C. $K_{sp}^{\ominus} = 2[Fe^{3+}] \times 3[S^{2-}]$

D. $K_{sp}^{\ominus} = [Fe^{3+}]^2[S^{2-}]^3$　　E. $K_{sp}^{\ominus} = [Fe^{3+}]^3[S^{2-}]^2$

（2）某难溶强电解质 A_2B（摩尔质量为 80g/mol），常温下在水中的溶解度为 2.4×10^{-3}g/L，则 A_2B 的溶度积为

A. 1.1×10^{-13}　B. 2.7×10^{-14}　C. 1.8×10^{-9}
D. 9.0×10^{-10}　E. 8.4×10^{-8}

（3）已知某 AB_2 型难溶强电解质在一定温度下，其溶度积常数为 1.08×10^{-13}，则其溶解度为
A. 4.76×10^{-5}mol/L　B. 3.00×10^{-5}mol/L　C. 3.29×10^{-7}mol/L
D. 1.64×10^{-7}mol/L　E. 2.57×10^{-6}mol/L

（4）$BaSO_4$ 在下列溶液中溶解度最大的是
A. 1mol/L NaCl　B. 1mol/L H_2SO_4　C. 2mol/L $BaCl_2$
D. 纯水　E. 乙醇

（5）某难溶强电解质 A_2B，在水溶液中达到溶解平衡时，$[A^+]=x$ mol/L，$[B^{2-}]=y$ mol/L，则 A_2B 的 $K_{sp}^{\ominus}$ 可表示为
A. $K_{sp}^{\ominus} = xy$　B. $K_{sp}^{\ominus} = (2x)^2y$　C. $K_{sp}^{\ominus} = x^2y$
D. $K_{sp}^{\ominus} = x^2y/2$　E. $K_{sp}^{\ominus} = 4x^3y$

（6）向含有相同离子浓度的混合溶液中，加入足量的 S^{2-} 离子，溶液中剩余离子最多的是 $[K_{sp}^{\ominus}(CuS) < K_{sp}^{\ominus}(CdS) < K_{sp}^{\ominus}(ZnS) < K_{sp}^{\ominus}(FeS)]$
A. Cd^{2+}　B. Zn^{2+}
C. Fe^{2+}　D. Cu^{2+}

（7）将 $BaSO_4$ 和 $CaSO_4$ 饱和溶液等体积混合，并加入足量固体 Na_2SO_4，其现象为
$(K_{sp}^{\ominus}(BaSO_4) = 1.08 \times 10^{-10}, K_{sp}^{\ominus}(CaSO_4) = 4.93 \times 10^{-6})$
A. 两种沉淀等量析出
B. 两者均沉淀，但以 $BaSO_4$ 为主
C. 两者均沉淀，但以 $CaSO_4$ 为主
D. 仅产生一种沉淀

（8）难溶强电解质 $BaCO_3$ 在下列溶液中溶解最多的是
A. HAc　B. H_2O
C. $BaCl_2$　D. Na_2CO_3

（9）Ag_3PO_4 的溶度积表达式是
A. $K_{sp}^{\ominus} = [Ag^+][PO_4^{3-}]$　B. $K_{sp}^{\ominus} = 3[Ag^+][PO_4^{3-}]$
C. $K_{sp}^{\ominus} = [Ag^+]^3[PO_4^{3-}]$　D. $K_{sp}^{\ominus} = [Ag^+][PO_4^{3-}]^3$

（10）在浓度均为 0.1mol/L 的 Cu^{2+}、Mg^{2+} 混合溶液中加入 $NH_3 \cdot H_2O$，要使两种离子分离，应控制 pH 为（Cu^{2+} 开始沉淀 pH 值为 5.21，完全沉淀 pH 值为 7.21；Mg^{2+} 开始沉淀 pH 值为 8.87，完全沉淀 pH 值为 10.87）
A. 5.21~7.21　B. 7.21~8.87
C. 8.87~10.87　D. 无法确定

3. 计算题

（1）分别计算 Ag_2CrO_4 在 0.10mol/L $AgNO_3$ 和 0.10mol/L Na_2CrO_4 溶液中的溶解度，已知：$K_{sp}^{\ominus}(Ag_2CrO_4) = 1.12 \times 10^{-12}$。

（2）已知：298.15K 时，$Zn(OH)_2$ 的溶度积为 3×10^{-17}，计算：
① $Zn(OH)_2$ 在水中的溶解度；
② $Zn(OH)_2$ 的饱和溶液中 $[Zn^{2+}]$、$[OH^-]$ 和 pH；
③ $Zn(OH)_2$ 在 0.10mol/L NaOH 中的溶解度；
④ $Zn(OH)_2$ 在 0.10mol/L $ZnSO_4$ 中的溶解度。

Chapter 6 Oxidation-Reduction Reaction and Electrochemistry
第六章 氧化－还原反应和电化学

学习目标

知识要求

1. **掌握** 用离子－电子法配平氧化还原反应方程式的方法；能斯特方程及电极电势的应用；判断氧化剂与还原剂的相对强弱和氧化还原反应的方向；氧化还原反应平衡常数的意义及其计算；判断氧化还原反应的程度。

2. **熟悉** 氧化还原反应实质、氧化值的概念；原电池的概念和符号书写；氧化还原电对、电极电势、电动势的概念。

3. **了解** 元素标准电极电势图的概念及其应用。

能力要求

学会设计原电池并进行相关应用

1. Oxidation-Reduction Reaction

PPT

Oxidation-reduction reactions or redox reactions (氧化－还原反应) for short are a part of the world around us. It is not only widely used in our daily production and life, but also plays an important role in medicine. Examples range from the rusting of iron, the metabolic breakdown of carbohydrates to the bleaching of hair. Many chemical reactions of drugs in vivo are redox reactions, and the quality, efficacy and stability of drugs are closely related to redox reactions.

In this chapter, we will learn the principles of oxidation-reduction reaction and an introduction to the principles of electrochemistry.

1.1 Oxidation Number

Oxidation-reduction or redox reactions are electron transfer reactions. In a redox reaction, electrons are transferred from one specie to another. The specie that loses electrons is oxidized and acts as a

reducing agent (reductant) (还原剂). The specie that gains electrons is reduced and acts as an oxidizing agent (oxidant) (氧化剂). For example, $Zn + Cu^{2+} \rightleftharpoons Zn^{2+} + Cu$, Zn loses electrons and acts as a reducing agent, and Cu^{2+} gains electrons and acts as an oxidizing agent.

In fact, there are other reactions, where there is no electron gain or loss, just electron migration. For example, $H_2 + Cl_2 \rightleftharpoons 2HCl$, in the reaction, H does not really lose an electron, and Cl does not really gain an electron. H and Cl just share the electron pairs which is in favor of Cl. But it is still a redox reaction.

Therefore, in the redox reaction, electron transfer or migration will occur, and this electron transfer or migration will cause the changes of the structure of the valence electron layer , thus changing the charged state of these atoms.

In order to tell whether a redox reaction has occurred or not, the best way to keep track of these electrons is by assigning oxidation numbers (or oxidation state) (氧化数、或氧化值) to the atoms or ions involved in the reaction. The oxidation number is the apparent charge that an atom has if all bonds between atoms of different elements are assumed to be completely ionic. The shared electrons in a bond are assigned to the more electronegative atom. Consequently, for elements and monatomic ions, the oxidation number is the same as the charge, while the oxidation number of a covalently bound element is the charge the element would carry if all the shared pairs of electrons were transferred to the more electronegative atom. Therefore oxidation number is hypothetical number assigned to an individual atom or ion present in substance. It should be noted that the oxidation number of an atom does not represent the "real" charge on that atom. For example, here are the oxidation numbers of the atoms on the species .

Oxidation numbers enable us to describe oxidation-reduction reactions, and balance redox chemical reactions. It is helpful and important to assign oxidation numbers to a variety of compounds and ions using a set of rules.

(1) The oxidation number of the atoms in any free, uncombined element is zero. This includes polyatomic elements such as H_2, O_2, P_4, and S_8.

(2) The oxidation number of an element in a simple (monatomic) ion is equal to the charge on the ion.

(3) The sum of the oxidation numbers of all atoms in a neutral compound is zero.

(4) In a polyatomic ion, the sum of the oxidation numbers of the constituent atoms is equal to the charge on the ion.

(5) Fluorine, the most electronegative element, has an oxidation number of –1 in all its compounds.

(6) Hydrogen has an oxidation number of +1 in compounds unless it is combined with metals, in which case it has an oxidation number of –1. Examples of these exceptions are NaH and CaH_2.

(7) Oxygen, the second most electronegative element, usually has an oxidation number of –2 in its compounds. There are just a few exceptions:

(a) Oxygen has an oxidation number of –1 in hydrogen peroxide, H_2O_2, and in peroxides, which contain the O_2^{2-} ion; examples are CaO_2, and Na_2O_2.

(b) Oxygen has an oxidation number of $\frac{1}{2}$ in superoxides, which contain the O_2^- ion; examples are KO_2, and RbO_2.

(c) When combined with fluorine in OF_2, oxygen has an oxidation number of +2.

(8) The position of the element in the periodic table helps to assign its oxidation number:

(a) Group IA elements have oxidation numbers of I in all of their compounds.

(b) Group IIA elements have oxidation numbers of +2 in all of their compounds.

(c) Group IIIA elements have oxidation numbers of +3 in all of their compounds with few rare exceptions.

Using the above rules, the oxidation numbers of various elements can be obtained.

Example 6-1 Determine the oxidation number of S atom in $NH_4S_2O_8$.

Solution Let x be the unknown oxidation number of sulfur. Therefore, for $NH_4S_2O_8$,

$$2 \times (+1) + 2x + 8 \times (-2) = 0$$

$$x = +7$$

So, the oxidation number of S is +7.

Example 6-2 Determine the oxidation number of Fe atom in Fe_3O_4.

Solution Let y be the unknown oxidation number of Fe. Therefore, for Fe_3O_4,

$$3y + 4 \times (-2) = 0$$

$$y = +\frac{8}{3}$$

So, the oxidation number of Fe is $+\frac{8}{3}$.

The difference between oxidation number and valence(氧化数与化合价的区别): the concept of valence was put forward in 1920, which refers to the proportional relationship between the number of atoms when combining components. From the perspective of molecular structure, valence reflects the ability to form chemical bonds between atoms. It can only be an integer corresponding to the electricity price and covalent number of ionic bonds and covalent bonds. Therefore, the valence is related to the microstructure of the substance. For some complex compounds, it is necessary to know the structure of the compound to determine the valence of the element. For example, the actual composition of Fe_3O_4 is $FeO \cdot Fe_2O_3$, so the combined valence of Fe is +2 and +3.

The oxidation number is a concept artificially introduced to explain the charged state of a substance. So the oxidation number of an element can be positive, negative, or fractional. When determining the oxidation number, it is not necessary to know the microstructure of the compound. Therefore, the application of oxidation number in the redox reaction is more convenient and practical than that of valence. For example, according to the chemical formula of Fe_3O_4, the oxidation number of Fe can be directly calculated as $+\frac{8}{3}$, without knowing the actual composition of Fe_3O_4. It should be noted that, in many cases, the oxidation numbers and valence numbers of the elements in a compound have the same number, but this does not imply that they are the same thing.

Redox reactions can be defined in terms of oxidation numbers: chemical reactions in which the oxidation numbers of elements change are called redox reactions (凡是有氧化值发生变化的反应称为氧化－还原反应). Among them, the process of increasing the oxidation number of elements is called the oxidation reaction (氧化反应), while the process of decreasing the oxidation number of elements is called the reduction reaction (还原反应). Substances with an increased oxidation number are called reducing agents(还原剂)and substances with a decreased oxidation number are called oxidizing agents(氧化剂).

1.2 Balance of Redox Reaction

Although some redox equations are relatively easy to balance by inspection, other more complex redox equations require systematic balancing procedures. Here we will discuss two such procedures, called the oxidation number method (氧化值法) and ion-electron method (离子－电子法). Whatever method you employed, any balanced equation must meet the following requirements.

(1) Atoms of each element must be conserved.

(2) Electrons must be conserved.

1.2.1 Oxidation Number Method

Oxidation number method (also called oxidation state method) is relatively an easy way to balance redox (including ionic redox) equations. The key principle is that the gain in the oxidation number in one reactant must equal to the loss in the oxidation number of the other reactant.

Example 6-3 Use the Oxidation Number Method to balance the equation for the reaction

$$KMnO_4+HCl \rightarrow MnCl_2+Cl_2+KCl$$

Solution

(1) Identify which elements undergo a change in the oxidation state.

$$K\overset{7}{Mn}O_4+2H\overset{-1}{Cl} \rightarrow \overset{2}{Mn}Cl_2+\overset{0}{Cl_2}+KCl$$

(2) Note the change in oxidation number of each element on each side of the equation. Draw a large bracket from the element in reactant side to that in product side, and write down the increase (+) and decrease (–) in oxidation number at the middle of the bracket. Multiply these changes by appropriate factors so that the total increase in oxidation number equals the total decrease in oxidation number.

$$\overbrace{KMnO_4 + HCl \rightarrow MnCl_2}^{(2-7=-5)\times 2} + Cl_2 + KCl$$
$$\underbrace{HCl \rightarrow MnCl_2 + Cl_2}_{[0-(-1)=+1]\times 2\times 5}$$

(3) Add coefficients to the formulas so as to obtain the correct ratio of the atoms whose oxidation numbers are changing (These coefficients are usually placed in front of the formulas).

$$2KMnO_4+10HCl \rightarrow 2MnCl_2+5Cl_2+KCl$$

(4) Balance the rest of the equation by final inspection.

$$2KMnO_4 + 16HCl = 2MnCl_2 + 5Cl_2 + 2KCl + 8H_2O$$

Firstly, balance all other atoms in the equation except those of hydrogen and oxygen. In this step, do not alter the coefficients determined in the previous step. Then balance the oxygen atoms by adding H_2O molecules to the side deficient to oxygen atoms. In this case, $8H_2O$ are needed on the products side to balance oxygen atoms. For reactions in acidic or basic solution, add H^+/H_2O (acidic) or OH^- / H_2O (basic) to balance the charges.

1.2.2 Ion-Electron Method (or Half-Reaction Method)

Ion-Electron Method (or Half-Reaction Method (半反应法)) is applied the best for redox reactions with ions and in aqueous solution. This method is to split an unbalanced redox equation into two parts called half-reactions. One half-reaction describes the oxidation, and the other describes the reduction. The balancing principle is that the total number of electrons gained by the oxidant must be equal to the total number of electrons lost by the reducing agent.

Example 6-4 Balance the following equation by Ion-Electron Method.

$$K_2Cr_2O_7 + KI + HCl \rightarrow CrCl_3 + I_2 + KCl$$

Solution

(1) Write the ion equation according to the given equation:

$$Cr_2O_7^{2-} + I^- \rightarrow Cr^{3+} + I_2$$

(2) The reaction is rewritten into two half reactions, namely the reduction half reaction of oxidant and the oxidation half reaction of reducing agent:

Reduction: $Cr_2O_7^{2-} \rightarrow Cr^{3+}$

Oxidation: $I^- \rightarrow I_2$

(3) Balance the number of atoms in the half-reaction formula.The principle of balancing the oxygen atoms is that in an acidic medium, add a certain amount of H_2O to the side with less O atoms, one O atom missing,plus one H_2O, and two H^+ to the other side;in an alkaline or neutral medium, add a certain number of OH^- on the less side of the O atom, one O atom is missing, add two OH^-, and get one H_2O on the other side.

Reduction: $Cr_2O_7^{2-} + 14H^+ \rightarrow 2Cr^{3+} + 7H_2O$

Oxidation: $2I^- \rightarrow I_2$

(4) Balance the charge in the half reaction equation:add a certain amount of electrons to both sides of the half reaction equation to balance the charges on both sides.

Reduction: $Cr_2O_7^{2-} + 14H^+ + 6e^- \rightarrow 2Cr^{3+} + 7H_2O$ ①

Oxidation: $2I^- \rightarrow I_2 + 2e^-$ ②

(5) In accordance with the principle that the number of electrons gained and lost must be equal for oxidant and reducing agent, the two half reactions are multiplied by the appropriate coefficient (determined by the least common multiple of electrons gained and lost), and then the two equations are added together to eliminate the number of electrons gained and lost, resulting in a balanced ion equation.

① ×1 $Cr_2O_7^{2-} + 14H^+ + 6e^- \rightarrow 2Cr^{3+} + 7H_2O$

② ×3 $2I^- \rightarrow 3I_2 + 6e^-$

$$Cr_2O_7^{2-} + 6I^- + 14H^+ \rightleftharpoons 2Cr^{3+} + 3I_2 + 7H_2O$$

(6) Fill the balanced ionic reaction equation with the positive or negative ions of the reactants and products that do not participate in the redox reaction, and write the corresponding molecular formula to get the balanced molecular reaction equation.

$$K_2Cr_2O_7 + 6KI + 14HCl \rightleftharpoons 2CrCl_3 + 3I_2 + 8KCl + 7H_2O$$

Example 6-5 Balance the following equation by Ion-Electron Method.

$$MnO_4^- + Fe^{2+} \rightarrow Mn^{2+} + Fe^{3+}$$

Solution Separate the equation into two half-reactions:

Oxidation: $Fe^{2+} \rightarrow Fe^{3+}$

Reduction: $MnO_4^- \rightarrow Mn^{2+}$

Balance each half-reaction:

$$2Fe^{2+} - 2e^- \rightarrow 2Fe^{3+}$$

$$MnO_4^- + 8H^+ + 5e^- \rightarrow Mn^{2+} + 4H_2O$$

Multiply each half-reaction by a coefficient,then add the two half reactions together.

$$2\times(MnO_4^- + 8H^+ + 5e^- \rightarrow Mn^{2+} + 4H_2O)$$
$$+5\times(2Fe^{2+} - 2e^- \rightarrow 2Fe^{3+})$$
$$= 2MnO_4^- + 16H^+ + 10Fe^{2+} \rightleftharpoons 2Mn^{2+} + 8H_2O + 10Fe^{3+}$$

PPT

2. Electrode Potential

2.1 Galvanic Cell

微课

When a strip of Zn is placed in a solution of copper(II) sulfate, the zinc is slowly dissolved and copper metal immediately begins to plate out on the zinc strip. This indicates that there is a redox reaction between zinc and copper sulfate, which is a spontaneous process. The reaction equation is as follows:

$$Zn + CuSO_4 \rightleftharpoons ZnSO_4 + Cu$$

In the above reaction, the thermal motion of the oxidant and the reducing agent results in effective collision and electron transfer. Since the thermal motion of molecules is not directional and does not form the directional motion of electrons, i.e., the electric current, we cannot directly observe the electron transfer phenomenon at the contact point between the metal and the solution. As the redox reaction proceeds, heat is released, indicating that chemical energy is converted into heat during the reaction.

However, when carried out under appropriate conditions, a redox reaction can proper device, concentration,temperature or pressure. It is possible to build such an apparatus where the two half-reactions take place at separate sites, with electrons being transferred indirectly. Such an apparatus which generates electricity through the use of a spontaneous redox reaction is known as a galvanic cell or voltaic cell (原电池). A wide variety of such cells may be constructed. The Cu-Zn cell also called the Daniel cell, which will illustrates the general principles involved.

In the device (Daniell cell) shown in Figure 6-1, the left hand compartment consists of a zinc strip or rod partly immersed in $ZnSO_4$ solution. The right hand compartment consists of a copper strip or rod partly immersed in $CuSO_4$ solution. A wire connects the two metal strips to allow indirect electron transfer.

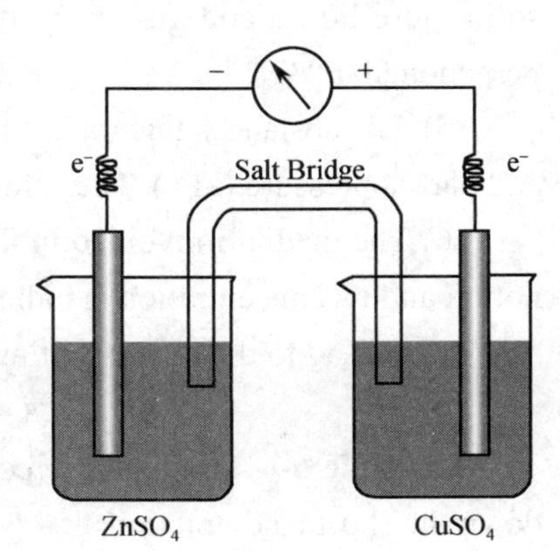

Figure 6-1 The Cu-Zn galvanic cell

As illustrated, the compartment on the left is the site for the oxidation half-reaction. Here zinc loses two electrons and oxidized to Zn^{2+} ions, which go into solution.

$$Zn(s) - 2e^- \rightarrow Zn^{2+}(aq)$$

The zinc strip is called the anode electrode(负极)where oxidation takes place, and the oxidation reaction that takes place at anode electrode is called Zn^{2+}/Zn half-cell reaction. The compartment on the right is the site for the reduction half-reaction. Here Cu^{2+} gains two electrons when they collide with the

surface of the copper strip.

$$Cu^{2+}(aq) + 2e^- \rightarrow Cu(s)$$

The copper strip is called the cathode electrode (正极) where reduction takes place, and the reduction reaction that takes place at cathode electrode is called Cu^{2+} /Cu half-cell reaction.

The electrons lost by the zinc electrode flow through the wire to the copper electrode where they are picked up by the Cu^{2+} ions. The overall cell reaction is the sum of the two half-cell reactions.

$$Zn(s) + Cu^{2+} \rightarrow Cu(s) + Zn^{2+}$$

The two half-cell reactions result in zinc cations (Zn^{2+}) being produced in the anode compartment, and sulphate anions (SO_4^{2-}) being left in the cathode compartment. A salt bridge (a glass tube filled with a jelly containing a suitable electrolyte such as saturated KCl or KNO_3) placed between the two half-cells (just like "liquid wire") connects the two solutions and allows the flow of cations and anions to maintain charge balance in each solution. Cations such as Zn^{2+} and K^+ move from anode to cathode, anions such as Cl^- and SO_4^{2-} move from cathode to anode along the salt bridge (盐桥).

2.1.1 Shorthand Notation (or Cell Diagram) for Galvanic Cells

In theory, any redox reaction can be designed into a galvanic cell. Galvanic devices may be represented by shorthand notations, which shall be written in accordance with the following provisions:

(1) The negative is written on the left, the positive is written on the right, and the outermost is marked with the symbols "(+)" and "(–)".

(2) Electrode materials are written on the left and right ends. If there is no conductive material in the electric pair, inert conductors (惰性导体)such as Pt and C can be used as electrode materials. The inert electrode does not participate in the electrode reaction, but only transports electrons.

(3) The double vertical line "‖" is used to represent the salt bridge, and the single vertical line "|" is used to represent the interface between different phases. Different substances in the same phase are separated by commas.

(4) Electrolyte solutions should be written close to the left and right sides of the salt bridge, pure solid, pure liquid and gas are written on one side of the electrode material and separated by a single perpendicular "|".

(5) All substances that make up a cell should be labeled with their states (s, l, g), concentrations (mol/L), and partial pressure (kPa). The state of the electrode material can be omitted.

(6) The medium involved in the electrode reaction, such as H^+ or OH^-, is also written in the battery notion and its concentration is indicated.

According to the above writing rules, the Cu-Zn cell can be represented as follows:

$$(-)\ Zn|\ ZnSO_4(c_1)\ \|\ CuSO_4(c_2)\ |\ Cu\ (+)$$

Example 6-6 Design the redox reaction $Cu + 2Fe^{3+} \rightleftharpoons Cu^{2+} + 2Fe^{2+}$ into a galvanic cell and write down the shorthand notation.

Solution The pairs that make up the reaction of the two electrodes are Fe^{3+}/Fe^{2+} and Cu^{2+}/Cu. However, neither the oxidized type nor the reduced type in the electric pair Fe^{3+}/Fe^{2+} is metal conductor, so platinum or carbon rods are usually required as conductors (solid conductors such as platinum and graphite do not participate in oxidation or reduction reactions, but only transport or receive electrons, so they are called "inert" electrodes) to form a half-cell. In this way, we insert the platinum plate into the beaker containing the mixed solution of Fe^{3+} and Fe^{2+}, insert the copper plate into the beaker containing

the solution of $CuSO_4$, connect the salt bridge and wire to form the galvanic cell, and the current will be generated. Electrode reaction is:

Cathode $Cu \rightarrow Cu^{2+} + 2e^-$ Oxidation

Anode $2Fe^{3+} + 2e^- \rightarrow 2Fe^{2+}$ Reduction

The shorthand notation is:

$$(-)Cu \mid Cu^{2+}(c_1) \parallel Fe^{3+}(c_2), Fe^{2+}(c_3) \mid Pt\ (+)$$

Where c_1, c_2 and c_3 represent the concentration of each substance respectively, (–) and (+) represent cathode and anode respectively, where Fe^{3+} and Fe^{2+} are in the same liquid phase, so they can be separated by commas, while salt bridge and inert electrode are in contact with both ions at the same time, so they are written in no particular order.

Example 6-7 Write down the shorthand notation for the reaction.

$$2MnO_4^- + 10Cl^- + 16H^+ \rightleftharpoons 2Mn^{2+} + Cl_2 + H_2O$$

Solution Decomposed the redox reaction into two halves:

Anode $MnO_4^- + 8H^+ + 5e^- \rightarrow Mn^{2+} + 4H_2O$ (the pair MnO_4^-/Mn^{2+})

Cathode $2Cl^- \rightarrow Cl_2 + 2e^-$ (the pair Cl_2/Cl^-)

The shorthand notation is

$$(-)Pt \mid Cl_2(p)|Cl^-(c_1) \parallel MnO_4^-\ (c_2), Mn^{2+}\ (c_3), H^+\ (c_4)|\ Pt(+)$$

In the galvanic cell, MnO_4^-, Mn^{2+} and H^+ are in the same phase. Write down these ions in no particular order and use commas to separate them. Cl^- and Cl_2 belong to different phase materials separated by a single vertical line, and Cl_2 is close to the electrode material, while Cl^- is close to the salt bridge. Since there is no metal conductor in the two electrode pairs, inert conductor Pt or C can be used as the electrode material.

Example 6-8 Known the galvanic cell(–)Pt | $Sn^{2+}(c_1)$,$Sn^{4+}(c_2)$||$Cl^-(c_3)$|$Cl_2(p)$| Pt(+), write the electrode reaction and the cell reaction.

Solution The electrode reaction:

Cathode: $Sn^{2+} \rightarrow Sn^{4+} + 2e^-$

Anode: $Cl_2 + 2e^- \rightarrow 2Cl^-$

The cell reaction: $Sn^{2+} + Cl_2 \rightleftharpoons Sn^{4+} + 2Cl^-$

2.1.2 Types of Electrodes

Electrodes are usually classified as four types according to the structure of the electrode and the characteristics of the reaction.

(1) Metal-Metal Ion Electrode

Metal-metal ion electrode is usually made by a metal electrode immersed in its salt solution. Electron transfer occurs between the metal atoms of the electrode and the metal ions in solution.

For example, $Cu\text{-}Cu^{2+}$ ion electrode is made by immersing a strip of Cu into a solution containing Cu^{2+} ion.

Electrode notation: $Cu \mid Cu^{2+}(c)$

Electrode reaction: $Cu^{2+} + 2e^- \rightleftharpoons Cu$

(2) Gas-ion Electrodes

Gas-ion electrode typically consists of a platinum (electrode) immersed in a solution containing the anion or cation of the gas. The gas is bubbled slowly over the surface of the platinum electrode. Platinum is a noble metal which is inert and only serves to establish the electric contact with the solution. Electron

transfer occurs between the gas and the ions in solution. For example, H^+/H_2 electrode and Cl_2/Cl^- electrode.

Chloride electrode:

Electrode notation: Pt | $Cl_2(p)$ | $Cl^-(c)$

Electrode reaction: $Cl_2 + 2e^- \rightleftharpoons 2Cl^-$

Hydrogen electrode:

Electrode notation: Pt | $H_2(p)$ | $H^+(c)$

Electrode reaction: $2H^+ + 2e^- \rightleftharpoons H_2$

(3) Redox electrodes

An inert electrode is formed by immersing it in a solution in which two ions with different oxidation numbers of the same element are dissolved.

For example, a platinum electrode is immersed in a solution containing both Sn^{4+} and Sn^{2+} ions.

Electrode notation: Pt | $Sn^{4+}(c_1)$, $Sn^{2+}(c_2)$

Electrode reaction: $Sn^{4+} + 2e^- \rightleftharpoons Sn^{2+}$

(4) Metal-Insoluble Salt -Anion Electrodes

Such electrodes are made by coating a metal surface with a layer of the metal's insoluble salt (or oxide) and then immersing it in an anion solution containing the corresponding insoluble salt. This kind of electrode is referred as metal-insoluble salt electrode.

For example, saturated calomel electrode (SCE) is made by solid calomel (mercury chloride) (甘汞) with mercury paste and potassium chloride solution (Figure 6-2).

Electrode notation: Hg(1)|Hg_2Cl_2(s)|$Cl^-(c)$

Electrode reaction: $Hg_2Cl_2(s) + 2e^- \rightleftharpoons 2Hg + 2Cl^-$

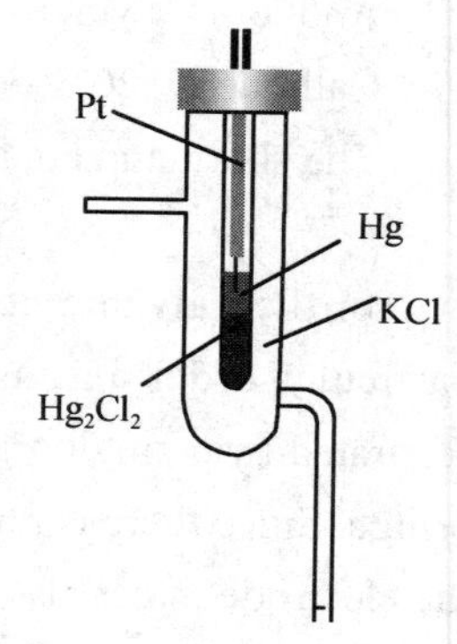

Figure 6-2 The saturated calomel electrode

2.2 Electrode Potential

2.2.1 The Generation of Electrode Potential

Connecting the two electrodes of the galvanic cell with a wire and a salt bridge can produce an electric current, indicating that there is a potential difference between the two electrodes. So, how does the electrode potential generat? Why the potentials of the two electrodes are different? In 1889, the German scientist Nernst (W.H.Nernst) put forward the "double layer theory"(双电层理论),which explained the mechanism of electrode potential generation. Take the metal electrode as an example to discuss the cause of the electrode potential.

The metal crystal is composed of metal atoms, metal ions and free electrons. When we immerse the metal in its salt solution, there are two tendencies. On one hand, one is that the metal ions in the metal crystal tend to leave the metal and enter the solution because of the thermal motion and the action of strong polar water molecules and the attraction of anions (the more active the metal or the smaller the concentration of metal ions in the solution, the stronger this trend). This process is called metal dissolution. On the other hand, due to the irregular thermal movement and the attraction of free electrons on the metal surface, the metal ions in the solution have a tendency to deposit from the solution to the metal surface (the more inactive the metal or the higher the concentration of metal ions in the solution,

the stronger this trend). This process is called metal ion deposition. In a certain concentration of solution, when the rate of metal dissolution equals to that of metal ion deposition, the dynamic equilibrium is reached:

$$Mn^{n+} + ne^- \rightleftharpoons M$$

If the trend of metal dissolution is greater than that of metal ion deposition, the metal surface is negatively charged and the solution is positively charged at equilibrium. Note that the metal ions in the solution are not uniformly distributed, and due to electrostatic attraction, more metal ions are concentrated near the metal surface, so an electric double layer is formed at the interface between the metal surface and the solution, and a potential difference occurs between the metal and the solution, as shown in Figure 6-3(a). If the trend of metal ion deposition is greater than that of metal dissolution, the metal surface is positively charged, the solution is negatively charged, and a double layer is formed at the interface between the metal and the solution, resulting in a potential difference, as shown in Figure 6-3(b). This potential difference between the metal electrode and the metal ion solution due to the establishment of the electric double layer is called the equilibrium electrode potential of the metal. The electrode potential (electrode potential) is denoted by the symbol $E^{\ominus}(M^{n+}/M)$ and the unit is V. For example, the electrode potential of zinc electrode is expressed by $E^{\ominus}(Zn^{2+}/Zn)$, and the electrode potential of copper electrode is expressed by $E^{\ominus}(Cu^{2+}/Cu)$.

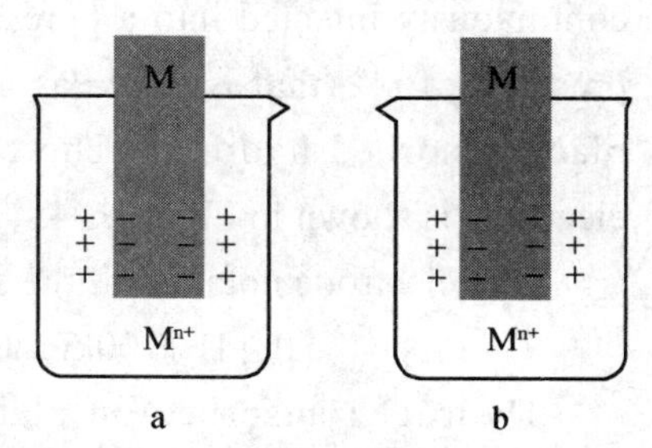

Figure 6-3 Double layer structure of metal electrode

The potential value of the electrode mainly depends on the nature of the electric pair and is affected by the temperature,concentration, pH value and ionic strength of the solution.

2.2.2 The Electromotive Force of the Galvanic Cell

The Galvanic cell is a device that converts chemical energy into electrical energy. Because the two electrodes of the galvanic cell produce different electrode potential, there exists potential difference between the two poles. If the potential difference between the two solutions is eliminated,the potential difference between the two electrodes of the galvanic cell is called the electromotive force (electromotive force, abbreviated as emf)(原电池的电动势), and expressed as E_{MF}. Namely

$$E_{MF} = E_+ - E_- \tag{6-1}$$

For example, in Cu-Zn galvanic cell, since Zn is more active than Cu, Zn is more likely to lose electrons into solution, and Zn sheet is more negatively charged than Cu sheet. Therefore, the electrode potential of Cu electrode is higher than that of Zn electrode, and the potential difference between the two electrodes results in electrons flowing from Zn electrode to Cu electrode.

2.2.3 Standard Electrode Potential

From the above discussion, it can be seen that the magnitude of the electrode potential expresses the trend and magnitude of the conversion from oxidizing electrons to reduced electrons, or from reducing lost electrons to oxidizing electrons. Therefore, the known electrode potential of the pair can directly compare the oxidation ability and reduction ability of the corresponding oxidation pair.

Because the thickness of the double layer is about 10^{-10} meters, so far, the absolute value of the electrode potential can not be measured by the general physical method. However, from the point of view of the actual need, we only need to know its relative value. In order to make practical use of the electrode potential, scientists have adopted a relative method, that is, select a reference electrode, specify that the

potential of the reference electrode is 0, and form a galvanic cell between the reference electrode and other electrodes to be tested. By measuring the electromotive force of the battery, the relative value of the electrode potential can be obtained.

(1) Standard hydrogen electrode (标准氢电极). According to IUPAC regulation in 1953, standard hydrogen electrode (abbreviated SHE) is used as the reference electrode. At 298K, the hydrogen electrode with a hydrogen partial pressure of 100kPa and a hydrogen ion concentration of 1mol/L is called a standard hydrogen electrode. The standard hydrogen electrode belongs to a gas-ion electrode.

The standard hydrogen electrode is a platinum electrode coated with active porous platinum black (with extremely high adsorption activity on hydrogen) immersed in an acid solution containing hydrogen ion concentration of 1mol/L, and continuously injected into a pure hydrogen flow with a standard pressure of $p^{\ominus}$ (that is 100kPa), so as to saturate the platinum black adsorbed hydrogen.The composition of the hydrogen electrode is shown in Figure 6-4.

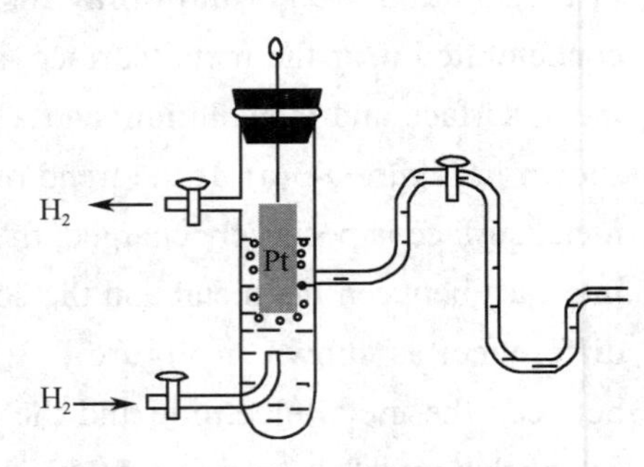

Figure 6-4 The standard hydrogen electrode

The electrode notation of the standard hydrogen electrode is:

$$Pt \mid H_2(100kPa) \mid H^+ (1mol/L)$$

Hydrogen adsorbed on platinum black and H^+ in solution establish the following dynamic equilibrium, namely, the electrode equilibrium formula of hydrogen electrode is:

$$2H^+ + 2e^- \rightleftharpoons H_2$$

The potential difference between the Pt tablet adsorbed with 100kPa saturated H_2 and the acid solution with a concentration of 1mol/L of H^+ is the electrode potential of the standard hydrogen electrode, whose value is defined as 0.0000V, namely:

$$E^{\ominus}(H^+/H_2) = 0.0000V$$

The "$\ominus$" in the upper right corner represents the standard state.

(2) Standard electrode potential and determination. If the electron-pair substance constituting the electrode is in a thermodynamic standard state, that is, the concentration of substance in solution is 1 mo/L, the partial pressure of gas is 100kPa, and solid or liquid is pure substance, then the electrode is called standard electrode. A galvanic cell is formed by a standard hydrogen electrode and a standard electrode to be measured, and the electromotive force of the galvanic cell is measured, so that the standard electrode potential of the electrode to be measured can be obtained, and the standard electrode potential is expressed by a symbol $E^{\ominus}$(oxidation type/reduced type), and the unit is V.

For example, to determine the standard electrode potential of a copper electrode, the standard copper electrode and the standard hydrogen electrode should form a galvanic cell. The current measured by galvanometer flows from the copper electrode to the hydrogen electrode, so the standard copper electrode is the positive electrode of the galvanic cell, and a reduction reaction occurs, while the standard hydrogen electrode is the negative electrode of the battery, and an oxidation reaction occurs. The electrode notation is:

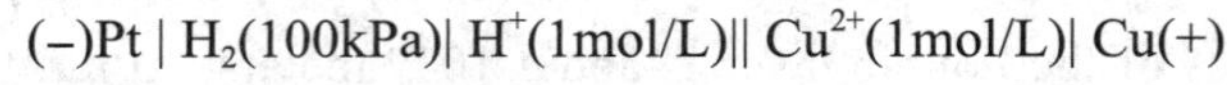

$$(-)Pt \mid H_2(100kPa) \mid H^+(1mol/L) \parallel Cu^{2+}(1mol/L) \mid Cu(+)$$

Anode Electrode reaction: $Cu^{2+} + 2e^- \rightarrow Cu$

Cathode Electrode reaction: $H_2 \rightarrow 2H^+ + 2e^-$

The cell reaction: $Cu^{2+} + H_2 \rightleftharpoons Cu + 2H^+$

At 298K, the electromotive force of the galvanic cell (the standard electromotive force) is measured, namely:

$$E^{\ominus}_{MF} = E^{\ominus}(Cu^{2+}/Cu) - E^{\ominus}(H^+/H_2) = 0.3419V$$

Because $E^{\ominus}(H^+/H_2) = 0.0000V$, so $E^{\ominus}(Cu^{2+}/Cu) = 0.3419V$

If the standard zinc electrode and the standard hydrogen electrode form a galvanic cell, the current measured by the galvanometer flows from the hydrogen electrode to the zinc electrode, so the standard hydrogen electrode is the positive electrode of the galvanic cell and the standard zinc electrode is the negative electrode of the galvanic cell. The notation of the cell is:

$(-)Zn|\ Zn^{2+}(1mol/L)||\ H^+(1mol/L)|\ H_2(100kPa)|\ Pt(+)$

Anode Electrode reaction: $2H^+ + 2e^- \rightarrow H_2$

Cathode Electrode reaction: $Zn \rightarrow Zn^{2+} + 2e^-$

The cell reaction: $Zn + 2H^+ \rightleftharpoons H_2 + Zn^{2+}$

At 298K, the standard electromotive force of the galvanic cell is determined, i.e

$$E^{\ominus}_{MF} = E^{\ominus}(H^+/H_2) - E^{\ominus}(Zn^{2+}/Zn) = 0.7618V$$

Because $E^{\ominus}(H^+/H_2) = 0.0000V$, so $E^{\ominus}(Zn^{2+}/Zn) = -0.7618V$

From the data measured above, it can be seen that, $E^{\ominus}(Zn^{2+}/Zn) < E^{\ominus}(H^+/H_2)$, indicating that the tendency of Zn to lose electrons is greater than that of H_2, or that of Zn^{2+} to gain electrons is less than that of H^+. $E^{\ominus}(Cu^{2+}/Cu) > E^{\ominus}(H^+/H_2)$ indicates that Cu has a lower tendency to lose electrons than H_2, or that Cu^{2+} has a greater tendency to gain electrons than H^+. Thus, Zn is more likely to lose electrons than Cu and convert to Zn^{2+} ions. On the other hand, it is much easier to convert electrons to Cu^{2+} than to Zn^{2+}.

The standard electrode potential of a series of electrodes can be measured in a similar way. However, the electrode potential of some pairs cannot be directly measured by experiments, such as Na^+/Na or F_2/F^-, and can only be calculated indirectly.

(3) Table of Standard Reduction Potentials (标准电极电势表). A standard electrode potential table is obtained by arranging the standard electrode potential values indirectly obtained by experimental measurement or calculation in the order from small to large. A complete list appears in Appendix 6. These available standard reduction potentials, often called half-cell potentials, could help us predict the cell potentials of voltaic cells created from any pair of electrodes, and predict comparative strengths of oxidizing and reducing agents.

It is important to know these points when using the values in the table.

① All of the half reactions in the table are written as reductions. If we reverse a half -reaction, then $E^{\ominus}$ changes sign.

② Electrode potentials are intensive properties. If we multiply a half-reaction by n, $E^{\ominus}$ is unchanged. The following two half-reactions have the same potential values.

$$Cl_2 + 2e^- \rightleftharpoons 2Cl^- \qquad E^{\ominus} = 1.3583V$$

$$\frac{1}{2}Cl_2 + e^- \rightleftharpoons Cl^- \qquad E^{\ominus} = 1.3583V$$

③ The bigger the standard reduction potential, the easier the specie to the left the arrow to be reduced. In other words, the specie is a stronger oxidizing agent, while the specie to the right of the arrow is a weaker reducing agent. The smaller the standard reduction potential, the easier the specie to the right

of the arrow to be oxidized. In other words, the specie is a stronger reducing agent, while the specie to the left of the arrow is a weaker oxidizing agent.

Since lithium at the top of the list has the most negative value, indicating that it is the strongest reducing agent and would rather undergo oxidation, Fluorine with the largest positive value for standard electrode potential is the most easily reduced species and the strongest oxidizing agent.

④ In the electrode equilibrium reaction: oxidation state + ne^- $\rightleftharpoons$ reduced state, the oxidation state includes oxidized substances and medium, while the reduced state includes reduced substances and other substances. For example, in $Cr_2O_7^{2-} + 14H^+ + 6e^- \rightleftharpoons 2Cr^{3+} + 7H_2O$, the oxidation state includes the oxidation type $Cr_2O_7^{2-}$ and H^+, and the reduced state includes the reduced Cr^{3+} and H_2O.

⑤ Standard electrode potential meter can be divided into acid table and base table. The acid table is the measured value in the medium with the concentration of H^+ 1mol/L, expressed by $E_A^\ominus$; The alkali table is the measured value in the medium with an OH^- concentration of 1mol/L, expressed by $E_B^\ominus$. When H^+ is involved in the electrode reaction, the acid table shall be checked. When OH^- is involved in the electrode reaction, the alkali table shall be checked; In the absence of H^+ or OH^-, we can think about it in terms of the state of existence of the electron pair.

For example, for the electrode $Fe^{3+} + e^- \rightleftharpoons Fe^{2+}$, Fe^{3+} and Fe^{2+} can only exist in acidic solution, so the pair check the acidic table.The electric pair of a amphoteric metal with its anion salt, such as ZnO_2^{2-}/Zn check the alkali table. In addition, the pairs of the medium not involved in the electrode reaction are usually listed in the acid table, such as $Cl_2(g) + 2e^- \rightleftharpoons 2Cl$.

2.3 Nernst Equation

2.3.1 Nernst Equation

The standard electrode potential is measured in a standard state, and changes in external conditions,such as temperature,concentration or pressure, will change the electrode potential of the pair. How to obtain the electrode potential in non-standard state?

The quantitative relationship between the electrode potential and the concentration and temperature of the pair of substances can be expressed by the Nernst equation (能斯特方程). For any electrode, its electrode balance reaction can be expressed in the following form:

$$\text{Oxidation state} + ne^- \rightleftharpoons \text{reduced state}$$

The electrode potential of the electrode in any state is:

$$E = E^\ominus + \frac{RT}{nF}\ln\frac{c(\text{oxidation state})}{c(\text{reduced state})} \tag{6-2}$$

The formula (6-2) is called the Nernst equation. Where, E is the electrode potential (V) of the electric pair in any state; $E^\ominus$ is the standard electrode potential (V) of the electric pair, R is the gas constant (8.314 J/K·mol), T is the absolute temperature (K), F is Faraday constant (96500 C/mol), n is the number of gain and loss electrons in the electrode balance, $\frac{c(\text{oxidation state})}{c(\text{reduced state})}$ represents the ratio of the product of the relative concentration or relative partial pressure powers of each substance (including oxidizing type and medium) on one side of the oxidation state to the product of the relative concentration or relative partial pressure powers of each substance (including reducing type and other substances) on the other side of the reduction state in the electrode equilibrium.

Since temperature has little effect on the electrode potential, the general chemical reaction takes place at the room temperature (298K). Then, at 298K, the natural logarithm is transformed into the common logarithm with base 10, and constants such as *R* and *F* are substituted, then the Nernst equation can be written as:

$$E = E^{\ominus} + \frac{0.0592}{n}\lg\frac{c(\text{oxidation state})}{c(\text{reduced state})} \tag{6-3}$$

The formula is the Nernst equation in common use. It should be noted that the rules for writing $\frac{c(\text{oxidation state})}{c(\text{reduced state})}$ in the Nernst equation are the same as the rules for writing the equilibrium constant, that is, the pure solid, the pure liquid and the solvent H_2O are not written in.

For example,the Nernst equation of $Cr_2O_7^{2-} + 14H^+ + 6e^- \rightleftharpoons 2Cr^{3+} + 7H_2O$ is expressed as:

$$E(Cr_2O_7^{2-}/Cr^{3+}) = E^{\ominus}(Cr_2O_7^{2-}/Cr^{3+}) + \frac{0.0592}{6}\lg\frac{c(Cr_2O_7^{2-})\cdot[c(H^+)]^{14}}{[c(Cr^{3+})]^2}$$

2.3.2 Examples of Nernst Equation Application

Example 6-9 Given that $Sn^{4+} + 2e^- \rightleftharpoons Sn^{2+}$ $E^{\ominus} = 0.151V$, calculate the electrode potential when ① $c(Sn^{4+})$=0.1mol/L, $c(Sn^{2+})$=1mol/L; ② $c(Sn^{4+})$=1mol/L, $c(Sn^{2+})$= 0.1mol/L.

Solution

(1) $E\ (Sn^{4+}/Sn^{2+}) = E^{\ominus}(Sn^{4+}/Sn^{2+}) + \frac{0.0592}{2}\lg\frac{c(Sn^{4+})}{c(Sn^{2+})} = 0.151V + \frac{0.0592}{2}\lg\frac{0.1}{1} = 0.121V$

(2) $E\ (Sn^{4+}/Sn^{2+}) = 0.151\ V + \frac{0.0592}{2}\lg\frac{1}{0.1} = 0.181V$

The above example illustrates the law of the influence of concentration on electrode potential: when the concentration of oxidized substance decreases or the concentration of reduced substance increases, the electrode potential decreases. On the contrary, the electrode potential value increases when the concentration of oxidized substance increases or decreases the concentration of reduced substance.

Example 6-10 Given that $MnO_2 + 4H^+ + 2e^- \rightleftharpoons Mn^{2+} + 2H_2O$ $E^{\ominus}$=1.224V, if $c(Mn^{2+})$=1mol/L, calculate the electrode potential when ① $c(H^+)$=0.1mol/L and ② $c(H^+)$=10mol/L at 298.15K.

Solution According to the Nernst equation:

$$E\ (MnO_2/Mn^{2+}) = E^{\ominus}(MnO_2/Mn^{2+}) + \frac{0.0592}{2}\lg\frac{[c(H^+)]^4}{c(Mn^{2+})}$$

(1) When $c(H^+)$= 0.1mol/L

$$E\ (MnO_2/Mn^{2+}) = 1.224V + \frac{0.0592}{2}\lg\frac{0.1^4}{1} = 1.106V$$

(2) When $c(H^+)$= 10mol/L

$$E(MnO_2/Mn^{2+}) = 1.224V + \frac{0.0592}{2}\lg\frac{10^4}{1} = 1.342V$$

The above example shows that the electrode potential of MnO_2 increases significantly with the increase of solution acidity, so does the oxidation of MnO_2.

In fact, the electrode potential of the reaction involving oxyacid salts (含氧酸盐) and some high

price oxygen-containing compounds is affected by the acidity of the solution. Some of the effects are even greater than the oxidation type and the concentration of the reduction itself because of the large H^+ or OH^- coefficients in these half-reactions.

For example, in $Cr_2O_7^{2-} + 14H^+ + 6e^- \rightleftharpoons 2Cr^{3+} + 7H_2O$, according to the Nernst equation:

$$E(Cr_2O_7^{2-}/Cr^{3+}) = E^{\ominus}(Cr_2O_7^{2-}/Cr^{3+}) + \frac{0.0592}{6}\lg\frac{c(Cr_2O_7^{2-})\cdot[c(H^+)]^{14}}{[c(Cr^{3+})]^2}$$

Thus, since the coefficient of H^+ is 14, the influence of the change of H^+ concentration on the electrode potential is much greater than that of $Cr_2O_7^{2-}$ and Cr^{3+}, which is the reason why some oxidants or reducing agents are often used in acidic solutions.

If some precipitant is added to the reaction system, the formation of precipitation will reduce the concentration of oxidized or reduced state, which will inevitably lead to the change of electrode potential.

Example 6-11 NaCl solution was added to the standard silver electrode solution so that its concentration finally reached to 1mol/L, and then calculate the electrode potential of the silver electrode.

Solution Given that

$$Ag^+ + e^- \rightleftharpoons Ag \quad E^{\ominus} = +0.7996V\text{, add } Cl^- \text{ to make AgCl} : Ag^+ + Cl^- \rightleftharpoons AgCl$$

According to the solubility product principle, the free Ag^+ concentration in the solution can be calculated when the precipitation-solution equilibrium is reached:

$$K^{\ominus}_{sp,AgCl} = [Ag^+][Cl^-]$$

$$[Ag^+] = K^{\ominus}_{sp,AgCl} / [Cl^-]$$

According to the Nernst equation:

$$E\,(Ag^+/Ag) = E^{\ominus}\,(Ag^+/Ag) + 0.0592\lg[Ag^+]$$

Substitute $[Ag^+]$ into it:

$$\begin{aligned} E\,(Ag^+/Ag) &= E^{\ominus}\,(Ag^+/Ag) + 0.0592\lg K^{\ominus}_{sp,\,AgCl} - 0.0592\lg[Cl^-] \\ &= 0.7996V + 0.0592\lg(1.77\times10^{-10}) - 0.0592\lg 1 \\ &= 0.222V \end{aligned}$$

Obviously, due to the formation of precipitation,the concentration of oxidized Ag^+ decreased greatly, and the electrode potential decreased greatly. Since AgCl precipitation is formed, the concentration of Ag^+ in the solution is very low, and the oxidized type basically exists in the form of AgCl. In fact, a new electrode has been formed in the solution, and the electric pair of the electrode is AgCl/Ag (a metal insoluble salt electrode). The electrode reaction formula of the pair is: $AgCl + e^- \rightleftharpoons Ag + Cl^-$. Since the concentration of Cl^- in the solution at equilibrium is 1mol/L, then $E\,(Ag^+/Ag) = E^{\ominus}\,(AgCl/Ag)$.

When a neutral oxidized substance or a reduced substance forms a complex with a ligand, the change of concentration will also cause the change of electrode potential. The calculation method is similar to Example 6-11, and the specific calculation is shown in the Chapter 9 (Coordination Compounds).

PPT

3. Applications of Electrode Potential

3.1 Compare the Relative Strength of Oxidant and Reducing Agent

The value of electrode potential reflects the relative strength of the oxidation type electron ability or the reduced electron loss ability of the electric pair. The larger the electrode potential, the stronger the oxidation ability and the weaker the reduction ability, and vice versa.

Example 6-12 Point out the strongest oxidants and reducing agents in the following pairs, and give the order of the oxidizing and reducing capacities of each type.

$$Fe^{3+}/Fe^{2+},\ Cu^{2+}/Cu,\ Fe^{2+}/Fe,\ Sn^{4+}/Sn^{2+},\ H_2O_2/H_2O,\ Cr_2O_7^{2-}/Cr^{3+}$$

Solution According to the table, the electrode potential of each pair is:

$$Fe^{3+} + e^- \rightleftharpoons Fe^{2+} \qquad E^\ominus = 0.771V$$

$$Cu^{2+} + 2e^- \rightleftharpoons Cu \qquad E^\ominus = 0.3419V$$

$$Fe^{2+} + 2e^- \rightleftharpoons Fe \qquad E^\ominus = -0.447V$$

$$Sn^{4+} + 2e^- \rightleftharpoons Sn^{2+} \qquad E^\ominus = 0.151V$$

$$H_2O_2 + 2H^+ + 2e^- \rightleftharpoons H_2O \qquad E^\ominus = 1.776V$$

$$Cr_2O_7^{2-} + 14^+ + 6e^- \rightleftharpoons 2Cr^{3+} + 7H_2O \qquad E^\ominus = 1.36V$$

The $E^\ominus$ value of the electric pair H_2O_2/H_2O is the largest, and the oxidizing type H_2O_2 is the strongest oxidant. The $E^\ominus$ value of electric pair Fe^{2+}/Fe is the smallest, and its reductive type Fe is the strongest reductant.

The order of oxidation ability of each oxidizing substance from strong to weak is:

$$H_2O_2 > Cr_2O_7^{2-} > Fe^{3+} > Cu^{2+} > Sn^{4+} > Fe^{2+}$$

The order of reducing ability of each reduced substance from strong to weak is:

$$Fe > Sn^{2+} > Cu > Fe^{2+} > Cr^{3+} > H_2O$$

The $E^\ominus$ (oxidation type/reduced type) value of strong oxidants commonly used in the laboratory is generally greater than 1.0V, such as $KMnO_4$, $K_2Cr_2O_7$, H_2O_2, etc. The $E^\ominus$ (oxidation type/reduced type) value of common reducing agents is generally less than zero or slightly greater than zero, such as the active metals Na, Zn, Fe, S^{2-}, I^-, Sn^{2+}, etc. It should be noted that the determination of redox capacity by $E^\ominus$ (oxidation type/reduced type) is carried out in the standard state. If the relative strength of oxidant and reducing agent is determined in the non-standard state, the value of E in the non-standard state must be calculated by the Nernst equation and then compared.

3.2 Determine the Direction of Redox Reaction

In principle, any redox reaction can be designed as a galvanic cell, and the reaction direction can be judged according to the value of electromotive force of the cell. If $E_{MF} > 0$, the galvanic cell reaction is progressing spontaneously. If E_{MF} is less than 0, the galvanic cell reaction proceeds spontaneously in reverse. If $E_{MF} = 0$, the galvanic cell reaction is in equilibrium. Of course, it is also possible to directly

compare the potentials of the two opposite electrodes to judge the direction of the redox reaction, i.e.

Example 6-13 Can copper powder interact with $FeCl_3$ solution in standard state?

Solution According to the table:

$$Cu^{2+} + 2e^- \rightleftharpoons Cu \qquad E^\ominus = 0.3419V$$
$$Fe^{3+} + e^- \rightleftharpoons Fe^{2+} \qquad E^\ominus = 0.771V$$
$$Fe^{3+} + 3e^- \rightleftharpoons Fe \qquad E^\ominus = -0.037V$$

If copper powder reacts with $FeCl_3$ solution, the following reaction may occur:

$$Cu + 2Fe^{3+} \rightleftharpoons Cu^{2+} + 2Fe^{2+} \qquad ①$$
$$Cu + 2Fe^{3+} \rightleftharpoons Cu^{2+} + 2Fe \qquad ②$$

Because $E^\ominus(Fe^{3+}/Fe^{2+}) > E^\ominus(Cu^{2+}/Cu)$, the $E^\ominus_{MF}$ of reaction ① is >0, and the reaction ① is positive and spontaneous. Because $E^\ominus(Fe^{3+}/Fe^{2+}) < E^\ominus(Cu^{2+}/Cu)$, the $E^\ominus_{MF}$ of the reaction ② is less than 0, the positive direction of the reaction ② can not be carried out spontaneously, and the reverse can be carried out spontaneously. Therefore, copper powder can interact with $FeCl_3$ solution to form Cu^{2+} and Fe^{2+}.

Example 6-14 (1) Determine whether the reaction

$$MnO_2(s) + HCl(aq) \rightleftharpoons MnCl_2(aq) + Cl_2(g) + 2H_2O(l)$$

can proceed to the right under the standard condition at 25°C?

(2) Why can we prepare chlorine by reacting MnO_2 with concentrated HCl ?

Solution

(1) Looking up the Appendix 6 and the potentials shows that:

$$MnO_2(s) + 4H^+ + 2e^- \rightleftharpoons Mn^{2+} + 2H_2O(l) \qquad E^\ominus = 1.224V$$
$$Cl_2 + 2e^- \rightleftharpoons 2\,Cl^- \qquad E^\ominus = 1.583V$$
$$E^\ominus_{MF} = E^\ominus_+ - E^\ominus_- = 1.224V - 1.3583V = -0.136V$$

Because $E^\ominus_{MF} < 0$, the above reaction cannot proceed spontaneously in standard condition.

(2) When preparing chlorine gas in laboratory, MnO_2 is used to react with concentrated HCl (12mol/L). According to the Nernst equation, the electrode potentials of the above two electrodes can be calculated respectively, and assuming that $c(Mn^{2+}) = 1.0$mol/L, $p(Cl_2) = 100$kPa, $c(H+) = c(Cl^-) = 12$mol/L, then

$$E(MnO_2/Mn^{2+}) = E^\ominus(MnO_2/Mn^{2+}) + \frac{0.0592}{2}\lg\frac{[c(H^+)]^4}{c(Mn^{2+})}$$

$$=1.224V + \frac{0.0592}{2}\lg\frac{12^4}{1} = 1.352V$$

$$E(Cl_2/Cl^-) = E^\ominus(Cl_2/Cl^-) + \frac{0.0592}{n}\lg\frac{p(Cl_2)/p^\ominus}{[c(Cl^-)]^2}$$

$$= 1.358V + \frac{0.0592}{2}\lg\frac{1}{12^2} = 1.294V$$

$$E(MnO_2/Mn^{2+}) > E(Cl_2/Cl^-)$$

Therefore, MnO_2 can react with concentrated HCl to prepare chlorine.

3.3 Determine the Degree of Redox Reaction

The limit of the redox reaction can be measured by the value of the standard equilibrium constant $K^{\ominus}$. If the redox reaction is designed as a galvanic cell, when the cell reaction reaches equilibrium, the electromotive force of the cell is zero, that is, the electrode potential of the cathode is equal to anode. From this, the relationship between the standard equilibrium constant of the cell reaction and the standard electromotive force can be deduced.

$$\ln K^{\ominus} = \frac{nFE^{\ominus}_{MF}}{RT} \tag{6-4}$$

At 298K, the R and F constants are substituted to obtain

$$\lg K^{\ominus} = \frac{nE^{\ominus}_{MF}}{0.0592} \tag{6-5}$$

In the formula, n is the total number of electrons gained and lost in the redox reaction. $E^{\ominus}_{MF}$ is the standard electromotive force of the galvanic cell.

From the equation (6-5), it can be seen that the equilibrium constant of redox reaction is only related to the nature of oxidant and reductant, but not to the concentration of reactant. The larger the standard electromotive force of the the galvanic cell, the larger the standard equilibrium constant of the cell reaction, and the more complete the positive reaction.

Example 6-15 Calculate the standard equilibrium constants of the following reactions and explain the degree of reaction progress.

$$Zn+Cu^{2+} \rightleftharpoons Cu^{2+} + Zn^{2+}$$

Solution Looking up the table:

Anode $E^{\ominus}(Cu^{2+}/Cu) = 0.3419V$

Cathode $E^{\ominus}(Zn^{2+}/Zn) = -0.7618V$

$$E^{\ominus}_{MF} = E^{\ominus}_{+} - E^{\ominus}_{-} = 0.3419 - (-0.7618) = 1.1037V$$

$$\lg K^{\ominus} = \frac{nE^{\ominus}_{MF}}{0.0592} = \frac{2\times1.1037}{0.0592} = 37.29$$

$$K^{\ominus} = 1.95\times10^{37}$$

The calculation results show that the $K^{\ominus}$ value is very large, indicating that the reaction is proceeding completely to the right. It should be pointed out that according to the size of the standard equilibrium constant, the reaction degree can be evaluated, but the speed of the reaction cannot be explained, that is, the reaction rate cannot be explained.

3.4 Reduction Potential Diagrams of Elements

3.4.1 Structure of Latimer Diagrams

Reduction potential diagrams (or Latimer diagrams) (元素电势图) show the standard reduction potentials connecting various oxidation states of an element. This is a convenient method of conveying large amounts of reduction potential values for half reactions involving all the oxidation states of a given element. In a Latimer diagram for an element, the formulas of species that represent each oxidation state of the element are written from left to right in order of decreasing oxidation number. That is the species

in the highest oxidation state is written on the left end, the species in the lowest oxidation state on the extreme right. The standard electrode potential value for the reduction half reaction involving any pair of the species connected by a horizontal line is shown above the line.

Since electrode potential differs between an acidic and a basic solution, different diagrams are required depending on the pH of the solution. Conventionally, Latimer diagrams are constructed for an element in two popular forms: (A) “In acidic medium”and (B) “In alkaline medium” for the two extremes of pH=0 and pH=14 (effective hydrogen ton concentrations of 1mol /L and 10^{-4}mol /L, respectively).

As examples, diagrams for iron and chlorine are shown below.

Acid medium($E_A^\ominus$/V) $FeO_4^{2-} \xrightarrow{+2.20V} Fe^{3+} \xrightarrow{+0.771V} Fe^{2+} \xrightarrow{-0.447V} Fe$

Alkaline medium ($E_B^\ominus$/V) $FeO_4^{2-} \xrightarrow{+0.9V} Fe(OH)_3 \xrightarrow{-0.56V} Fe(OH)_2 \xrightarrow{-0.88V} Fe$

Acid medium($E_A^\ominus$/V)

$$ClO_4^- \xrightarrow{+1.19V} ClO_3^- \xrightarrow{+1.21V} ClO_2^- \xrightarrow{+1.64V} HClO \xrightarrow{+1.63V} Cl_2 \xrightarrow{+1.35V} Cl^-$$

Alkaline medium($E_B^\ominus$/V)

$$ClO_4^- \xrightarrow{+0.36V} ClO_3^- \xrightarrow{+0.33V} ClO_2^- \xrightarrow{+0.66V} ClO^- \xrightarrow{+0.40V} Cl_2 \xrightarrow{+1.35V} Cl^-$$

Any pair of the species in diagram could easily be converted back to the redox reaction by writing the predominant species present first and then adding the other characteristic species in the acidic ($H^+$and H_2O) and in alkaline (OH^- and H_2O) solutions respectively. Charge balancing is affected using the appropriate number of electrons. Thus, the redox reaction for the ClO_4^-/ClO_3^- couple in acidic solution is

$$ClO_4^-(aq) + 2H^+(aq) + 2e^- \rightleftharpoons ClO_3^-(aq) + H_2O(l) \qquad E^\ominus = +1.19V$$

and in basic solution will be

$$ClO_4^-(aq) + H_2O(l) + e^- \rightleftharpoons ClO_3^-(aq) + OH^-(aq) \qquad E^\ominus = +0.36V$$

Variations arise in $E^\ominus$ values in the two media, due to the involvement of H^+ or H^- in the steps. If the involvement is not there, the values remain the same ($E^\ominus(Cl_2/Cl^-)$ is almost the same in the two diagrams).

3.4.2 Application of Latimer Diagrams

(1) Determine whether disproportionation reaction can take place or not. A disproportionation reaction is one in which a single element is simultaneously oxidized and reduced.

For example, the three oxidation values of an element form two pairs, which are arranged from left to right according to their oxidation values from high to low:

$$A \xrightarrow{E^\ominus_{left}} B \xrightarrow{E^\ominus_{right}} C$$

$$\xrightarrow{\text{Oxidation state reduction}}$$

Assuming that B can undergo disproportionation reaction, B→A+C should proceed spontaneously in the forward direction. As the oxidation value of B to C decreases, a reduction reaction occurs,and the electric pair B/C is the positive electrode of the galvanic cell. B becomes A, the oxidation increases, and the oxidation reaction occurs, and the electric pair A / B is the negative electrode of the cell, so

$$E^\ominus_{MF} = E^\ominus(B/C) - E^\ominus(A/B) = E^\ominus_{right} - E^\ominus_{left} > 0$$

That is, when $E^{\ominus}_{right} > E^{\ominus}_{left}$, B can have the disproportionation reaction.

If B cannot produce disproportionation reaction, similarly:

$$E^{\ominus}_{MF}=E^{\ominus}_{B/C} - E^{\ominus}_{A/B} = E^{\ominus}_{right} - E^{\ominus}_{left} < 0$$

That is, when $E^{\ominus}_{right} < E^{\ominus}_{left}$, B cannot produce disproportionation reaction, A+C→B can proceed spontaneously.

Example 6-16 Determine whether Fe^{2+} will disproportionate in acid solution.

Solution:

The Latimer diagram for iron in acid solution is shown below:

$$Fe^{3+} \xrightarrow{+0.771\ V} Fe^{2+} \xrightarrow{-0.447\ V} Fe$$

Let us consider both of these processes:

$$Fe^{2+}(aq) + 2e^- \rightleftharpoons Fe(aq) \qquad E^{\ominus}(Fe^{2+}/Fe) = -0.447\ V$$

$$-)2Fe^{3+}(aq) + 2e^- \rightleftharpoons 2Fe^{2+}(aq) \qquad E^{\ominus}(Fe^{3+}/Fe^{2+}) = 0.771\ V$$

$$3Fe^{2+}(aq) \rightleftharpoons 2Fe^{3+}(aq) + Fe(s) \qquad E^{\ominus}_{cell} = -1.218\ V$$

A negative $E^{\ominus}_{MF}$ value indicates that the final reaction is nonspontaneous, i.e., Fe^{2+} will not disproportionate in acid solution.

Example 6-17 When pH = 7 is known, the potential diagram of oxygen element is: $O_2 \xrightarrow{-0.33\ V} \cdot O_2^- \xrightarrow{+0.87\ V} O_2^{2-}$

Judge whether $\cdot O_2^-$ can exist stably under this condition? If there is a reaction, write down the ionic equation.

Solution Because of $E^{\ominus}_{right} > E^{\ominus}_{left}$, $\cdot O_2^-$ will undergo disproportionation reaction in aqueous solution:

$$2 \cdot O_2^- + 2H^+ \rightleftharpoons H_2O_2 + O_2$$

$\cdot O_2^-$ is the superoxide free radical. From the element potential diagram, $E^{\ominus}_{right} \gg E^{\ominus}_{left}$, the superoxide free radical has a great disproportionation trend. It is unstable in aqueous solution. However, in the physiological pH range (pH=7), $\cdot O_2^-$ mainly exists in the form of dissociation. As the disproportionation reaction occurs between two $\cdot O_2^-$, it is difficult for the two $\cdot O_2^-$ to approach due to the repulsion between negative charges, so the reaction rate is not high. But superoxide dismutase (SOD; an enzyme containing Cu and Zn) can catalyze this reaction, resulting in a very high rate of disproportionation.

(2) Calculate the $E^{\ominus}$ value of the unknown power pair from the $E^{\ominus}$ value of the known power pair. Using the element potential diagram, the standard electrode potentials of other pairs can be found from the known standard electrode potentials of several adjacent pairs.

Suppose the potential diagram of an element:

$$A \xrightarrow[n_1]{E_1^{\ominus}} B \xrightarrow[n_2]{E_2^{\ominus}} C \xrightarrow[n_3]{E_3^{\ominus}} D$$

$$A \xrightarrow[n_1+n_2+n_3]{E_x^{\ominus}} D$$

The following formula can be derived from thermodynamic method:

$$E_x^{\ominus} = \frac{n_1E_1^{\ominus} + n_2E_2^{\ominus} + n_3E_3^{\ominus}}{n_1 + n_2 + n_3}$$

If there are *i* adjacent pairs, then:

$$E_x^{\ominus}=\frac{n_1E_1^{\ominus}+n_2E_2^{\ominus}+\cdots+n_iE_i^{\ominus}}{n_1+n_2+\cdots+n_i} \tag{6-6}$$

In the formula (6-6), n_1、n_2…n_i are the electron transfer numbers of the corresponding pairs respectively. According to it, the $E^{\ominus}$ value of the unknown pair can be calculated from the $E^{\ominus}$ value of the known power pair.

Example 6-18 The potential diagram of manganese in acid solution is known,

$E^{\ominus}$/V:

$$BrO_3^- \xrightarrow{?} BrO^- \xrightarrow{+0.4556V} Br_2 \xrightarrow{+1.066V} Br^-$$

($BrO_3^- \to Br^-$: +0.61V; $BrO^- \to Br^-$: ?)

(1) Try to find $E^{\ominus}(BrO_3^-/BrO^-)$=?, $E^{\ominus}(BrO^-/Br^-)$=?

(2) Determine whether BrO^- can undergo disproportionation.

Solution

(1) According to the formula (6-6), there are

$$0.61V=\frac{E^{\ominus}(BrO_3^-/BrO^-)\times4+0.4556\ V\times1+1.066V\times1}{6}$$

$$E^{\ominus}(BrO_3^-/BrO^-)=\frac{0.61V\times6-0.4556V\times1-1.066V\times1}{4}=0.535V$$

$$E^{\ominus}(BrO^-/Br^-)=\frac{0.4556V\times1+1.066V\times1}{2}=+0.7608V$$

(2) Because of $E^{\ominus}(BrO^-/Br^-)>E^{\ominus}(BrO_3^-/BrO^-)$, BrO^- can disproportionate in acid solution.

重 点 小 结

1. 氧化值 氧化值是元素的一个原子所带的形式电荷。氧化还原反应实质是电子的转移或者电子对的偏移，形式上是元素氧化值发生了变化。

2. 反应方程式配平 常用配平氧化还原反应方程式的方法有：氧化值法、离子电子半反应法。

3. 电极电势、电池电动势 电极电势用符号 E 表示。通过标准电极电势的大小，可以判断氧化态物质和还原态物质的相对强弱。电池电动势用符号 E_{MF} 表示。通过电池电动势的正负，可以判断氧化还原反应进行的方向。

4. Nernst 方程式 对于任一电对：氧化态 + ne^- ⇌ 还原态，常用能斯特方程为：

$$E=E^{\ominus}+\frac{0.0592}{n}\lg\frac{c(\text{氧化态})}{c(\text{还原态})}$$

5. 电极电势的应用

（1）利用电极电势的大小判断氧化态物质和还原态物质的相对强弱。

（2）判断氧化还原反应进行的方向。

（3）标准电势图的构成；元素电势图的应用：判断中间态物质能否发生歧化反应；通过已知电对求解未知电对的标准电极电势。

Exercises

1. Indicate the oxidation numbers of elements in the following compounds:

$H_3\underline{P}O_4$; $Na_2\underline{S}_2O_3$; $K_2\underline{Cr}O_4$; $\underline{N}_2O_5$; $Na\underline{H}$; $K_2\underline{O}_2$; $\underline{Cl}O_2$; $\underline{P}_4O_6$

2. Balance the following aqueous skeleton reactions with ion-electron method:

(1) $Cr_2O_7^{2-} + Cl^- \rightarrow Cr^{3+} + Cl_2$ (acidic medium)

(2) $As_2S_3 + ClO_3^- \rightarrow Cl^- + H_3AsO_4 + SO_4^{2-}$ (acidic medium)

(3) $MnO_4^- + SO_3^{2-} \rightarrow Mn^{2+} + SO_4^{2-}$ (acidic medium)

(4) $PbO_2 + Cl^- \rightarrow Pb^{2+} + Cl_2$ (acidic medium)

(5) $HgS + NO_3^- + Cl^- \rightarrow HgCl_4^{2-} + NO_2 + S$(acidic medium)

(6) $CrO_4^{2-} + HSnO_2^- \rightarrow HSnO_3^- + CrO_2^-$ (basic medium)

3. Design the following redox reaction as a galvanic cell, mark it with a galvanic cell symbol, and write down the electrode equation:

(1) $Zn + 2Ag^+ \rightleftharpoons Zn^{2+} + 2Ag$

(2) $Sn^{2+} + 2Fe^{3+} \rightleftharpoons Sn^{4+} + 2Fe^{2+}$

(3) $Cl_2 + 2Fe^{2+} \rightleftharpoons 2Cl^- + 2Fe^{3+}$

(4) $Zn + 2H^+ \rightleftharpoons Zn^{2+} + H_2$

4. Explain the following phenomena according to the electrode potential:

(1) Why should Sn particles be added to the preparation of $SnCl_2$ solution?

(2) Why do Na_2SO_3 or $FeSO_4$ solutions fail after long periods of exposure?

5. Answer the following questions according to the standard electrode potential of Zn^{2+}/Zn, Ce^{4+}/Ce^{3+}, Fe^{2+}/Fe, Ag^+/Ag and MnO_4^-/MnO_2.

(1) Which is the strongest reducing agent? Which is the strongest oxidizing agent?

(2) Which can reduce Fe^{2+} to Fe?

(3) Which can oxidize Ag to Ag^+?

6. Write a balanced equation for the overall cell reaction and calculate their electromotive force at 298K.

(1) (−) Ni(s) | Ni^{2+}(0.01 mol/L) || Sn^{2+}(0.10 mol/L) | Sn(s) (+)

(2) (−) Pt | $Cl_2(p^{\ominus})$ | Cl^-(10 mol/L) || Mn^{2+}(1 mol/L), H^+(10 mol/L) | MnO_2(s)| Pt (+)

7. If the electromotive force of the following galvanic cells at 298K is 0.3884V, What's the concentration of Zn^{2+}?

$$(-)\ Zn(s)\ |\ Zn^{2+}(x\ mol/L)\ ||\ Cd^{2+}(0.20\ mol/L)\ |\ Cd(s)\ (+)$$

8. If the electromotive force of the following galvanic cell at 298K is 0.16V, What's the concentration of H^+?

$$(-)\ Pt(s)|H_2(100\ kPa)|H^+(x\ mol/L)\ ||\ H^+(pH{=}1.0)|H_2(100\ kPa)|Pt(s)\ (+)$$

9. If the electromotive force of the following galvanic cell at 298K is 0.480V. What is the pH of the solution for which $E^{\ominus}(Cu^{2+}/Cu)=0.3419V$.

$$(-)\ Pt(s)\ |\ H_2(100\ kPa)\ |\ H^+(x\ mol/L)\ ||\ Cu^{2+}(1.0\ mol/L)\ |\ Cu(s)\ (+)$$

10. Use the following standard reduction potentials:

$$Pb^{2+}+2e^- \rightleftharpoons Pb(s) \qquad E^{\ominus}(Pb^{2+}/Pb)= -0.1262V$$

$$PbSO_4(s)+2e^- \rightleftharpoons Pb(s)+SO_4^{2-} \qquad E^{\ominus}(PbSO_4/Pb)= -0.3588V$$

Calculate the solubility product constant of $PbSO_4$?

11. The equilibrium constant is 0.334 for the reaction: $Ag(s)+Fe^{3+} \rightleftharpoons Fe^{2+}+Ag^+$. Calculate the $E^{\ominus}(Ag^+/Ag)$ at 298K when $E^{\ominus}(Fe^{3+}/Fe^{2+})=0.771V$.

12. Use the following standard reduction potential and the solubility product constant of $Ag_2C_2O_4$ (3.5×10^{-11}):

$$Ag^+ + e^- \rightleftharpoons Ag(s)\ ,\ E^{\ominus} = +0.799V$$

Calculate the $E^{\ominus}(Ag_2C_2O_4/Ag)$.

13. Use the following standard reduction potentials:

$$Hg_2SO_4(s)+2e^- \rightleftharpoons 2Hg(l)+SO_4^{2-}(aq) \qquad E^{\ominus}(Hg_2SO_4/Hg)=0.6125V$$

$$Hg_2^{2+}(aq)+2e^- \rightleftharpoons 2Hg(l) \qquad E^{\ominus}(Hg_2^{2+}/Hg)=0.7973V$$

(1) Suppose the above pairs are combined into a galvanic cell. Calculate the electromotive force $E^{\ominus}_{MF}$.

(2) Calculate the solubility product constant ($K^{\ominus}_{sp}$) of Hg_2SO_4?

14. The copper plate was inserted into 0.1mol/L $CuSO_4$ solution and the silver plate was inserted into 0.1mol/L $AgNO_3$ solution to form the galvanic cell. Answer the following question according the standard reduction potentials of the two half-reactions obtained from Appendix 6.

(1) Calculate the electrode potential of each-half electrode and write down the galvanic cell symbol.

(2) Write the electrode reaction and the Galvanic reaction.

(3) Calculate the equilibrium constant of the battery reaction.

15. The standard reduction potentials of MnO_2/Mn^{2+} and Cl_2/Cl^- is 1.224V and 1.358V, repectively. Answer the questions about the following redox reaction:

$$MnO_2+4HCl \rightleftharpoons MnCl_2+Cl_2+2H_2O$$

(1) Judge the direction of the reaction in the standard state.

(2) Judge the direction of the reaction with the following initial concentrations: $[H^+]=[Cl^-]=12mol/L$; $[Mn^{2+}]=1mol/L$; $p(Cl_2)=p^{\ominus}$.

16. Calculate the equilibrium constant for the redox reaction

$$3CuS+2NO_3^-+8H^+ \rightleftharpoons 3S+2NO+3Cu^{2+}+4H_2O$$

The standard reduction potentials of NO_3^-/NO and S/S^{2-} is 0.957V and −0.4763V, repectively, while the solubility product constant of CuS is 6.3×10^{-36}.

17. Calculate the $E^{\ominus}(ClO_3^-/ClO_2^-)$ according to the potential diagram of chlorine in a basic solution:

$$E_B^\ominus/V \quad ClO_3^- \xrightarrow{?} ClO_2^- \xrightarrow{+0.66V} ClO^- \quad (ClO_3^- \to ClO^-: +0.50V)$$

18. Answer the following questions based on the potential diagram of oxygen:

$$E_A^\ominus/V \quad O_2 \xrightarrow{+0.695V} H_2O_2 \xrightarrow{\quad} H_2O \quad (O_2 \to H_2O: +1.23V) \qquad E_B^\ominus/V \quad O_2 \xrightarrow{\quad} HO_2^- \xrightarrow{+0.88V} OH^- \quad (O_2 \to OH^-: +0.40V)$$

(1) Clarify the strength of oxidization of H_2O_2 in acidic medium and the strength of reducibility of H_2O_2 in alkaline medium through the calculation.

(2) Judge the stability of H_2O_2 in acidic and alkaline medium.

目标检测

1. 指出下列化合物中画线元素的氧化值

$H_3\underline{P}O_4$；$Na_2\underline{S_2}O_3$；$K_2\underline{Cr}O_4$；$\underline{N}_2O_5$；$Na\underline{H}$；$K_2\underline{O_2}$；$\underline{Cl}O_2$；$\underline{P}_4O_6$

2. 利用离子－电子法配平下列各反应式

(1) $Cr_2O_7^{2-} + Cl^- \rightarrow Cr^{3+} + Cl_2$（酸性介质）

(2) $As_2S_3 + ClO_3^- \rightarrow Cl^- + H_3AsO_4 + SO_4^{2-}$（酸性介质）

(3) $MnO_4^- + SO_3^{2-} \rightarrow Mn^{2+} + SO_4^{2-}$（酸性介质）

(4) $PbO_2 + Cl^- \rightarrow Pb^{2+} + Cl_2$（酸性介质）

(5) $HgS + NO_3^- + Cl^- \rightarrow HgCl_4^{2-} + NO_2 + S$（酸性介质）

(6) $CrO_4^{2-} + HSnO_2^- \rightarrow HSnO_3^- + CrO_2^-$（碱性介质）

3. 将下列反应设计为原电池，用电池符号标示，并写出电极反应式。

(1) $Zn + 2Ag^+ \rightleftharpoons Zn^{2+} + 2Ag$

(2) $Sn^{2+} + 2Fe^{3+} \rightleftharpoons Sn^{4+} + 2Fe^{2+}$

(3) $Cl_2 + 2Fe^{2+} \rightleftharpoons 2Cl^- + 2Fe^{3+}$

(4) $Zn + 2H^+ \rightleftharpoons Zn^{2+} + H_2$

4. 根据电极电势解释下列现象

(1) 配制 $SnCl_2$ 溶液时，常需加入 Sn 粒。

(2) Na_2SO_3 或 $FeSO_4$ 溶液久置后失效。

5. 查出下列电对的 $E^\ominus$：Zn^{2+}/Zn，Ce^{4+}/Ce^{3+}，Fe^{2+}/Fe，Ag^+/Ag，MnO_4^-/MnO_2。回答下列问题：

(1) 上述物质中，哪个是最强的还原剂？哪个是最强的氧化剂？

(2) 上述物质中，哪些能把 Fe^{2+} 还原成 Fe。

(3) 上述物质中，哪些能把 Ag 氧化成 Ag^+。

6. 写出下列原电池的电池反应式，并计算 298K 时它们的电动势

(1) $(-)\ Ni(s) \mid Ni^{2+}(0.01mol/L) \parallel Sn^{2+}(0.10mol/L) \mid Sn(s)\ (+)$

(2) $(-)\ Pt \mid Cl_2(p^\ominus) \mid Cl^-(10mol/L) \parallel Mn^{2+}(1mol/L), H^+(10mol/L) \mid MnO_2(s) \mid Pt\ (+)$

7. 已知 298K 时，下列原电池的电动势为 0.3884V：

$$(-)\ Zn(s)\mid Zn^{2+}(x\ mol/L)\parallel Cd^{2+}(0.20\ mol/L)\mid Cd(s)\ (+)$$

则 Zn^{2+} 的浓度应该是多少？

8. 已知 298K 时，下列原电池的电动势为 0.16V：

$$(-)\ Pt(s)|H_2(100\ kPa)|H^+(x\ mol/L)\parallel H^+(pH=1.0)|H_2(100\ kPa)|Pt(s)\ (+)$$

则 H^+ 离子的未知浓度是多少？

9. 已知 298K 时，下列原电池的电动势为 0.480V：

$$(-)\ Pt(s)\mid H_2(100\ KPa)\mid H^+(x\ mol/L)\parallel Cu^{2+}(1.0\ mol/L)\mid Cu(s)\ (+)$$

已知 $E^{\ominus}(Cu^{2+}/Cu)=0.3419V$，求溶液的 pH 值。

10. 用下列数据：$Pb^{2+}+2e^- \rightleftharpoons Pb(s)$　　$E^{\ominus}(Pb^{2+}/Pb)=-0.1262V$

$PbSO_4(s)+2e^- \rightleftharpoons Pb(s)+SO_4^{2-}$　　$E^{\ominus}(PbSO_4/Pb)=-0.3588V$

计算 $PbSO_4$ 的溶度积常数。

11. 在 298K 时反应 $Ag(s)+Fe^{3+} \rightleftharpoons Fe^{2+}+Ag^+$ 的平衡常数为 0.334。已知 $E^{\ominus}(Fe^{3+}/Fe^{2+})=0.771V$，计算 $E^{\ominus}(Ag^+/Ag)$。

12. 已知电对 $Ag^++e^- \rightleftharpoons Ag(s)$，$E^{\ominus}=+0.799V$；$Ag_2C_2O_4$ 的溶度积为：3.5×10^{-11}。求算电对 $Ag_2C_2O_4+2e^- \rightleftharpoons 2Ag(s)+C_2O_4^{2-}$ 的标准电极电势。

13. 已知：$Hg_2SO_4(s)+2e^- \rightleftharpoons 2Hg(l)+SO_4^{2-}(aq)$　　$E^{\ominus}(Hg_2SO_4/Hg)=0.6125V$

$Hg_2^{2+}(aq)+2e^- \rightleftharpoons 2Hg(l)$　　$E^{\ominus}(Hg_2^{2+}/Hg)=0.7973V$

（1）若将上述电对组成原电池，计算该电池的 $E^{\ominus}_{MF}$。

（2）求 Hg_2SO_4 的溶度积常数 K_{sp}。

14. 已知 $E^{\ominus}(Cu^{2+}/Cu)=0.3419V$，$E^{\ominus}(Ag^+/Ag)=0.7991V$，将铜片插入 0.1mol/L $CuSO_4$ 溶液中和银片插入 0.1mol/L $AgNO_3$ 溶液中组成原电池。

(1) 计算各个电极的电极电势，写出原电池符号。

(2) 写出电极反应和电池反应。

(3) 计算电池反应的平衡常数。

15. 反应 $MnO_2+4HCl \rightleftharpoons MnCl_2+Cl_2+2H_2O$

（1）判断标准状态时反应进行的方向。

（2）当使用浓盐酸 $[H^+]=[Cl^-]=12mol/L$，$[Mn^{2+}]$、$p(Cl_2)$ 均为标准状态时，判断反应自发进行的方向。

（已知：$E^{\ominus}(MnO_2/Mn^{2+})=1.224V$；$E^{\ominus}(Cl_2/Cl^-)=1.358V$）

16. 已知：$E^{\ominus}(NO_3^-/NO)=0.957V$，$E^{\ominus}(S/S^{2-})=-0.4763V$，$K_{sp}(CuS)=6.3\times10^{-36}$，计算反应：$3CuS+2NO_3^-+8H^+ \rightleftharpoons 3S+2NO+3Cu^{2+}+4H_2O$ 的平衡常数。

17. 已知在碱性溶液中：

$$E_B^{\ominus}/V \quad ClO_3^- \xrightarrow{\ ?\ } ClO_2^- \xrightarrow{+0.66V} ClO^- \quad (ClO_3^- \to ClO^-:\ +0.50V)$$

试求 $E^{\ominus}(ClO_3^-/ClO_2^-)$。

18. 下面是氧元素的电势图，根据此图回答下列问题：

$E_A^{\ominus}/V$ $O_2 \xrightarrow{+0.695V} H_2O_2$ —— H_2O；$O_2 \to H_2O$：+1.23V

$E_B^{\ominus}/V$ O_2 —— $HO_2^- \xrightarrow{+0.88V} OH^-$；$O_2 \to OH^-$：+0.40V

(1) 通过计算说明 H_2O_2 在酸性介质中的氧化性的强弱，在碱性介质中还原性的强弱。

(2) 判断 H_2O_2 在酸性介质和碱性介质中的稳定性。

Chapter 7 The Atomic Structure and Periodic Law of Elements
第七章 原子结构与元素周期律

学习目标

知识要求

1．**掌握** 核外电子运动特征；原子轨道的概念；四个量子数的名称、取值和意义；原子核外电子排布规则；基态原子的电子层结构与元素周期系。

2．**熟悉** 薛定谔方程和波函数；原子轨道和电子云的角度分布图、径向分布图；鲍林近似能级图；屏蔽效应和钻穿效应；元素性质的周期性变化规律。

3．**了解** 氢原子光谱；原子模型的建立。

能力要求

通过对原子核外电子排布规则的学习，掌握元素基态原子的电子层结构及元素周期系，理解元素性质的周期性变化原因。

1. Atomic Structure

PPT

1.1 The Structure of an Atom

1.1.1 Discovery of Electrons

In 1879, the British physicist W. Crookes (English, 1832—1919) discovered that the cathode ray (阴极射线) would be deflected in a magnetic or electric field. It indicated that the cathode ray was composed of negatively charged particles called electrons (电子). In 1897, Thomson (English, 1856—1940) experimentally determined that the charge-to-mass ratio of the electrons was 1.759×10^{11}C/kg. It is therefore that electrons exist in all substances in the same form, irrelevant to where they come from. In 1909, Michigan (US, 1868~1953) used the "oil drop test" to determine the electric quantity of the electron as 1.602×10^{-19} C, hence the mass of the electron can be calculated as 9.109×10^{-31}kg.

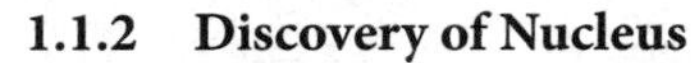

1.1.2 Discovery of Nucleus

In 1910, Rutherford (English, 1871—1937) and his students bombarded a very thin piece of gold foil with α-particles (alpha particles are known to be He^{2+} ions, α粒子) from a radioactive source. See Figure 7-1.

It was found that most of the α-particles appeared to hit the fluorescent zinc sulfide screen through the gold foil unhindered. Only a few of them bounced back or refracted as if they encountered impenetrable barriers. By measuring the number of α-particles that bounced back, Rutherford pointed out that there is a positively charged particle in the atom with the mass being equal to that of a gold atom but the diameter being no more than $\frac{1}{10000}$ that of an atom. Many experiments with foils of different metals yielded similar results. Realizing that these observations were inconsistent with previous theories about atomic structure, Rutherford discarded the old theory. By 1911, he had an explanation that each atom contains a tiny, positively charged, massive center that he called the atomic nucleus (原子核). Subsequent experiments gave him a hint, and he predicted the existence of protons (质子). In 1920, Rutherford predicted the existence of neutrons (中子). In 1932, J. Chadwick (English, 1891—1974) confirmed the existence of neutrons by experiments and determined that the mass of neutrons was almost equal that of protons.

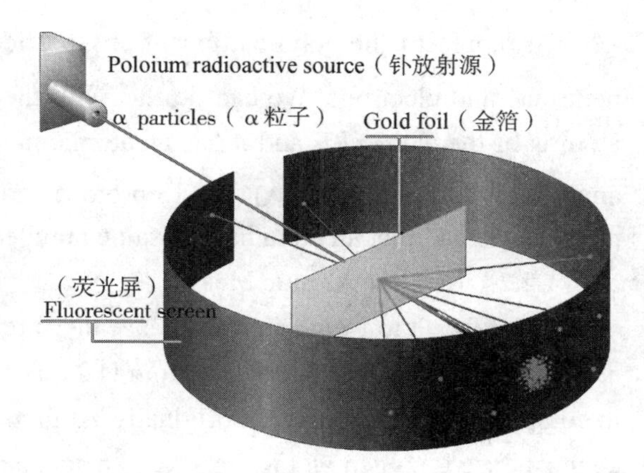

Figure 7-1 Rutherford's α-particle scattering experiment

Atoms are composed of nucleus and electrons. The nucleus is composed of protons and neutrons. Atoms are tiny, but the nucleus of an atom is far tinier still. If the diameter of an atom were the size of a football stadium, the nucleus would only be the size of a pea. This means that most of the atom is empty space! Electrons are even smaller than protons and neutrons.

1.2 The Relative Atomic Mass

1.2.1 Element and Atomic Number

Element (元素) refers to a class of atoms with the same number of protons. For example, tritium (氕), deuterium (氘), and tritium (氚) are all hydrogen elements (氢元素).

What makes an atom of one element different from an atom of another element? The atoms of each element have a characteristic number of protons. The number of protons in an atom of any particular element is called that element's atomic number (原子序数), which is also called nuclear charge (核电荷数, Z). Because an atom has no net electrical charge, the number of electrons it contains must equal the number of protons. All atoms of carbon, for example, have six protons and six electrons, whereas all atoms of oxygen have eight protons and eight electrons. Thus, carbon has atomic number 6, and oxygen has atomic number 8. It is obviously for an atom that

Atomic number = nuclear charge = number of protons = number of electrons

(原子序数 = 核电荷数 = 质子数 = 电子数)

The total number of protons and neutrons in an atom is called the mass number (质量数, A). Atom that have the same atomic number (Z) but different mass number (A) are called isotopes (同位素). Atoms with the same number of protons and different numbers of neutrons are at the same position in the periodic table (周期表).

To represent the composition of any particular atom, we need to specify its number of protons, neutrons, and electrons. We can do this with the symbolism of $^{A}_{Z}E$. This symbolism indicates that the atom is of the element E and it has an atomic number Z and a mass number A. For example, an atom of aluminum represented as $^{27}_{13}Al$ has 13 protons and 14 neutrons in its nucleus and 13 electrons outside the nucleus. (Recall that an atom has the same number of electrons as protons.)

1.2.2 Relative Atomic Mass

It is difficult to weigh atoms because they are very slight. For example, the actual mass of a hydrogen atom is 1.673×10^{-27}kg, an oxygen atom is 2.657×10^{-26}kg, and a ^{12}C atom is 1.993×10^{-26}kg. Chemists had to compare the masses of atom originally. At first, the mass of a hydrogen atom was taken as 1 unit. Later, chemists used oxygen as the stand as the standard for comparison. Finally, in 1961, chemists took one isotope of carbon as the standard. The isotope is $^{12}_{6}C$ (called carbon–12, a carbon atom with 6 neutrons in the nucleus). The mass of one atom ^{12}C is taken as exactly 12 units. As a result of action taken by the International Union of Pure and Applied Chemistry (IUPAC) in 1962, the mass of microscopic particles is generally expressed in atomic mass unit (amu or u), 1 u = 1.6605837×10^{-27}kg, which is exactly $\frac{1}{12}$ of the mass of a carbon–12 atom. The relative atomic mass (相对原子质量) or atomic mass (原子质量) of an element is the ratio of the average mass of one atom in the element (the weighted average of the masses of its isotopes.) to $\frac{1}{12}$ of the mass of ^{12}C atom. The numbers at the top left of each element in the periodic table are their relative atomic mass.

2. Motion Characteristics of Electrons

PPT

Rutherford's atomic model (卢瑟福原子模型) can explain the scattering phenomenon of the α-particles, so it can be as the foundation for modern atomic structure theory. However, if the negatively charged electrons in the atom do not move, they will be attracted by the positively charged nucleus. If the electron moves around the nucleus, there will be energy radiation, then the electrons' speed will slow down and finally fall onto the nucleus spirally. As a result, the atom will no longer exist. Apparently, this is not true. How to explain these problems? In the early 20th century, Planck (Germany, 1858—1947) proposed the quantum theory (量子论), and Einstein (Germany, 1885—1962) proposed the photon theory (光子学说). Bohr (Denmark, 1885—1962) put forward his assumption in 1913 based on the above theories. Bohrs theory (玻尔理论) successfully solved the above problems, and explained the hydrogen atom spectrum, therefore promoted the development of atomic structure theory.

2.1 The Particularity of the Motion of Electrons

2.1.1 Quantum Theory

In 1900, Max Planck made a revolutionary proposal: Energy, like matter, is discontinuous. Here, then, is the essential difference between the classical physics of Planck's time and the new quantum theory

that he proposed: Classical physics places no limitations on the amount of energy a system may possess, whereas quantum theory limits this energy to a discrete set of specific values. The difference between two of the allowed energies of a system also has specific value, called a quantum of energy (能量的量子化). This means that when the energy increases from one allowed value to the next, it increases by a tiny jump, or quantum. Here is a way of thinking about a quantum of energy: It bears a similar relationship to the total energy of a system as a single atom does to an entire sample of matter. Plank's assumption was that the group of atoms, the oscillator, must have an energy corresponding to the equation: $E = nh\nu$. Where E is the energy, n is a positive integer, ν is the oscillator frequency, and h is a constant that had to be determined by experiment. Using his theory and experimental data for the distribution of frequencies with temperature, Planck established the following value for the constant h. We now call it Planck's constant (普朗克常数), and it has the value $h = 6.62607 \times 10^{-34}$ Planck's postulate can be rephrased in this more general way: The energy of a quantum of electromagnetic radiation is proportional to the frequency of the radiation—the higher the frequency, the greater the energy. This is summarized by what we now call Planck's equation:

$$E = h\nu \tag{7-1}$$

Because the energy can be released only in specific amounts, we say that the allowed energies are quantized—their values are restricted to certain quantities. Planck's revolutionary proposal that energy is quantized was proved correct. And he was awarded the 1918 Nobel Prize in Physics for his work on quantum theory.

2.1.2 Wave-Particle Duality

As we know, light has wave-particle duality (波粒二象性). Just like the wave-particle duality of light, in 1924, Louis de Broglie (France, 1892—1987) assumed that all physical particles have wave-particle duality. And it is believed that the relation of the wave-particle duality of light can also be applied to physical particles such as electrons.

$$\lambda = \frac{h}{p} = \frac{h}{m\nu} \tag{7-2}$$

In formula (7-2), m, ν, p, and λ are the mass, velocity, momentum, and wavelength of the electron, respectively. h is the Planck's constant. By Planck's constant, the fluctuations (波动性) is quantitatively related to the particle-like nature (粒子性) of electrons.

In 1927, When Davidson (US, 1881—1958) and Germer (US, 1896—1971) bombarded a nickel crystal with high-energy electron beams instead of X-rays at Bell Labs in New York (Figure 7-2), the obtained electron diffraction photo is very similar to the X-ray image. The electron diffraction photograph has a series of light and dark diffraction ring patterns. This is due to the mutual interference between the waves. The wavelength of the electron wave obtained from the diffraction pattern confirms De Broglie's prediction.

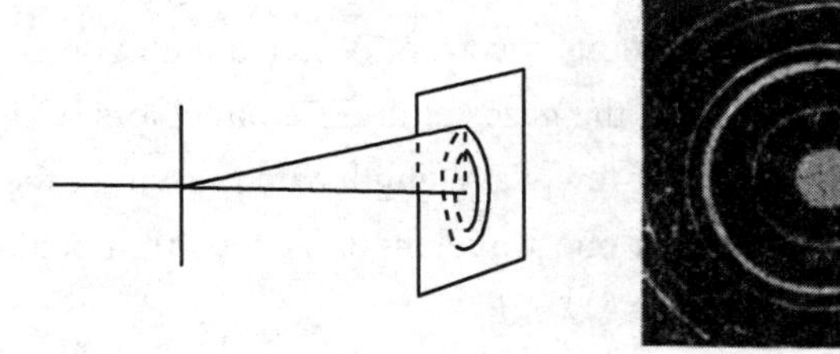

Figure 7-2 Schematic diagram of electron diffraction

Wave-particle duality is a universal phenomenon of substances. The fluctuation of microscopic particles is a characteristic of each moving particle itself, and it follows the statistical law.

2.1.3 Heisenberg's Uncertainty Principle

In classical physics, the movement of a macroscopic object has a fixed orbital. They have a fixed position and momentum (or speed) at a specific instant. During the movement, the macro objects, such as the bullets, cannonballs, and planets, have not only a certain speed, but also a position that can be accurately determined at any time. For microscopic particles with wave-particle duality, is this also right? The answer is no. For a microscopic particle moving at a high speed, if its position can be determined at a moment, its momentum cannot be determined accurately, and vice versa.

In 1927, Heisenberg (Germany, 1901—1976) proposed the famous uncertainty principle (测不准原理，或称作不确定原理). In a quantum mechanical system, the position of a particle and its momentum cannot be determined simultaneously. That is, it is impossible for us to know simultaneously both the exact momentum of an electron and its exact location in space. The mathematical formula for the uncertainty principle is

$$\Delta x \cdot \Delta p \geqslant \frac{h}{4\pi} \tag{7-3}$$

A brief calculation illustrates the dramatic implications of the uncertainty principle. For the ground state electrons (基态电子) of hydrogen atom, the electron's velocity is 2.18×10^7 m/s and its mass is 9.1×10^{-31}kg. Let's assume that we know the speed to an uncertainty of 1%, then

$$\Delta p = \Delta mv = 9.1\times10^{-31}\times2.18\times10^{7}\times0.01 = 2.0\times10^{-25}\ \text{kg}\cdot\text{m/s}$$

And the measurement deviation of the motion coordinates of the electron is

$$\Delta x = \frac{h}{4\pi\Delta mv} = \frac{6.63\times10^{-34}}{4\times3.14\times2\times10^{-25}} = 2.64\times10^{-10}\,\text{m} = 264\ \text{pm}$$

The covalent radius of the hydrogen atom is only 37 pm. The uncertainty in the position (264 pm) of the electron in the atom is much greater than the size of the atom itself. Thus, we have essentially no idea where the electron is located in the atom. The uncertainty principle states that there is a relationship between the inherent uncertainties in the location and momentum of an electron.

2.2 The Bohr Theory

2.2.1 The Spectrum of Hydrogen Atom

Continuous spectrum (连续光谱): After being refracted by a prism, the light of sunlight or incandescent lamps can be divided into seven colored light bands. This rainbow of colors, containing light of all wavelengths, is called a continuous spectrum. The most familiar example of a continuous spectrum is the rainbow produced when raindrops or mist acts as a prism for sunlight.

Not all radiation sources produce a continuous spectrum. When a high voltage is applied to tubes that contain different gases under reduced pressure, the gases emit different colors of light. When light coming from such tubes passes through a prism, only a few wavelengths are present in the resultant spectra. Each colored line in such spectra represents light of one wavelength. A spectrum containing radiation of only specific wavelengths is called a line spectrum (线状光谱).

When scientists first detected the line spectrum of hydrogen in the mid-1800s, they were fascinated

by its simplicity. At that time, only five lines at wavelengths of 410 nm, 410 nm, 434 nm, 486 nm, and 656 nm were observed (Figure 7-3).

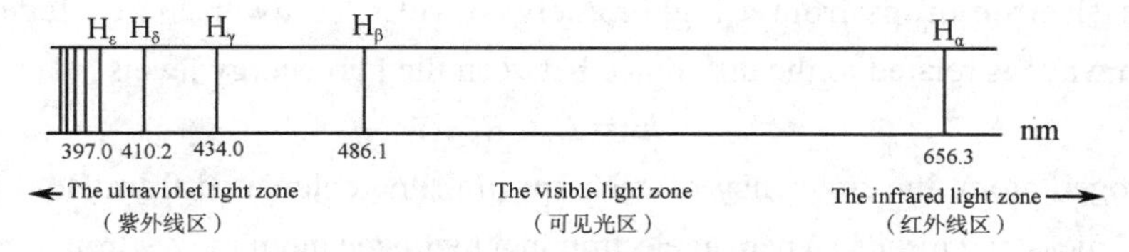

Figure 7-3 Line spectrum of hydrogen atom

In 1883, Balmer (Sweden, 1825~1898) showed that the wavelengths of these four lines fit an intriguingly simple formula that relates the wavelengths to integers.

$$\lambda = \mathrm{B}\frac{n^2}{n^2-4} \tag{7-4}$$

In the formula, λ is the wavelength and B is a constant. When n is equal to 3, 4, 5, 6, 7, respectively, the wavelengths of these 5 spectral lines will be calculated respectively. Therefore, these spectral lines in the visible light are also called the Balmer line.

In 1913, Rydberg (Sweden, 1854~1919) finds a more general equation, which allows us to calculate the wave lengths of all the spectral lines of hydrogen:

$$\sigma = R_H\left(\frac{1}{n_1^2}-\frac{1}{n_2^2}\right) \tag{7-5}$$

In formula (7-5), σ is the wave number, R_H is called the Rydberg's constant, its value is $1.097\times10^5\text{cm}^{-1}$, n_1 and n_2 are positive integers, and $n_2 > n_1$.

2.2.2 The Bohr's Model of the Hydrogen Atom

To explain the line spectrum of hydrogen, Niels Bohr (Denmark, 1885~1962) assumed that electrons in hydrogen atoms move in circular orbits around the nucleus. According to classical physics, a charged particle (such as an electron) moving in a circular path should continuously lose energy. As an electron loses energy, therefore, it should spiral into the positively charged nucleus. This behavior, however, does not happen—hydrogen atoms are stable. So how can we explain this apparent violation of the laws of physics? Bohr assumed that the prevailing laws of physics were inadequate to describe all aspects of atoms.

Bohr based his model on three postulates:

(1) The electrons in the hydrogen atom can only move in some specific circular orbitals (stable orbitals), and the angular momentum P of these orbitals must be an integer multiple of h / 2π,

$$P = mvr = nh/2\pi$$

Where m and v are the mass and velocity of the electron respectively, r is the radius of the circular orbital, h is the Planck's constant, n is the quantum number, which is 1, 2, 3, ... Electrons neither absorb nor release energy when moving in these stable orbitals. Bohr deduced that the energy of various specific orbitals in the hydrogen atom obeys the following formula:

$$E = -\frac{13.6}{n^2}(eV) = -\frac{2.18\times 10^{-18}}{n^2}(J) \tag{7-6}$$

(2) The electrons in the atom have different energy in different orbitals, that is, they have different energy levels (能级). Electrons neither absorb nor radiate energy when orbiting a nucleus in orbital. The electrons in the orbitals are as close to the nucleus as possible, and the atom is now in the ground state.

When the electrons are excited by external energy, it moves to an outer orbital with higher energy. Then the atoms and electrons are in an excited state (激发态).

(3) When an electron drops from a higher energy level to a lower energy level, energy will be emitted. Radiation wave is related to the difference between the two energy levels

$$h\nu = E_2 - E_1 \tag{7-7}$$

For the hydrogen atom, the frequency of each spectral line calculated from Bohr theory is in good agreement with the measured results. When an electron in a hydrogen atom moves from a higher energy level of $n = 3$ to a lower energy level of $n = 2$, it emits a photon with a frequency of 4.57×10^{14} (s^{-1}) (wavelength of 656 nm), which is equivalent to the H_α line (See Figure 7-4). According to the Bohr theory, the orbital radius of the ground state hydrogen atom is 53 pm (0.53 Å). This value is the Bohr radius (玻尔半径, a_0). a_0 is often used as the unit of length for atoms and molecules.

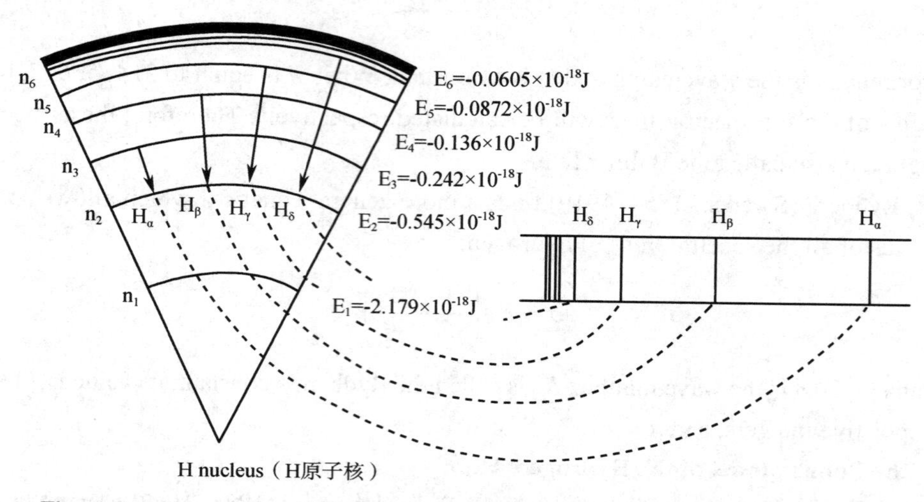

Figure 7-4 Generation of hydrogen atom spectrum and hydrogen atom structure

Bohr theory is very successful in explaining the relationship between the spectrum of hydrogen atoms (or hydrogen-like ions, 类氢原子, such as He^{2+}, Li^{3+}, etc.) and the energy levels of atomic orbitals. It also points out the quantized characteristics of the motion of electrons.

Despite its great success in predicting the spectral lines of hydrogen atoms, the Bohr's model failed with other atoms. The reason is that it is a mono-electron model: it works beautifully for the H atom and for mono-electron ions. It fails completely for species with more than one electron because electron-electron repulsion and additional nucleus-electron attractions create much more complex interactions. Even more fundamentally, electrons do not move in fixed, defined orbitals.

3. The Quantum Mechanical Model of the Atom

PPT

3.1 The Schrödinger Equation

The most basic equation in the theoretical model for processing electrons outside the nucleus of hydrogen atom by quantum mechanics (量子力学) was proposed by Schrödinger (Austria, 1887—1961) in 1926, which is a wave equation describing the motion of microscopic particles, that is, the Schrödinger equation (薛定谔方程).

$$\frac{\partial^2\psi}{\partial x^2}+\frac{\partial^2\psi}{\partial y^2}+\frac{\partial^2\psi}{\partial z^2}+\frac{8\pi^2 m}{h^2}(E-V)\psi=0 \tag{7-8}$$

In the formula, ψ is wave function (波函数), which is a function of the space coordinates x, y, and z. E is the total energy, V is the potential energy, m is the mass of the electron, and h is the Planck's constant.

By solving the Schrödinger equation of a system (such as a hydrogen atom system), it can obtain a series of wave functions ψ_1, ψ_2, ψ_3, ..., ψ_n, and a series of corresponding energy E_1, E_2, E_3, …, E_n. Each reasonable solution of the Schrödinger equation represents the possible motion state of the electrons in the system. Therefore, in quantum mechanics, the wave function and its corresponding energy are used to describe the motion state of microscopic particles. The wave function ψ is not only a mathematical function describing the motion state of electrons, but also a function of spatial coordinates. Its spatial image can be viewed as the spatial range of electron motion, so it is also called "atomic orbital" (原子轨道). For the motion of electrons in a hydrogen atom, an accurate solution can be obtained by solving the wave function ψ from the Schrödinger equation, but for electrons in other multi-electron atoms, only approximate solutions can be obtained.

When solving Schrödinger equation, three parameters (n, l, m) must be introduced for getting a reasonable solution. Since each solution is constrained by the three parameters n, l, and m, a wave function (a state of motion or an atomic orbital) can be simply described by a set of quantum numbers.

3.2 Quantum Numbers

3.2.1 The Principal Quantum Number *n*

The first number to be fixed is the principal quantum number, n (主量子数), which may have a only positive, nonzero integral value, n = 1, 2, 3,… All orbitals with the same value of n are in the same principal electronic shell (电子层). As n increases, the orbital becomes larger, and the electron spends more time farther from the nucleus. An increase in n also means that the electron has a higher energy and is therefore less tightly bound to the nucleus.

The value of n can also be represented by the corresponding electronic shell symbol.

n value	1	2	3	4	5	6	7
Electronic shell	1	2	3	4	5	6	7
Electronic shell spectrum symbol	K	L	M	N	O	P	Q
Average distance from nucleus	near ⟶						far

3.2.2 The Angular Momentum Quantum Number *l*

The second number to be fixed is the orbital angular momentum quantum number, l (角量子数), which may be zero or a positive integer, but not larger than n–1 (where n is the principle quantum number), l = 0, 1, 2, 3, …, n–1. All orbitals with the same n and l values are in the same electronic subshell (电子亚层). The value of l for a particular orbital is generally designated by the electronic spectrum symbol *s*, *p*, *d*, and *f*, corresponding to l values of 0, 1, 2, and 3. In the first shell, the maximum value of l is zero, which tells us that there is only an *s* subshell and no *p* subshell. In the second shell, the permissible values of l are 0 and 1, which tells us that there are only *s* and *p* subshells.

The orbital angular momentum quantum number designates a specific shape of atomic orbital that an electron may occupy. For different l values, the shapes of the corresponding orbitals are different. When $l = 0$, the orbital symbol is s, and the shape is spherical (球形); when $l = 1$, the orbital symbol is p, and the shape is a dumbbell without handle (无柄哑铃形); when $l = 2$, the orbital symbol is d, and the shape is a plum petal (梅花瓣形).

Note that the orbital energy level of hydrogen atom or hydrogen-like ions is determined only by n (because there is only one electron outside their nucleus). However, For multi-electron atoms (多电子原子), the orbital energy levels are determined by both n and l.

3.2.3 The Magnetic Quantum Number m

The third number to be fixed is the magnetic quantum number, m (磁量子数). It may be a negative or positive integer, including zero, and ranging from $-l$ to $+l$ (where l is the orbital angular momentum quantum number), $m = 0, \pm1, \pm2, ..., \pm l$. The magnetic quantum number designates a specific orbital within a subshell, prescribing the three-dimensional orientation of the orbital in the space (轨道的空间取向) around the nucleus. The orbitals with the same n and l, and different m basically have the identical energy. We call these orbitals degenerate orbitals (等价轨道，或简并轨道). Orbitals with different l values are known by the following labels. When $l = 0$, $m = 0$, it can be labeled by s. When $l = 1$, $m = +1, 0, -1$, the spectral symbols are p_x, p_y, p_z. When $l = 2$, $m = +2, +1, 0, -1, -2$, the spectral symbols are d_{xz}, d_{yz}, d_{xy}, $d_{x^2-y^2}$, d_{z^2}.

Each atomic orbital refers to a wave function $\psi_{n,l,m}$. It represents the motion state of an electron, for example:

Quantum numbers	$\psi_{n,l,m}$	Motion state
$n = 2, l = 0, m = 0$	$\psi_{2,0,0}$ or ψ_{2s}	$2s$
$n = 2, l = 1, m = 0$	$\psi_{2,1,0}$ or ψ_{2p_z}	$2p_z$
$n = 3, l = 2, m = 0$	$\psi_{3,2,0}$ or $\psi_{3d_{z^2}}$	$3d_{z^2}$

3.2.4 The Spin Quantum Number m_s

The three quantum numbers n, l, and m describe the size, shape, and orientation of the orbital, respectively. An additional quantum number, called the spin quantum number, m_s (自旋量子数), describes the spin property of the electron. Corresponding to the two directions of the electron's field, the m_s has two possible values, $+\frac{1}{2}$ or $-\frac{1}{2}$. Thus, each electron in an atom is described completely by a set of four quantum numbers (n, l, m, and m_s): the first three describe its orbital, and the fourth describes its spin. The possible motion states of electrons are summarized in Table 7-1.

3.3 Interpreting and Representing the Wave Function

It is the solution of the Schrödinger equation (Table 7-2). After converting rectangular Cartesian coordinate system into spherical polar coordinate system, the orbitals can be expressed in terms of one function $R_{n,l}(r)$ that depends only on r, and a second function $Y_{l,m}(\theta, \varphi)$ that depends on θ and φ. That is,

$$\psi_{n,l,m}(r, \theta, \varphi) = R_{n,l}(r) \cdot Y_{l,m}(\theta, \varphi) \tag{7-9}$$

Among them, $R_{n,l}(r)$ is called the radial function (径向波函数), and $Y_{l,m}(\theta, \varphi)$ is called the angular function (角度波函数).

Table 7-1 The possible motion states of electrons

n	l ($l<n$)	Orbital symbol (Energy level)	m ($\lvert m\rvert \leqslant l$)	Orbital number	Orbital number of each electron shell (n^2)	Number of electrons that can be accommodated ($2n^2$)
1	0	1*s*	0	1	1	2
2	0	2*s*	0	1	4	8
	1	2*p*	+1, 0, -1	3		
	0	3*s*	0	1		
3	1	3*p*	+1, 0, -1	3	9	18
	2	3*d*	+2, +1, 0, -1, -2	5		
	0	4*s*	0	1		
4	1	4*p*	+1, 0, -1	3	16	32
	2	4*d*	+2, +1, 0, -1, -2	5		
	3	4*f*	+3, +2, +1, 0, -1, -2, -3	7		

Table 7-2 The angular and radial wave functions of a hydrogen-like atom (a_0/bohr radius)

Orbital	ψ (r, θ, φ)	R (r)	Y (θ, φ)
1*s*	$\sqrt{\frac{1}{\pi a_0^3}}e^{-r/a_0}$	$2\sqrt{\frac{1}{a_0^3}}e^{-r/a_0}$	$\sqrt{\frac{1}{4\pi}}$
2*s*	$\frac{1}{4}\sqrt{\frac{1}{2\ \ a_0}}(2\ \ \frac{}{a_0})\ \ ^{r/2a}$	$\sqrt{\frac{1}{8a_0^3}}\left(2-\frac{r}{a_0}\right)e^{-r/2a_0}$	$\sqrt{\frac{1}{4\pi}}$
2p_x	$\frac{1}{4}\sqrt{\frac{1}{2\pi a_0^3}}(\frac{r}{a_0})e^{-r/2a_0}\sin\theta\cos\varphi$		$\sqrt{\frac{3}{4\pi}}\sin\theta\cos\varphi$
2p_y	$\frac{1}{4}\sqrt{\frac{1}{2\pi a_0^3}}(\frac{r}{a_0})e^{-r/2a_0}\sin\theta\sin\varphi$	$\sqrt{\frac{1}{24a_0^3}}\left(\frac{r}{a_0}\right)e^{-r/2a_0}$	$\sqrt{\frac{3}{4\pi}}\sin\theta\sin\varphi$
2p_z	$\frac{1}{4}\sqrt{\frac{1}{2\pi a_0^3}}(\frac{r}{a_0})e^{-r/2a_0}\cos\theta$		$\sqrt{\frac{3}{4\pi}}\cos\theta$

3.3.1 Angular Distribution Plot of Wave Function

If the $Y(\theta, \varphi)$ of the wave function is plotted with the angle θ and φ, the angular distribution of the atomic orbital can be obtained. Figure 7-5 is the angular distribution of the *s*, *p*, and *d* orbitals of a hydrogen atom.

The angular distribution plot of the wave function (波函数角度分布图) describes the angular distribution of the atomic orbital. Its shape is independent of the number of energy levels. For the *s* orbitals in each shell, their angular distribution images are the same because they have same *Y* values. The angle distribution diagram of *p* orbital of each shell is also the same. There are three *p* orbitals (called p_x, p_y, and p_z respectively), five *d* orbitals and seven *f* orbitals (images are omitted due to the complexity).

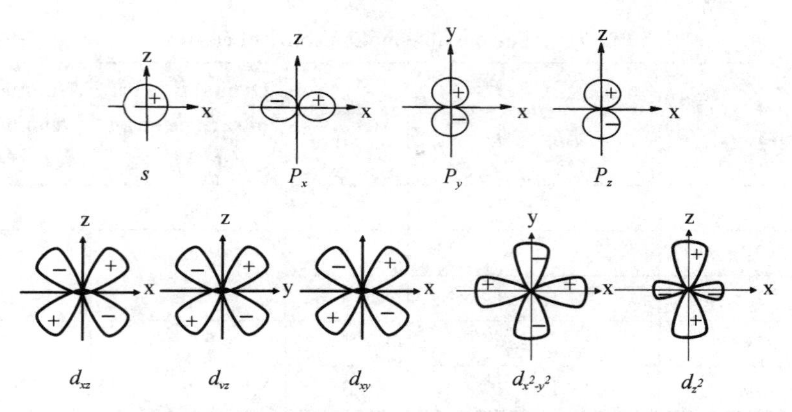

Figure 7-5　Angle distribution of the hydrogen atom wave function

3.3.2　Angle probability Distribution Plot of Electron Cloud

While we cannot know exactly where the electron is at any moment, we can know where it probably is, that is, where it spends most of its time. We get this information by squaring the wave function. Thus, even though ψ has no physical meaning, $|\psi|^2$ does and is called the probability density (概率密度), a measure of the probability of finding the electron in some tiny volume of the atom. In order to visualize the probability density distribution of electrons, the small black dots is often used to represent the probability density in space. Because the nucleus is enveloped by the negatively charged electrons, so it was called "electron cloud" (电子云). A place with a high density of electron clouds indicates that electrons are more likely to appear there. $|\psi|^2$, which is a quantity related to probabilities, is an approximate image of the electron cloud. Figure 7-6 is a schematic diagram of an electron cloud of a hydrogen atom. The 1*s* electron cloud of a hydrogen atom is spherically symmetric.

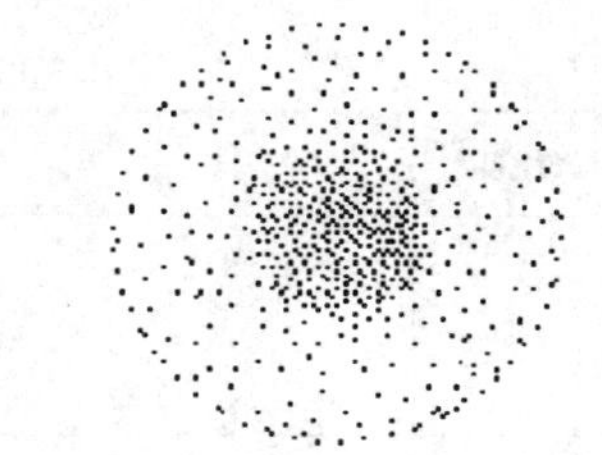

Figure 7-6　1*s* electron cloud depictions of hydrogen atom

If the $|Y(\theta, \varphi)|^2$ is plotted with the angle θ and φ, the angle probability distribution plot of the electron cloud (电子云的角度分布图) is obtained. Figure 7-7 is a schematic diagram of the angular distribution of the hydrogen atom electron cloud. The angle distribution of the electron cloud can indicate the probability density of electrons appearing at different angles in space, but it cannot indicate the probability density of electrons appearing along the radius of the atom. They are similar in shape to the angular distribution of the wave function, except that the angular distribution of the wave function has a plus and minus sign.

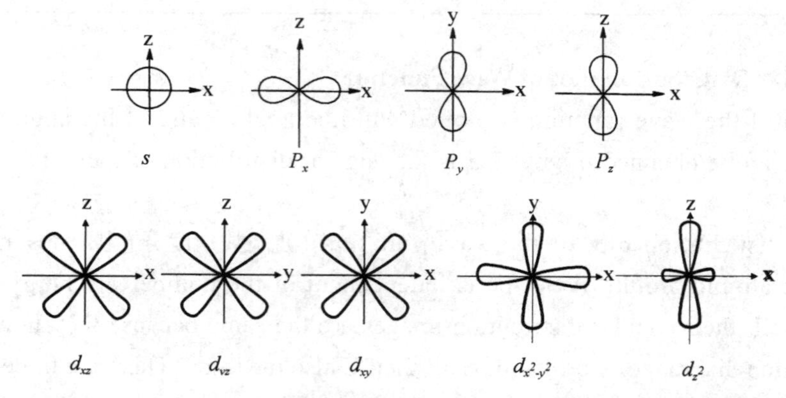

Figure 7-7　Angle probability distribution plot of the electron cloud of hydrogen atom

The angular distribution of the electron cloud is "thinner" than the angular distribution of the atomic orbital. This is because $|Y|$ is less than "1", so $|Y|^2$ is smaller.

It is worth noting that neither the atomic orbital angle distribution diagram nor the electron cloud angular distribution diagram is the orbital of electron movement. They are not the results of experiments, but two kinds of function graphics and each graphic represents different meanings.

3.3.3 Radial Distribution Plot of Wave Function

If the $R_{n,l}(r)$ of the wave function is plotted with r, the radial distribution of the atomic orbital can be obtained (Figure 7-8).

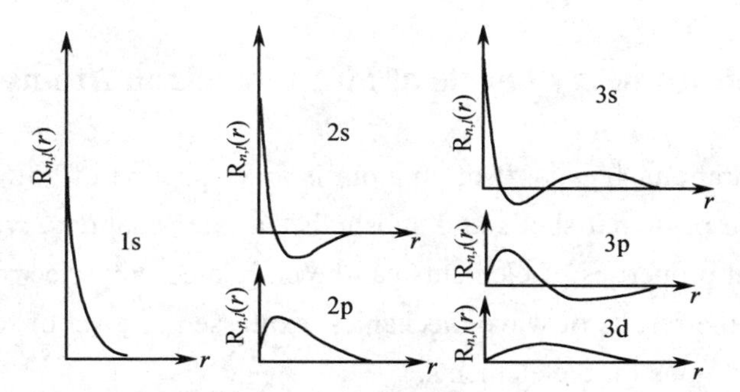

Figure 7-8 Radial distribution of hydrogen atom orbitals

Because the wave function $R_{n,l}(r)$ is only a mathematical formula describing the state of electron movement without physical meanings, so the above picture is not described in details.

3.3.4 Radial Probability Distribution Plot of Electron Cloud

We mentioned earlier that the wave function $|\psi|^2$ represents the probability that an electron appears in space. How to describe the probability of the electrons appearing in a certain spatial region?

Probability = probability density×microscopic volume

（概率 = 概率密度×微体积）

The distance from an electron to the nucleus is r, the thickness of the thin spherical shell is d_r. Thus, the volume of the spherical shell is $4\pi r^2 dr$ ($dr = 1$). The probability of an electron appearing in a spherical shell should be $|R(r)|^2 \times 4\pi r^2 = 4\pi r^2 R^2$, where $4\pi r^2 R^2$ is called the D function. If the D function is plotted with r, the radial probability distribution plot of the electron cloud (电子云的径向分布图) can be obtained (Figure 7-9). In comparing the radial probability curves for the 1*s*, 2*s*, and 3*s* orbitals, we find that a 1*s* electron has a greater probability of being close to the nucleus than a 2*s* electron does, which in turn has a greater probability than does a 3*s* electron. In comparing 2*s* and 2*p* orbitals, a 2*s* electron has a greater chance

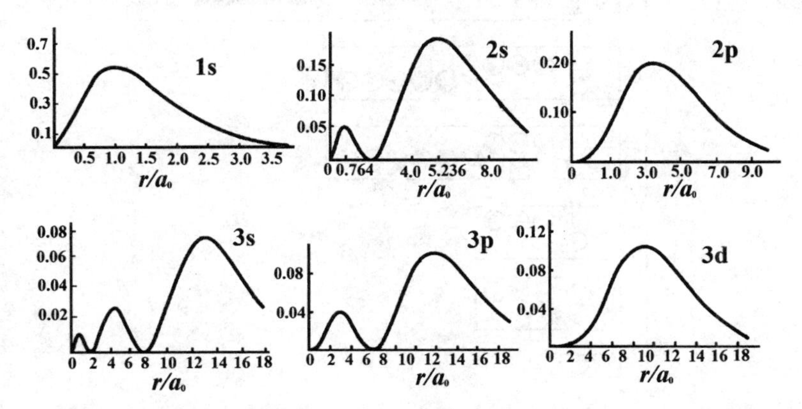

Figure 7-9 Radial probability distribution plot of the electron cloud

of being close to the nucleus than a $2p$ electron does. The $2s$ electron exhibits greater penetration than the $2p$ electron. Electrons having a high degree of penetration effectively block the view of an electron in an outer orbital looking for the nucleus.

4. Electron Configurations of Multi-electron Atoms

4.1 Atomic Orbital Energy Levels of Multi-electron Atoms

The electron configuration (电子构型) of an atom is a designation of how electrons are distributed among various orbitals in principal shells and subshells. In later chapters, we will find that many of the physical and chemical properties of elements can be correlated with electron configurations. In this section, we will see how the results of wave mechanics, expressed as a set of rules, can help us to write probable electron configurations for the elements.

The electron configuration of multi-electron atoms (all atoms except hydrogen and hydrogen-like ions) is based on the order of the energy levels of the atomic orbitals. Due to the penetration effect (钻穿效应) and the shielding effect (屏蔽效应), the orbital energy is not only determined by n and l , but also related to the electron configuration.

4.1.1 Pauling's Atomic Orbital Approximate Energy Level Diagram

According to the results of the spectral experiments, Linus Carl Pauling (US, 1901—1994) summarized the order of the energy levels of the atomic orbitals in the multi-electron atoms (Figure 7-10):

(1) According to the order of energy from low to high, atomic orbitals with similar energy levels are arranged in one group, and each small circle represents an atomic orbital. It is currently divided into

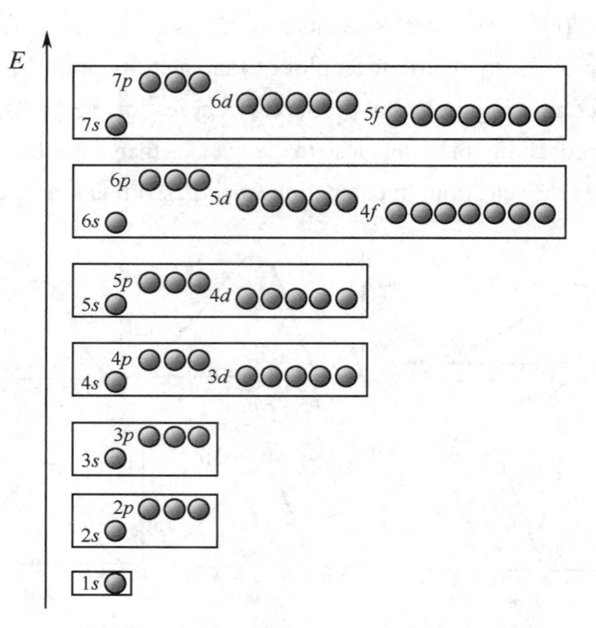

Figure 7-10 Pauling's atomic orbital approximate energy level diagram

seven energy level groups: (1*s*), (2*s*, 2*p*), (3*s*, 3*p*), (4*s*, 3*d*, 4*p*), (5*s*, 4*d*, 5*p*), (6*s*, 4*f*, 5*d*, 6*p*), (7*s*, 5*f*, 6*d*, 7*p*). Note that, the energy difference between the energy levels group is larger, while the energy levels in each group is smaller.

(2) In the same atom, the energy states of the atomic orbital is mainly determined by the principal quantum number *n* and the angular momentum quantum number *l*, such as: $E_{1s} < E_{2s} < E_{3s} < E_{4s}$, $E_{4s} < E_{4p} < E_{4d} < E_{4f}$. This phenomenon of different energy states of atomic orbital in the same shell (atomic orbital with same *n*) is called "Energy-level splitting" (能级分裂). In the fourth energy level group, *n* is large, but the energy is low, such as: $E_{4s} < E_{3d}$. This phenomenon is called "energy level overlap" (能级交错). The changes in the energy levels of these atomic orbitals can usually be explained by the "shielding effect" and "penetration effect".

4.1.2 Shielding Effect

In multi-electron atoms, each electron 'feels' not only the attraction to the nucleus, but also repulsion from other electrons. Repulsion counteracts the nuclear attraction somewhat, making each electron easier to remove by helping to push it away. We speak of each electron "shielding" the other electrons to some extent from the nuclear charge (*Z*). Shielding effect (also called screening effect) reduces the full nuclear charge to an effective nuclear charge (有效核电荷), which is usually represented by the symbol Z^*. The relationship between the effective nucleus charge number and the nucleus charge number is:

$$Z^* = Z - \sigma \tag{7-10}$$

In the formula, σ is called the shielding constant (屏蔽常数), which represents the repulsion of other electrons against the specified electron. We can approximate the multi-electron atom system as a single-electron system with Z^*. Thus, the energy level of each orbital in a multi-electron atom can be described as follows:

$$E = -2.18\times 10^{-18} \frac{(Z^*)^2}{n^2} = -2.18 \quad 10^{-18} \frac{(Z-\sigma)^2}{n^2} (J) \tag{7-11}$$

The value of σ is related to both *n* and *l*. Slater (US, 1900~1976) formulated a set of simple rules to estimate the shielding constant σ (called the Slater rules, 斯莱特规则). The σ values of various electrons provided by this rule are listed in Table 7-3.

Table 7-3 Slater shielding constant of each electron in multi-electron atom orbitals

Shielded electron	Shielding electron							
	1*s*	2*s*, 2*p*	3*s*, 3*p*	3*d*	4*s*, 4*p*	4*d*	4*f*	5*s*, 5*p*
1*s*	0.3							
2*s*, 2*p*	0.85	0.35						
3*s*, 3*p*	1.00	0.85	0.35					
3*d*	1.00	1.00	1.00	0.35				
4*s*, 4*p*	1.00	1.00	0.85	0.85	0.35			
4*d*	1.00	1.00	1.00	1.00	1.00	0.35		
4*f*	1.00	1.00	1.00	1.00	1.00	0.85	0.35	
5*s*, 5*p*	1.00	1.00	1.00	1.00	0.85	0.85	0.85	0.35

According to the above table, the effective nucleus charge of electrons outside the nucleus can be calculated. For example, the electron configuration of element 23Vanadium is $1s^22s^22p^63s^23p^63d^34s^2$,

For 1*s* electronics: $\sigma_{1s} = 0.30$, $Z^* = 23 - 0.30 = 22.70$,

For 2*s*, 2*p* electrons: $\sigma_{2s} = \sigma_{2p} = 0.35\times7 + 0.85\times2 = 4.15$, $Z^* = 23 - 4.15 = 18.85$,

For 3*s*, 3*p* electronics: $\sigma_{3s} = \sigma_{3p} = 0.35\times7 + 0.85\times8+1.00\times2 =11.25$, $Z^* = 23 - 11.25= 11.75$,

For 3*d* electronics: $\sigma_{3d} = 0.35\times2 + 1.00\times18 =18.70$, $Z^* = 23 - 18.70 = 4.30$,

For 4*s* electronics: $\sigma_{4s} = 0.35\times1 + 0.85\times11 + 1.00\times10 = 19.70$, $Z^* = 23 - 19.70 = 3.30$,

From the above calculations, it can be seen that the farther the electrons are from the nucleus, the smaller the attraction of the electron to the nucleus is, and hence the higher the energy of the orbital is.

4.1.3 Penetration Effect

The magnitude of the reduction of the nuclear charge depends on the types of orbitals the inner electrons are in and the type of orbital that the screened electron is in. From the probability distribution plot of electron cloud (Figure 7-9), we have seen that electrons in *s* orbitals have a high probability density at the nucleus, whereas *p* and *d* orbitals have low probability densities near the nucleus. Thus, electrons in *s* orbitals are more effective at screening the nucleus from outer electrons than are electrons in *p* or *d* orbitals. This ability of electrons in *s* orbitals that allows them to get close to the nucleus is called penetration (钻穿). An electron in an orbital with good penetration is better at screening than one with low penetration.

(1) For a given *l* value, the larger *n* value indicates a much stronger shielding effect of the electrons in outer energy levels by electrons in inner energy levels. Therefore, $E_{1s} < E_{2s} < E_{3s} < E_{4s}$.

(2) For a given *n* value, a lower *l* value indicates a much stronger penetration effect. Therefore, the penetrate ability is *ns* > *np* > *nd* > *nf*, and the attraction by nucleus is *ns* > *np* > *nd* > *nf*. Thus, energy-level splitting occurs, such as: $E_{4s} < E_{4p} < E_{4d} < E_{4f}$.

(3) For energy levels with different *n* and *l*, the order of the energy levels is complicated. For example, the energy level of 4*s* is lower than 3*d*. This is because the penetration of a 4*s* electron is much stronger than that of a 3*d* electron, causing energy levels to overlap. The 4*s* energy level is below the 3*d* level despite its higher principal quantum number *n*. So $E_{4s} < E_{3d}$. It also can explain the energy level overlap phenomenon: $E_{5s} < E_{4d}$, $E_{6s} < E_{4f} < E_{5d}$.

4.2 Electron Configurations of the Ground State Atom

4.2.1 Rules for Assigning Electrons to Orbitals

(1) The lowest energy principle (能量最低原理) Electrons occupy orbitals in a way that minimizes the energy of the atom. So the electrons are filled in the atomic orbits from the lowest level 1*s* orbital, and then are filled in order in the higher energy levels.

According to Pauling's atomic orbital approximate energy level diagram (Figure 7-10), the order in which orbitals fill is 1*s*, 2*s*, 2*p*, 3*s*, 3*p*, 4*s*, 3*d*, 4*p*, 5*s*, 4*d*, 5*p*, 6*s*, 4*f*, 5*d*, 6*p*, 7*s*, 5*f*, 6*d*, 7*p*.

(2) The pauli exclusion principle (Pauli不相容原理) In 1926, Wolfgang Pauli (Austria, 1900~1958) explained complex features of emission spectra associated with atoms in magnetic fields by proposing that no two electrons in an atom can have all four quantum numbers alike. The first three quantum numbers *n*, *l* and *m* determine a specific orbital. Two electrons may have these three quantum numbers alike; but if they do, they must have different values of *ms*, the spin quantum number. Another way to state this result

is that only two electrons may occupy the same orbital and these electrons must have opposing spins.

(3) Hund's Rule Hund (Germany, 1896—1997) summarized the Hund's rule (洪特规则) according to a large amount of spectral experimental data that electrons occupy as many orbitals as possible in the degenerate orbitals with same n and l. When orbitals of equal energy are available, the electron configuration of lowest energy has the maximum number of unpaired electrons with parallel spins. This is understandable, as it takes energy to pair up electrons. The requirement of parallel spins for electrons that do occupy different orbitals is a consequence of a quantum mechanical effect called spin correlation, the tendency for two electrons with parallel spins to stay apart from one another and hence to repel each other less.

In addition, quantum mechanics theory also states that the system is under the most stable state when the electron configuration is full filled (全充满), half-filled (半充满), or full empty (全空) in degenerate orbitals. It can also be called a special case of Hund's rule.

Full filled: p^6, d^{10}, f^{14}

Half-filled: p^3, d^5, f^7

Full empty: p^0, d^0, f^0

4.2.2 Electron Configurations of the Ground State Atom

According to the rules for assigning electrons to orbitals, several examples of electron configurations of ground state atom are discussed as below.

Example 7-1 According to the rules for assigning electrons to orbitals, write the symbol and electron configurations of the elements with atomic number of 6, 17, and 27.

Solution

^{6}C $1s^22s^22p^2$

^{17}Cl $1s^22s^22p^63s^23p^5$ or $[Ne]3s^23p^5$

^{24}Cr $1s^22s^22p^63s^23p^63d^54s^1$ or $[Ar]3d^54s^1$

In determining the ground-state electron configuration of Cr, we have to choose between $[Ar]3d^44s^2$ and $[Ar]3d^54s^1$, because the energies of the 3d and 4s orbitals are very similar. In this case, we can use Hund's rule to decide that the most stable electron configuration is that with the most unpaired electrons, that is $[Ar]3d^54s^1$.

The electrons involved in forming compounds are called the valence electrons (价电子). For main group elements (主族元素), the valence electrons are the outermost electrons. For transition elements (过渡元素), in addition to the outermost ns electrons, the $(n\text{-}1)d$ electrons are also valence electrons. According to the results of the spectroscopic experiments, the valence electrons of each element have been listed in the periodic table.

It should be pointed out that not all the valence electron configurations in the periodic table can be explained by the three rules for assigning electrons mentioned above. As the number of nuclear charges increases, the interactions between the nucleus and electrons, and between electrons and electrons, become more complicated. From the fifth to the seventh period, there are "exceptions" in the electron configurations, such as Nb, Ru, Rh, Pd, W, Pt and so on. Therefore, the electron configurations are finally determined by the results of spectral experiments.

5. Structure of the Periodic Table

In 1870, Dmitri Mendeleev (Russia, 1834—1907) arranged the 65 elements known at the time into a table and summarized their behavior in the periodic law: when arranged by atomic mass, the elements exhibit a periodic recurrence of similar properties. Mendeleev left blank spaces in his table and was even able to predict the properties of several elements. Today's periodic table commonly used includes 117 kinds of elements, of which there are 52 elements not known in 1870. Most importantly, it arranges the elements by atomic number (number of protons) not atomic mass. By 2019, the number of elements has increased to 118, and the newly added element is synthetic element.

5.1 Period

In the periodic table, each horizontal rows is a period (周期). With the increase of the atomic number, the electrons enter the next energy level group after filling one electron level group, so new electronic shells are constantly appearing. The electrons of each period are filled from ns^1 to ns^2np^6 (except for the first period). The number of elements in a complete period is exactly twice the number of orbitals in the energy level group. So, the period starts at the beginning of a new electron shell, and the period number is the n value of the highest energy level. The elements known can be divided into periods, including 1 ultra-short period, 2 short periods, 2 long periods, and 2 extra-long periods. The relationship between the number of periods and the number of energy level groups, and other relevant topics are shown in Table 7-4.

Table 7-4 Relationship between the number of periods and the number of energy level groups, and other relevant topics

Energy level groups	Periods	Characteristics of periods	Number of elements	orbitals
1	1	Ultra-short period	2	1
2	2	short period	8	4
3	3	short period	8	4
4	4	Long period	18	9
5	5	Long period	18	9
6	6	Extra-long Period	32	16
7	7	Extra-long Period	32	16

The electron configuration, the division of the energy level group and the division of the period of an element are related as follows:

Number of periods = number of electron shells = number of energy level groups

(周期数 = 电子层数 = 能级组数)

5.2 Group

The labeling of the groups (族) has been a matter of some debate among chemists. Group labels previously used in the United States consisted of a letter and a number, closely following the method adopted by Mendeleev, the discoverer of the table. The International Union of Pure and Applied Chemistry (IUPAC) recommended the simple 1-18 numbering scheme in order to avoid confusion between the American number and letter system and that used in Europe, where some of the A and B designations were switched! Currently, the IUPAC system is officially recommended by the American Chemical Society (ACS) and chemical societies in other nations. Because both numbering systems are in use, we show both in Appendix 12, where the periodic table is inside. However, except for an occasional reminder of the newer system, we will use the earlier American numbering system in this textbook.

In the long periods of periodic table, there are 18 columns from left to right. In earlier American numbering system, all elements are divided into 16groups, including 8 main groups (known as A groups, 主族) and 8 subgroups (known as B groups, 副族). Generally, the subgroup elements are also called a transition elements (过渡元素). Because of all of them are metals, they are also called the transition metals (过渡金属). Like the main group metals, the transition metals form positive ions, but the number of electrons lost is not related in any simple way to the group number, mostly because transition metals can form two or more ions of differing charge. We will discuss transition metal ions in several later chapters.

5.2.1 Main Groups

For the main group elements, the last electron of atoms is filled in the *ns* or *np* orbital (where *n* is the number of electron shells of the atom), occupying 8 columns. Among these 8 columns, Ⅷ A group (or called zero group) is the last column, where elements of this group are also called rare gases or noble gas, and their electron configuration is extremely stable. The $ns^{1\sim2}np^{0\sim6}$ is known as the valence electron configuration of the main group elements. The electron configuration and the division of main groups of an element are related as follows:

Number of main groups = number of valence electrons = the outermost electrons = the total number of *ns* and *np* electrons

[主族族数 = 价电子数 = 最外层电子数 = (*ns* + *np*) 电子数]

5.2.2 Subgroups

For the subgroup elements, the last electron of atoms is filled in the (*n*-1)*d* orbital while the *ns* orbital is filled with 1~2 electrons at the same time (where *n* is the number of electron shells of the atom), occupying 10 columns. Ⅰ B~ Ⅱ B each occupies one column, Ⅲ B~ Ⅶ B each occupies one column, and Ⅷ B group (or called Ⅷ group) occupies 3 columns. The $(n\text{-}1)d^{1\sim10}ns^{1\sim2}$ is known as the valence electron configuration of the subgroup elements.

For the subgroup elements, the group number equals the number of electrons on the valence electron shell except Ⅰ B, Ⅱ B and Ⅷ B. The subgroups are divided according to the following points:

(1) When (*n*-1)*d* orbital is not full filled and the number of valence electrons ranges from 3 to 7, the number of subgroup equals the number of valence electrons (This is the case for Ⅲ B~ Ⅶ B).

(2) When (*n*-1)*d* orbital is not full filled and the number of valence electrons ranges from 8 to 10, the

number of subgroup equals Ⅷ (This is the case for ⅧB).

(3) When $(n\text{-}1)d$ orbital is full filled, the number of subgroup equals the number of ns electrons (This is the case for ⅠB and ⅡB).

Because of the similar configuration of their valence electron shells, the chemical properties for the elements located in the same group are also very similar.

5.3 Block

According to the orbital filled by the last electron, the periodic table is divided into five blocks (区).

5.3.1 *s* block Element with the last electron filled in *s* orbital is a *s*-block element. Elements of the *s*-block all have configuration $ns^{1\sim2}$, and are located in the leftmost two columns in the periodic table. They are active metal elements, including IA group elements of alkali metals (碱金属) and ⅡA group elements of alkaline earth metals (碱土金属).

5.3.2 *p* block Element with the last electron filled in the *p* orbital is a *p*-block element. Elements of the *p*-block all have configuration $ns^2np^{1\sim6}$, and are located in the rightmost 6 columns of the periodic table. They include boron group elements (ⅢA), carbon group elements (ⅣA), nitrogen group elements (ⅤA), oxygen group elements (ⅥA), halogen elements (ⅦA) and "zero group" rare gas elements (ⅧA).

5.3.3 *d* block Element with the last electron filled in the *d* orbital is an *d*-block element. Elements of the *d*-block all have configuration $(n\text{-}1)d^{1\sim9}ns^{1\sim2}$, and are located in the middle of the periodic table. They include elements of ⅢB~ⅧB group. It is worth noting that the valence electron configuration of the elements in this block is written $(n\text{-}1)d$ first and then ns. But the order of electron filling is ns first and then the $(n\text{-}1)d$.

5.3.4 *ds* block The last electron of the *ds* block elements is filled in the *d* orbital, and the *d* orbital reaches the full-filled state. Elements of the *ds* block all have configuration $(n\text{-}1)d^{10}\ ns^{1\sim2}$, including elements of IB and IIB group.

5.3.5 *f* block Element whose last electron is filled in the *f* orbital is called *f* block element, and its valence electron configuration is $(n\text{-}2)f^{1\sim14}(n\text{-}1)d^{0\sim2}ns^2$, including lanthanide elements (57~71) and actinides (89~103) elements, which are at the bottom of the periodic table. The difference in properties between the elements of the same period in this block is small.

Example 7-2 For an element with an atomic number of 28, write its electron configuration, and point out which period of the periodic table the element is on? Which group? Which block? And write the name and chemical symbol of the element.

Solution the electron configuration is: $1s^22s^22p^63s^23p^63d^84s^2$ or abbreviated as $[Ar]3d^84s^2$

This element belongs to the 4th period, VIII Group, and *d* block. The element name is nickel and the chemical symbol is Ni.

6. Periodic Properties of the Elements

In this section, we focus on four atomic properties that reflect the central importance of electron configuration and effective nuclear charge: atomic radius, ionization energy, electron affinity, and electronegativity. Most notably, these properties are periodic, which means they generally exhibit consistent changes, or trends, within a group or period.

6.1 Periodic Trends in Atomic Radii

To understand certain physical and chemical properties, we need to know something about atomic size. Unfortunately, atomic radius is hard to define. The probability of finding an electron decreases with increasing distance from the nucleus, but nowhere does the probability fall to zero. There is no precise outer boundary to an atom. We might describe an effective atomic radius as, say, the distance from the nucleus within which 90% of all the electron charge density is found, but, in fact, all that we can measure is the distance between the nuclei of adjacent atoms (internuclear distance). Even though it varies, depending on whether atoms are chemically bonded or merely in contact without forming a bond, we define atomic radius in terms of internuclear distance. Because we are primarily interested in bonded atoms, we will emphasize an atomic radius based on the distance between the nuclei of two atoms joined by chemical bond. Atoms do not have sharply defined boundaries, so the radius of an individual atom cannot be accurately determined. Nevertheless, we can define atomic radius in several ways, based on the distances between atoms in various situations. Atomic radius (原子半径) usually refers to half of the distance between adjacent two nuclei measured by experimental methods (such as the X-ray diffraction).

(1) Covalent radius (共价半径): used for elements occurring as molecules, mostly nonmetals, it is one-half the shortest distance between nuclei of two identical atoms joined by a single covalent bond.

(2) Metallic radius (金属半径): used mostly for metals, it is one-half the shortest distance between the nuclei of two atoms in contact in the crystalline solid metal.

(3) Van der Waals radius (范德华半径): in a solid sample of a noble gas the distance between the centers of neighboring atoms is called the van der Waals radius. Figure 7-11 shows the relationship between the covalent radius and the van der Waals radius, where r_c (one-half the length of the solid line in the figure) is the covalent radius and r_v (one-half the length of the dotted line in the figure) is the van der Waals radius.

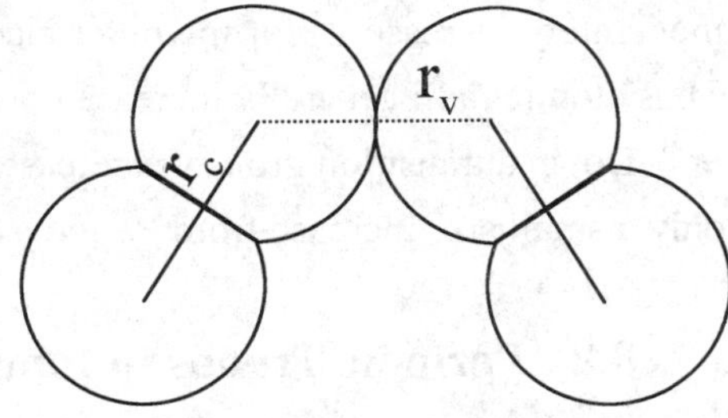

Figure 7-11 Relationship between covalent radius and van der Waals radius

Generally, the covalent radius is the smallest, the metallic radius is the larger, and the van der Waals radius is the largest. When comparing certain properties of elements, it is best to use the same set of values for atomic radii. The atomic radius of each element in the periodic table is listed in Table 7-5.

Table 7-5 Atomic radii of elements (pm)

H																	He
37																	122
Li	Be											B	C	N	O	F	Ne
152	111											88*	77	70	66	64	160
Na	Mg											Al	Si	P	S	Cl	Ar
186	160											143*	117	110	104	99	191
K	Ca	Sc	Ti	V	Cr	Mn	Fe	Co	Ni	Cu	Zn	Ga	Ge	As	Se	Br	Kr
227	197	161	145	132	125	124	124	125	125	128	133	122	122	121	117	114	198*
Rb	Sr	Y	Zr	Nb	Mo	Tc	Ru	Rh	Pd	Ag	Cd	In	Sn	Sb	Te	I	Xe
248	215	181	160	143	139	136	135	138	137	144	149	163	141	141	137	133	217
Cs	Ba	La	Hf	Ta	W	Re	Os	Ir	Pt	Au	Hg	Tl	Pb	Bi	Po	At	Rn
265	217	188	159	143	137	137	134	136	136	144	160	170	175	155	153		
			Ce	Pr	Nd	Pm	Sm	Eu	Gd	Tb	Dy	Ho	Er	Tm	Yb	Lu	
			183	183	182	181	180	204	180	178	177	177	176	175	194	173	

6.1.1 Varitation of Atom Radii Within a Period

Among the main-group elements, across a period, Z^* dominates. Across a period from left to right, electrons are added to the same outer level, so the shielding by inner electrons does not change. Despite greater electron repulsion, outer electrons shield each other only slightly, so Z^* rises significantly. And the outer electrons are pulled closer to the nucleus. Thus, atomic radii generally decrease across a period.

Across a transition series, atomic size shrinks through the first two or three elements because of the increasing nuclear charge. But, from then on, size remains relatively constant because shielding by the inner *d* electrons counteracts the increase in Z^*.

6.1.2 Variation of Atomic Radii Within a Group

Among the main-group elements, as we move down a main group, each member has one more level of inner electrons that shield the outer electrons very effectively. Even though additional protons do moderately increase Z^* for the outer electrons, the atoms get larger as a result of the increasing n value. Thus atomic radii generally increase down a group.

Down a transition group, n increases, but shielding by an additional level of inner electrons results in only a small size increase from Period 4 to 5 and none from 5 to 6.

6.2 Periodic Trends in Ionization Energy

The ionization energy (I, 电离能) is the energy required for the complete removal of 1mol of electrons from 1mol of gaseous atoms or ions. Electrons are attracted to the positive charge on the nucleus of an atom, and energy is needed to overcome that attraction. Because energy flows into the system, the ionization energy is always positive.

Multi-electron atoms can lose more than one electron. The first ionization energy (I_1) removes an outermost electron (highest energy sublevel) from a gaseous atom. The second ionization energy (I_2) removes a second electron. Further ionization energies are I_3, I_4 and so on. Invariably, we find that succeeding ionization energy each is larger than the preceding one. The more easily its electrons are lost, the more metallic an atom is considered to be.

E.g: $Li(g) - e^- \rightarrow Li^+(g)$ $I_1 = 520$ kJ/mol

$Li^+(g) - e^- \rightarrow Li^{2+}(g)$ $I_2 = 7298$ kJ/mol

$Li^{2+}(g) - e^- \rightarrow Li^{3+}(g)$ $I_3 = 11815$ kJ/mol

Ionization energies are usually measured through the emission spectrum experiment of the element. Usually, the I_1 is used. The ionization energies of the elements exhibit obvious periodicity in the periodic table. Figures 7-12 shows the trend in first ionization energy with atomic number.

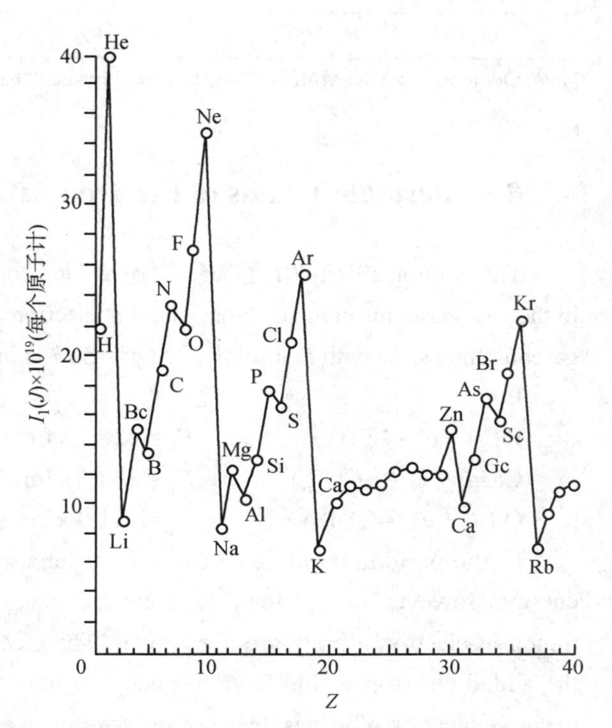

Figure 7-12 First ionization energies as a function of atomic number

6.2.1 Variation of Ionization Energy Within a Period

As we move left to right across a period, Z^* increases and atomic size decreases. The attraction between nucleus and outer electron increases, so the electron is harder to remove: Ionization energy generally increases across a period.

There are two exceptions to the otherwise smooth increase in I_1 across periods: in Periods 2 and 3, there are dips at the Group 3A elements, boron (B) and aluminum (Al). These elements have the first *np* electrons, which are removed more easily because the resulting ion has a filled (stable) *ns* sublevel. In Periods 2 and 3, once again, there are dips at the Group 6A elements, O and S. These elements have a fourth *np* electron, the first to pair up with another *np* electron, and electron-electron repulsions raise the orbital energy. The fourth *np* electron is easier to remove because doing so relieves the repulsions and leaves a half-filled (stable) *np* sublevel.

6.2.2 Variation of Ionization Energy Within a Group

The *n* value increases down a main group, so atomic size does as well. As the distance from nucleus to outer electron increases, their attraction lessens, so the electron is easier to remove: ionization energy generally decreases down a group.

The only significant exception occurs in Group 3A: I_1 decreases from B to Al, but not for the rest of the group. Filling the transition series in Periods 4, 5, and 6 causes a much higher Z^* and an unusually small change in size, so outer electrons in the larger Group 3A members are held tighter.

In addition, the ionization energy can also be used to explain the oxidation number of the elements. Table 7-6 lists the ionization energies of Na, Mg, and Al. It can be seen that the second ionization energy of Na, the third ionization energy of Mg, and the fourth ionization energy of Al increases rapidly. It indicates that it is difficult for Na, Mg, and Al to lose the second, third, and fourth electrons, respectively. So the oxidation number of them usually presented as +1, +2, and +3, respectively. Therefore, the ionization energy of an element is one of the important properties of the elements.

Table 7-6 Ionization energies of Na, Mg, and Al (kJ/mol)

Element	Ionization energy				
	I_1	I_2	I_3	I_4	I_5
Na	494	4560	6940	9540	13400
Mg	736	1450	7740	10500	13600
Al	577	1820	2740	11600	14800

Note: Quoted from Mac Millian. Chemical and Physical Date (1992)

6.3 Periodic Trends in Electron Affinity

The electron affinity (电子亲和能) of an element is the energy change upon adding an electron to an atom in the gas phase, forming an anion. The first electron affinity E_1 refers to the formation of 1mol of monovalent gaseous anions. As with ionization energy, there is a first electron affinity, a second, and so forth.

E.g:

$F(g) + e^- \rightarrow F^-(g)$ $E_1 = 328.2$ kJ/mol

$Cl(g) + e^- \rightarrow Cl^-(g)$ $E_1 = 348.6$ kJ/mol

$O(g) + e^- \rightarrow O^-(g)$ $E_1 = 141.0$ kJ/mol

In the periodic table, electron affinity energy changes in a similar way to that of the ionization energy. However, except for Z^* and atomic size, other factors will the affect electron affinities, so the trends of electron affinity are not regular. The electron affinities of the noble gases are positive because the added electron would have to occupy a new, higher-energy subshell. The halogens have the most negative electron affinities, because by gaining an electron, a halogen atom forms a stable anion that has a noble-gas configuration.

Because the electronic affinity is difficult to measure directly, and the indirect method is inaccurate, the data of electronic affinity is incomplete, so this book does not list the data of electronic affinity.

6.4 Periodic Trends in Electronegativity

Ionization Energy and electron affinity relate to the ability of atoms to lose or gain electrons. While electronegativity (电负性) describes an atom's ability to compete for electrons with other atoms to which it is bonded. The greater the electronegativity of an atom is, the greater its ability to attract electrons. The American chemist Linus Pauling developed the first and most widely used electronegativity scale, which is based on thermochemical data (Table 7-7). As Table 7-7 shows, because the nucleus of a smaller atom is closer to the shared pair than that of a larger atom, it attracts the bonding electrons more strongly. So, in general, electronegativity is inversely related to atomic size. Thus, for the main group elements, electronegativity generally increases up a group and across a period. With some exceptions (especially in the transition metals), electronegativity decreases with increasing atomic number in a group.

Pauling's electronegativity values range from about 0.7 to 4.0. In general, the lower its electronegativity, the more metallic the element is. And the higher the electronegativity, the more nonmetallic it is.

Table 7-7 Electronegativity values based on Pauling's thermochemical data

H 2.18																
Li 0.98	Be 1.57											B 2.04	C 2.55	N 3.04	O 3.44	F 3.98
Na 0.93	Mg 1.31											Al 1.61	Si 1.90	P 2.19	S 2.58	Cl 3.16
K 0.82	Ca 1.00	Sc 1.36	Ti 1.54	V 1.63	Cr 1.66	Mn 1.55	Fe 1.8	Co 1.88	Ni 1.91	Cu 1.90	Zn 1.65	Ga 1.81	Ge 2.01	As 2.18	Se 2.55	Br 2.96
Rb 0.82	Sr 0.95	Y 1.22	Zr 1.33	Nb 1.60	Mo 2.16	Tc 1.9	Ru 2.28	Rh 2.2	Pd 2.2	Ag 1.93	Cd 1.69	In 1.73	Sn 1.96	Sb 2.05	Te 2.1	I 2.66
Cs 0.79	Ba 0.89	La 1.10	Hf 1.3	Ta 1.5	W 2.36	Re 1.9	Os 2.2	Ir 2.2	Pt 2.28	Au 2.54	Hg 2.00	Tl 2.04	Pb 2.33	Bi 2.02	Po 2.0	At 2.2

Note: Quoted from Mac Millian. Chemistry and Physical Date (1992).

重点小结

1. 微观粒子的运动特征 量子化特征、波粒二象性、测不准关系。

2. 原子轨道 描述核外电子运动状态的概率波函数称为原子轨道。每一个原子轨道是指 *n*，*l*，*m* 组合一定时的波函数 $\psi_{n,l,m}$。

3. 四个量子数

n，主量子数，取 1，2，3，4 等正整数。*n* 表示电子云几率密度最大区域离核的平均距离，是决定多电子原子原子轨道能量的主要因素。*n* 越大，表示电子离核越远，能量越高。

l，角量子数，取值受到主量子数 *n* 的限制，当 *n* 一定时，*l* 可取 0，1，2，3，…(*n*-1)。*l* 表示原子轨道形状，是决定多电子原子原子轨道能量的次要因素。

m，磁量子数，取值受 *l* 限制，可取 0，±1，±2，…± *l*。*m* 可描述原子轨道在空间的不同伸展方向。

m_s，自旋量子数，只能取 $+\frac{1}{2}$ 或 $-\frac{1}{2}$。m_s 表示电子自旋的方向。核外电子只有两种相反的自旋方向，分别是顺时针和逆时针方向。

4. 电子云 电子云是 $|\psi|^2$ 空间图像的形象化描述。小黑点密集的地方，表明电子在该处出现的概率密度大；小黑点稀疏的地方，表明电子在该处出现的概率密度小。

5. 多电子原子轨道的能级 多电子原子的原子轨道能级由 *n* 和 *l* 的共同决定。能级顺序可以用屏蔽效应和钻穿效应来解释，也可以用 Pauling 原子轨道近似能级图来表示。

6. 核外电子排布（填充）原则

（1）能量最低原理 电子优先占据能级最低的 1*s* 轨道，然后依次占据能级较高的轨道。能量最低原理是核外电子排布需要遵守的第一个原则，该原则保证原子体系处于能量最低的状态，即基态。

（2）Pauli 不相容原理 在同一个原子中不可能存在四个量子数完全相同的两个电子。Pauli 不相容原理是核外电子排布需要遵守的第二个原则，还可以表述为：每个原子轨道最多排布 2 个

电子，且自旋方向相反。

（3）Hund 规则　电子排布在简并轨道时，尽可能分占各等价轨道，且自旋方向相同。Hund 规则是核外电子排布需要遵守的第三个原则，还可以表述为：电子排布在简并轨道时，电子以相同的自旋分占各等价轨道。

（4）Hund 规则特例　当等价轨道处于全充满、半充满和全空状态时，体系能量最低、最稳定。

7. 元素周期表和元素周期律　元素周期表可分为 8 个周期、16 个族和 5 个区。原子半径、电离能、电子亲和能、电负性等元素性质呈现周期性变化，是原子的电子层结构周期性变化所导致。电离能、电子亲和能、电负性等可以衡量原子得失电子的能力，可以说明元素周期表中元素金属性和非金属性的递变规律。

Exercises

题库

1. Judgement question

(1) The wave function represents the possible motion states of an electron with definite energy.

(2) For the ground state electron of hydrogen, its momentum and position can be determined accurately at a given moment.

(3) Three quantum numbers are need for describing an atomic orbital. However, four quantum numbers are need for describing the motion states of an electron.

(4) The energy, shape and size of the 5 degenerate *d* orbitals are the same, except the spatial orientation.

(5) The black dot of the electron cloud indicates the possible location of the electron, and the density indicates the probability density in space.

(6) Cl^- has the same electron configurations as Ar, so the 3*p* energy levels of them are the same.

(7) The larger the ionization energy of element is, the larger the electron affinity is.

(8) The electronegative reflects the ability of the atoms to attract bonding pair of electrons.

2. Single-Choice question

(1) Which group of quantum numbers has the highest energy?

A. (3, 2, 2, +1/2)　B. (3, 1, -1, +1/2)　C. (3, 1, 1, +1/2)　D. (2, 1, -1, +1/2)

(2) Which group of quantum numbers has the lowest energy?

A. (3, 2, 2, +1/2)　B. (3, 1, -1, +1/2)　C. (3, 1, 1, +1/2)　D. (2, 1, -1, +1/2)

(3) Of the following sets of quantum numbers, which is unreasonable?

A. (3, 2, 1, +1/2)　B. (3, 1, 0, +1/2)　C. (3, 0, 0, +1/2)　D. (2, 2, -1, +1/2)

(4) The valence electron configuration of an element is $4d^{10}5s^1$. Its atomic number is

A. 25　B. 39　C. 41

D. 47　E. 57

(5) If the electron motion state of the 2*p* orbital of the ground state nitrogen atom is described as: (2, 1, 0, +1/2), (2, 1, 0, -1/2), (2, 1, +1, -1/2), it violates

A. Minimum energy principle　B. The same principle of symmetry

C. Hund's rule　D. Pauli exclusion principle

E. Raoult's law

(6) How many columns do *d* block elements include

A. 2 B. 4 C. 6

D. 8 E. 10

(7) Of the following elements, which has the least electronegativity

A. H B. Ca C. F

D. Cr E. Cs

(8) The quantum numbers that determine the energy of electron of the ground state hydrogen atom are

A. n, l, m B. n, m C. n, l

D. n, l, m, m_s E. n

(9) When there are only two electrons in the fifth electron shell of the atom in ground state, the number of the electrons on the fourth electron shell is

A. 8 B. 18 C. 8~18

D. 32 E. 8~32

(10) The magnetic quantum number of electrons in the 4*d* orbital may be

A. 0, 1, 2 B. 1, 2, 3 C. 0, ±1

D. 0, ±1, ±2 E. ±1/2

(11) When the angular momentum quantum number is 2, the possible number of degenerate orbitals is

A. 1 B. 3 C. 5

D. 7 E. 9

3. Multiple - choice question

(1) What does the wave function ψ describe

A. The spatial motion state of the electrons

B. Motion orbital of the electrons

C. The amplitude of fluctuation equation of probability density

D. Atomic orbital

E. The amplitude of fluctuation equation

(2) The schematic diagram of an electron cloud can be used to describe

A. The energy of the electron

B. Probability of the electron appearing in a certain region

C. The motion orbital of the electron

D. Probability density of the electron appearing in a certain region

E. The strength index of the wave function of the electron

(3) The quantum number *n*, *l* and *m* cannot determine

A. The energy of the atomic orbital

B. The shape of the atomic orbital

C. The spatial orientation of the atomic orbital

D. The number of the electrons

E. The motion state of the electrons

(4) In the following description of the orbital energy of multi-electronic atoms, which is correct

A. $E_{3s} = E_{3p} = E_{3d}$ B. $E_{3s} < E_{3p} < E_{3d}$

C. $E_{3s} > E_{3p} > E_{3d}$　　D. K atom: $E_{3d} > E_{4s}$

E. Ba atom: $E_{6s} < E_{4f} < E_{5d}$

(5) Which of the following quantum numbers can be used to describe the valence electrons of elements whose atomic number is 21

A. 2, 1, 0, -1/2　　B. 3, 2, 1, +1/2　　C. 4, 0, 0, -1/2

D. 4, 0, 0, +1/2　　E. 4, 1, 0, +1/2

(6) The principles to be observed in the arrangement of the electrons outside the nucleus are

A. Minimum energy principle

B. the Principle of Maximum Overlap of Atomic orbitals

C. Pauli exclusion principle

D. Hund's rule

E. Symmetry-adapted principle

(7) For the second period B, C, N, O, F, they are

A. The element with the smallest atomic radius in the same group

B. The highest electronegativity element in the same group

C. The highest density of electron cloud in the outermost shell

D. The element with the smallest first ionization energy in the same group

E. The element with the largest first electron affinity in the same group

(8) Valence shell structure of Carbon atomic in ground state is $2s^2 2p^2$, the following sets of quantum numbers can be used to describe the electron motion state are

A. 2, 0, 0, +1/2　　B. 2, 0, 0, –1/2　　C. 2, 1, 0, +1/2

D. 2, 1, 1, +1/2　　E. 2, 1, –1, –1/2

(9) Of the following atomic orbital, which belong to the fourth energy level

A. 3*p*　　B. 3*d*　　C. 4*s*

D. 4*p*　　E. 4*d*

4. Question

(1) How many types of orbitals are there in the electron shells when atomic quantum number is *n*? How many orbitals are there in electron shell *n*? How many orbitals are contained in the subshell with an angular quantum number *l*?

(2) Which quantum numbers are related to the wave functionψ?

(3) Write out the electron configurations for the following elements:

F　　Mg　　Cu　　Ar　　Co

(4) The +3 ions of an element have the same electron configuration as Ar. Please point out which period of the periodic table the element is on? Which group? Which block? And write the chemical symbol of the element.

(5) Why does the I_1 of the second period elements show a dip between Be and B, N and O?

目 标 检 测

1. 判断题

（1）原子的核外电子的波函数，代表了该电子可能存在的运动状态，该运动状态具有确定的能量与之对应。

（2）基态氢原子核外电子的动量一定时，其位置也就确定。

（3）每一个原子轨道需要有 3 个量子数才能确定，而每一个电子则需要 4 个量子数才能具体确定。

（4）5 个 *d* 轨道的能量、形状、大小都相同，不同的是在空间的取向。

（5）电子云的黑点表示电子可能出现的位置，疏密程度表示电子出现在该范围的机会大小。

（6）Cl 离子与 Ar 原子具有相同的电子层结构，故其 3*p* 能级与 Ar 原子的 3*p* 能级具有相同的能量。

（7）电离能大的元素，其电子亲和能也大。

（8）电负性反映了化合态原子吸引电于能力的大小。

2. 单项选择题

（1）下列各组量子数中，能量最高的是

A.（3, 2, 2, +1/2） B.（3, 1, −1, +1/2）
C.（3, 1, 1, +1/2） D.（2, 1, −1, +1/2）

（2）下列各组量子数中，能量最低的是

A.（3, 2, 2, +1/2） B.（3, 1, −1, +1/2）
C.（3, 1, 1, +1/2） D.（2, 1, −1, +1/2）

（3）下列各组量子数中，不合理的是

A.（3, 2, 1, +1/2） B.（3, 1, 0, +1/2）
C.（3, 0, 0, +1/2） D.（2, 2, -1, +1/2）

（4）价电子层构型为 $4d^{10}5s^{1}$ 的元素，其原子序数为

A. 25 B. 39 C. 41
D. 47 E. 57

（5）如果将基态氮原子的 2*p* 轨道的电子运动状态描述为：（2, 1, 0, +1/2）、（2, 1, 0, −1/2）、（2, 1, +1, −1/2），则违背了

A. 能量最低原理 B. 对称性原则 C. 洪特规则
D. 泡利不相容原理 E. 拉乌尔定律

（6）*d* 区元素包括几个纵列

A. 2 B. 4 C. 6
D. 8 E. 10

（7）下列元素中电负性最小的是

A. H B. Ca C. F
D. Cr E. Cs

（8）决定氢原子核外电子能量的量子数为

A. n, l, m B. n, m C. n, l
D. n, l, m, m_s E. n

（9）基态原子的第五电子层只有 2 个电子时，则原子的第四电子层的电子数为

A. 8 B. 18 C. 8~18
D. 32 E. 8~32

（10）4*d* 轨道的磁量子数可能有

A. 0, 1, 2 B. 1, 2, 3 C. 0, ±1
D. 0, ±1, ±2 E. ±1/2

（11）当角量子数为 2 时，可能的简并轨道数是

A. 1　　B. 3　　C. 5

D. 7　　E. 9

3. 多项选择题

（1）波函数 ψ 描述的是

A. 核外电子的空间运动状态　　B. 核外电子的运动轨迹

C. 概率密度波动方程的振幅　　D. 原子轨道

E. 波动方程的振幅

（2）电子云图可以描述

A. 电子的能量　　B. 电子在某空间出现的概率大小

C. 电子运动的轨道　　D. 电子在空间某处出现的概率密度大小

E. 电子波函数的强弱

（3）量子数 n, l 和 m 不能决定

A. 原子轨道的能量　　B. 原子轨道的形状

C. 原子轨道在空间的伸展方向　　D. 电子的数目

E. 电子的运动状态

（4）下列对多电子原子轨道能量的描述中，正确的是

A. $E_{3s} = E_{3p} = E_{3d}$　　B. $E_{3s}<E_{3p}<E_{3d}$

C. $E_{3s}>E_{3p}>E_{3d}$　　D. K 原子的 $E_{3d}>E_{4s}$;

E. Ba 原子的 $E_{6s}<E_{4f}<E_{5d}$

（5）下列哪些量子数的组合可以用来描述原子序数 21 的元素的价电子

A. 2, 1, 0, -1/2　　B. 3, 2, 1, +1/2

C. 4, 0, 0, -1/2　　D. 4, 0, 0, +1/2

E. 4, 1, 0, +1/2

（6）核外电子排布遵守的原则是

A. 能量最低原理　　B. 原子轨道最大重叠原理　　C. 泡利不相容原理

D. 洪特规则　　D. 对称性相同原理

（7）对于第二周期 B、C、N、O、F 来说，它们是

A. 同一族中原子半径最小的元素

B. 同一族中电负性最大的元素

C. 同族中最外层电子云密度最大的

D. 同一族中元素的第一电离能最小的元素

E. 同一族中元素的第一电子亲和能最大的元素

（8）基态 C 原子价层结构为 $2s^2 2p^2$，下列各组量子数可用来描述其电子运动状态的是

A. 2, 0, 0, +1/2　　B. 2, 0, 0, −1/2　　C. 2, 1, 0, +1/2

D. 2, 1, 1, +1/2　　E. 2, 1, −1, −1/2

（9）下列原子轨道中，属第四能级的有

A. $3p$　　B. $3d$　　C. $4s$

D. $4p$　　E. $4d$

4. 简答题

（1）原子的主量子数为 n 的电子层中有几种类型的原子轨道？第 n 层中共有多少个原子轨道？角量子数为 l 的亚层中含有几个原子轨道？

（2）波函数 ψ 和哪些量子数有关？

（3）请写出下列原子的电子排布式

F Mg Cu Ar Co

（4）某元素 +3 价离子和氩原子的电子构型相同，该元素属哪个元素周期表中的哪个周期、哪个族、哪个区？并写出其元素符号。

（5）第二周期元素的第一电离能为什么在 Be 和 B 以及 N 和 O 之间出现转折？

Chapter 8　Chemical Bonds and Molecular Structures
第八章　化学键与分子结构

学习目标

知识要求

1．掌握　价键理论（价键理论、价层电子对互斥理论和杂化轨道理论）的基本要点；分子间作用力和氢键。

2．熟悉　分子轨道理论的要点，运用分子轨道理论处理第一、二周期同核双原子分子；共价键的键参数；离子键的形成与特征；离子极化的概念。

3．了解　分子间作用力对物质性质的影响；离子极化对化合物性质的影响。

能力要求

能应用价层电子对互斥理论和杂化轨道理论解释多原子分子的空间构型及成键情况。

The structure of matter determines the properties of matter, while molecule is the smallest particle that can exist independently and keep its chemical properties. Therefore, the chemical properties of substances mainly depend on the properties of molecules. To study the structure of molecules is to master the properties of molecules and ultimately the properties of substances.

Molecules are made up of atoms. The strong interaction between neighboring atoms (or ions) in molecules or crystals is called chemical bond (化学键). The chemical bond is a kind of binding force. Chemical bond can be divided into ionic bond, covalent bond, and metallic bond. Covalent bond theory includes valence bond theory and molecular orbital theory, in which valence bond theory includes the concept of electron pair, hybrid orbital theory and valence shell electron pair repulsion theory. It is of great significance to study the characteristics of chemical bonds between atoms in molecules for exploring the properties, structures and functions of substances. This chapter focuses on covalent bond, ionic bond and intermolecular force and other issues.

1. Ionic Bond

1.1 Formation of Ionic Bond

W. Kossel (Germany, 1888—1956) put forward the concept of ionic bond in 1916. He thought that the ionic bond is a chemical bond formed by the electrostatic interactions between the anion and cation after the atoms gain and lose electrons.

In the model of ionic bond, the anion and cation can be regarded as spherical charge approximately. In this way, according to Coulomb's law, the electrostatic attraction F between two kinds of ions with opposite charges (q^+ and q^-) is directly proportional to the product of ion charges as follows:

$$F = \frac{q^+ \cdot q^-}{d^2} \tag{8-1}$$

It can be seen that the larger the charge of the ion is, and the smaller the distance d between the charge centers of the ion is, thus the stronger the attraction between the ions is.

Under certain conditions, when the atom of the active metal element with lower electronegativity is close to the atom of the active non-metal element with higher electronegativity, the active metal atom loses the outermost electrons and forms a cation with positive charge and stable electron shell configuration, while the active non-metal atom gains electrons and forms a anion with negative charge and stable electron shell configuration. Anions and cations can attract each other by electrostatic attraction, however, when they are close to each other, the repulsion between the nucleus and between the electrons of the ions increase. When the repulsion and the attraction are balanced, the energy of the system is reduced to the lowest. The chemical bond formed by the electrostatic interaction between anion and cation is called ionic bond (离子键).

Taking NaCl as an example, the process of ionic bond formation can be simply expressed as follows:

$$\left.\begin{aligned} nNa(3s^{\square}) &\xrightarrow{-ne^-} nNa^+(2s\ \ 2p\ \ ) \\ nCl(3s^2 3p^5) &\xrightarrow{+ne^-} nCl^-(3s^2 3p^6) \end{aligned}\right\} \longrightarrow nNa^+Cl^- \tag{8-2}$$

Compounds formed by ionic bonds are called ionic compounds (离子型化合物).

1.2 Characteristics of Ionic Bond

The ionic bond is characterized by no saturation and no directionality (无饱和性、无方向性). Ion is a charged sphere, and its electrostatic effects are the same in all directions of space. Anion and cation can attract ions with opposite charges in any direction of space, so the ionic bond has no directionality. As long as space allows, an anion or a cation can combine with more ions with opposite charges, which is not limited by the charge of the ion itself, so the ionic bond is not saturated.

In ionic crystals, the number of opposite charged ions arranged around each anion and cation is fixed. For example, in NaCl crystal, there are 6 Cl^- around each Na^+ and 6 Na^+ around each Cl^-; in CsCl

crystal, there are 8 Cl^- around each Cs^+ and 8 Cs^+ around each Cl^-.

1.3 Characteristics of Ion

1.3.1 Ionic Radius

Ionic radius (离子半径) is one of the important characteristics of ions. Similar to atoms, there is no clear interface for a single ion. The so-called ionic radius refers to the contact radius of the ions in the crystal. According to the nuclear distance of anion and cation in ionic crystal, it is assumed that the nuclear distance of anion and cation is the sum of the radii of anion and cation. The average nuclear distance between anion and cation can be determined by X-ray diffraction. If the radius of anion is known, the radius of cation can be deduced. Table 8-1 lists the ionic radii of some common ions.

Table 8-1 Ionic radic of some common ions (r/pm)

Ions	Radius /pm	Ions	Radius /pm	Ions	Radius /pm
Li^+	60	Cr^{3+}	64	Hg^{2+}	110
Na^+	95	Mn^{2+}	80	Al^{3+}	50
K^+	133	Fe^{2+}	76	Sn^{2+}	102
Rb^+	148	Fe^{3+}	64	Sn^{4+}	71
Cs^+	169	Co^{2+}	74	Pb^{2+}	120
Be^{2+}	31	Ni^{2+}	72	O^{2-}	140
Mg^{2+}	65	Cu^+	96	S^{2-}	184
Ca^{2+}	99	Cu^{2+}	72	F^-	136
Sr^{2+}	113	Ag^+	126	Cl^-	181
Ba^{2+}	135	Zn^{2+}	74	Br^-	196
Ti^{4+}	68	Cd^{2+}	97	I^-	216

The change of ionic radius has the following rules.

(1) For the main group elements in the same period, with the increase of group number, the charge number of positive charge increases, and the ionic radius decreases in turn, for example:

$$r(Na^+) > r(Mg^{2+}) > r(Al^{3+})$$

(2) Among the main group elements in the periodic table, the radii of the same group ions with the same number of charges increase in turn due to the increase of the number of electron shells from top to bottom, for example:

$$r(Li^+) < r(Na^+) < r(K^+) < r(Rb^+) < r(Cs^+)$$

$$r(F^-) < r(Cl^-) < r(Br^-) < r(I^-)$$

(3) If the same element can form several cations with different charges, the radius of high valence ion is smaller than that of low valence ion, for example:

$$r(Fe^{3+})(60pm) < r(Fe^{2+})(75pm)$$

(4) Relatively speaking, the radius of anion is larger, about 130~250pm, while the radius of cation is smaller, about 10~170pm.

The ionic radius has a great influence on the bond strength. Generally speaking, when the ionic charges are the same, the smaller the ionic radius is, the greater the attraction between the ions is, and the stronger the ionic bond is, thus, the more energy is needed to break them, and the higher the melting and boiling points of the ionic compounds are.

1.3.2 Ionic Charge

It can be seen from the formation process of the ionic bond that the charge of the cation is the number of electrons lost by the corresponding atom (or group); the charge of the anion is the number of electrons gained by the corresponding atom (or group).

Ionic charge is also an important factor affecting the bond strength. The more ionic charges, the more attractive they are to the opposite charged ions, and the higher melting point of the formed ionic compounds. For example, the melting points of most salts of alkaline earth metal ion M^{2+} are higher than those of alkali metal ion M^+.

1.3.3 Electron Shell Configuration of Ions

When an atom forms an ion, the number of electrons lost or gained is related to the electron shell configuration of the atom. Generally, after an atom gains or loses electrons, it makes the electron shell of an ion reach a more stable configuration, that is, the electronic subshell is full filled.

The configuration of the outer electron shell of simple anions (for example, F^-, S^{2-}, Cl^-, etc.) is a noble gas configuration with 8 electrons of ns^2np^6. However, the electron shell configuration of simple cations is relatively complex, as follows:

(1) 2 electron shell configuration: the outer shell is $1s^2$ arrangement, such as Li^+, Be^{2+}, etc.

(2) 8 electron shell configuration: the outer shell is ns^2np^6 arrangement, such as Na^+, Ca^{2+}, etc.

(3) 18 electron shell configuration: the outer shell is $ns^2np^6nd^{10}$ arrangement, such as Ag^+, Zn^{2+}, etc.

(4) 18+2 electron shell configuration: there are 18 electrons in the subouter shell and 2 electrons in the outermost shell, namely $(n\text{-}1)s^2(n\text{-}1)p^6(n\text{-}1)d^{10}ns^2$, such as Sn^{2+}, Pb^{2+}, etc.

(5) 9-17 electron shell configuration: it belongs to irregular electron shell configuration. The outermost shell has 9-17 electrons, namely $ns^2np^6nd^{1-9}$, such as Fe^{2+}, Cr^{3+}, etc.

The configuration of the outer electron shell of an ion has an effect on the interaction between the ions, thus changing the properties of the chemical bond. For example, Na^+ and Cu^+ have the same charge and nearly the same ionic radius, but NaCl is easy to dissolve in water, while CuCl is difficult to dissolve in water. Obviously, this is due to the different electron shell configurations of Na^+ and Cu^+, which will be discussed in "ionic polarization".

2. Covalent Bond Theory

PPT

The theory of ionic bond can well explain the formation of ionic compounds, but it cannot explain the bonding between two identical atoms or atoms with little difference in electronegativity. An American chemist, Lewis (US, 1875—1946), put forward the classical theory of covalent bond in 1916. He thought that such atoms are bound by sharing electron pairs. The classical covalent bond theory reveals the difference between covalent bond and ionic bond, but it cannot explain the nature of covalent bond. It cannot explain why the two negatively charged electrons do not repel each other but they pair to make the

two atoms join together. In 1927, Heitler (Germany, 1904—1981) and London (Germany, 1900—1954) treated the H_2 structure with quantum mechanics, which initially explained the nature of covalent bond in theory. In 1931, Pauling (US, 1901—1994) proposed the hybrid orbital theory and further developed the valence bond theory; in 1932, Hund (Germany, 1896—1997) and Mulliken (US, 1896—1986) proposed the molecular orbital theory and further pointed out that the bonding electrons can move through the entire molecular region.

2.1 Valence Bond Theory

2.1.1 Covalent Bond Formation

Heitler and London treated two H atoms to form H_2 molecule by quantum mechanics, and obtained the curve of the relationship between the energy of H_2 molecule and the distance between atomic nuclei, as shown in Figure 8-1:

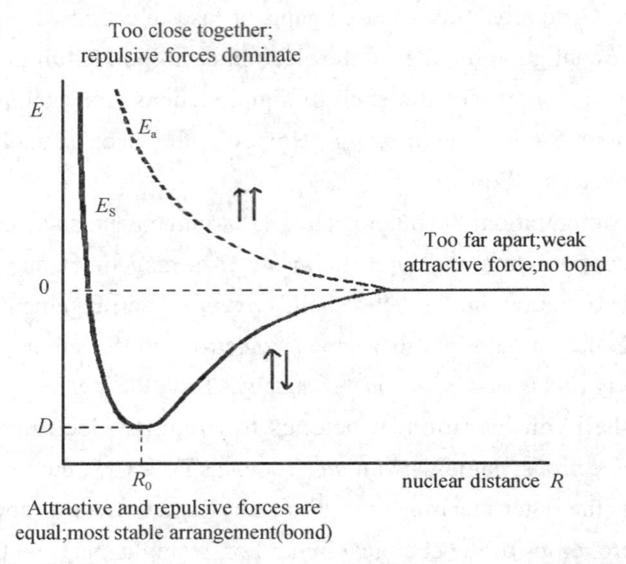

Figure 8-1 Schematic diagram of energy change with nuclear distance during H_2 molecular formation

Let's look at the simplest case of covalent bonding, the reaction of two hydrogen atoms to form the diatomic molecule H_2. As you recall, an isolated hydrogen atom has the ground state electron configuration $1s^1$, with the probability density for this one electron spherically distributed about the hydrogen nucleus. As two hydrogen atoms approach each other from large distance, the electron of each hydrogen atom is attracted by the nucleus of the other hydrogen atom as well as by its own nucleus. If these two electrons have opposite spins so that they can occupy the same region (orbital), both electrons can now preferentially occupy the region between the two nuclei, where they are strongly attracted by both nuclei. A pair of electrons is shared between the two hydrogen atoms, and a single covalent bond is formed. We say that the $1s$ orbitals overlap so that both electrons are now in the orbitals of both hydrogen atoms. The close together the atoms come, the more nearly this is true. In that sense, each hydrogen atom then has helium configuration $1s^2$.

Let's see Figure 8-1. When two H atoms approach each other from a distance, there are two situations: If the $1s$ electron spin direction of the two H atoms is opposite, with the decrease of the distance R between the two atoms, the system energy (E_s) gradually decreases. When the distance R =

R_0 (about 0.74 Å), the energy E_s drops to the lowest value D (about 436 kJ/mol), the density of electron cloud between two nuclei is relatively dense, which indicates that the two hydrogen atoms are attracted to each other. If the 1*s* electron spin direction of the two H atoms is the same, the system energy (E_a) increases with the decreasing of distance *R* between the two nuclei, which means that the two hydrogen atoms tend to separate and cannot bond. It can be seen that two H atoms with opposite spin direction combine at the nuclear distance of R_0 to form stable H_2 molecule. This state is called the ground state of hydrogen molecule. At this time, the energy of the system is lower than that of the two H atoms when they are not combined. On the contrary, if the 1*s* electrons of the two H atoms spin in the same direction, the energy of the system increases with the decreasing of *R*, and the probability density of 1*s* electrons between the nuclei is very small, which means that the two H atoms tend to separate rather than bond.

Therefore, according to the basic principles of quantum mechanics, the reason why the ground state of hydrogen molecule can be bonded is that when the 1*s* atomic orbitals of two hydrogen atoms overlap each other, ψ_{1s} is positive, and the electron cloud density between the two nuclei increases after the addition. In the region with high density of electron cloud between the two nuclei, on the one hand, the positive repulsion between the two nuclei is reduced, on the other hand, the attraction of the region with high density of electron cloud between the two nuclei is increased, which is conducive to the reduction of potential energy and the formation of stable chemical bond. The chemical bond formed by the attraction of two nuclei in the region with high density of electron cloud between atoms is called covalent bond (共价键). See Figure 8-2:

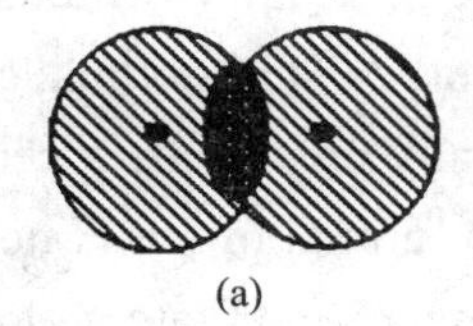

(a)

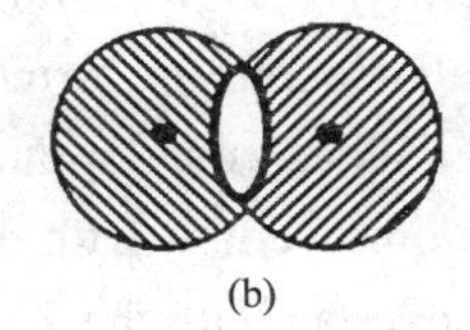

(b)

Figure 8-2 The two states of H_2 molecule

(a) When the electron spin direction is opposite, the electron cloud density between the two nuclei is large so as to form a stable chemical bond. (b) When the electron spin direction is the same, the electron cloud density between the two nuclei is zero, so the chemical bond cannot form.

In a word, the valence bond theory inherits the concept of Lewis "shared electron pair". Moreover, on the basis of quantum mechanics theory, it is pointed out that the nature of covalent bond is due to the overlapping of atomic orbitals and the high probability density of electrons between two atomic nuclei attracting atomic nuclei so as to form chemical bonds.

Other pairs of nonmetal atoms share electron pairs to form covalent bonds. The result of this sharing is that each atom attains a more stable electron configuration–frequently the same as that of the nearest noble gas. This results in a more stable arrangement for the bonded atoms. Most covalent bonds involve sharing of two, four, or six electrons—that is, one, two, or three pairs of electrons. Two atoms form a single covalent bond (共价单键) when they share one pair of electrons, a double covalent bond (共价双键) when they share two electron pairs, and a triple covalent bond (共价三键) when they share three electron pairs. These are usually called simply single, double, and triple bonds. Covalent bonds that involve sharing of one and three electrons are known but are relatively rare.

2.1.2 Basic Points of Valence Bond Theory

(1) When two atoms are close to each other, unpaired electrons with opposite spin direction can pair (be shared) to form covalent bond.

(2) When electrons pair (is shared), the more the atomic orbitals overlap, the stronger the covalent bond is, the more energy is released, and the more stable the chemical bond is. For example, forming a C—C bond releases 346 kJ/mol energy, and forming an H-F bond releases 570 kJ/mol energy.

2.1.3 Characteristics of Covalent Bond

(1) Saturation of covalent bond The saturation of covalent bond (共价键的饱和性) means that an atom with several unpaired single electrons can pair with several single electrons with opposite spin to form several covalent bonds. That is to say, the number of covalent bonds formed by an atom is not arbitrary but is generally limited by the number of single electrons. If atom A and B each has one, two or three single electrons, and with opposite spin, they can pair with each other to form covalent single bond, double bond or triple bond (for example, H—H, O=O, and N≡N). If atom A has one single electron and atom B has two single electrons, and if the spin is opposite, one atom B can combine with two atoms A to form A_2B molecule. For example, two atoms H and one atom O can form H_2O molecule.

(2) Directionality of covalent bond According to the principle of the maximum overlap of atomic orbitals, the formation of covalent bonds will follow the direction of the maximum overlap of atomic orbitals, so that the denser the electron cloud between the two nuclei, the stronger the covalent bond formed. This is the directionality of the covalent bond (共价键的方向性). Except for the spherical symmetry and non-directionality of the *s* orbital, the *p*, *d* and *f* orbitals all have certain extension directions in space. When covalent bonds are formed, *s* orbital and *s* orbital can overlap maximally in any direction, while *p*, *d* and *f* orbitals can overlap maximally only along a certain direction. For example, when the 1*s* orbital of the H atom overlaps with the $2p_x$ orbital of the F atom to form HF molecule, the 1*s* orbital of the H atom must overlap with the $2p_x$ orbital of the F atom along the X axis to form stable covalent bond (Figure 8-3(a)); While the overlaps in other directions, the atomic orbitals cannot overlap (Figure 8-3(b)) or there are few overlaps (Figure 8-3(c)), therefore, they cannot form chemical bonds or the chemical bonds formed are unstable.

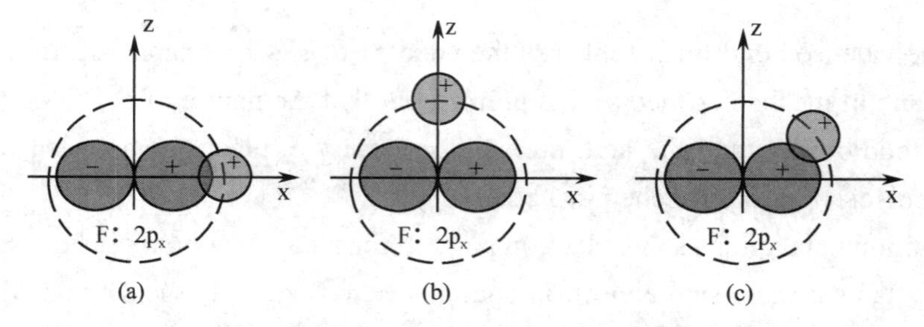

Figure 8-3 Schematic diagram of atomic orbital overlapping between H and F

2.1.4 Types of Covalent Bond

There are many types of covalent bonds. Here, we will introduce two main types: σ bond and π bond.

(1) σ bond The atomic orbital is in the direction of the bond axis (the line between two atoms) [the orbital overlaps in the way of head-on ("end-to-end", also visualized as "head to head", 头碰头), and the overlapped part of the orbital is distributed in a cylindrical shape along the bond axis. This bond is called σ bond (σ键)], and the electrons that form the σ bond are called σ electrons. When the σ bond is formed,

the overlapped part of the atomic orbitals is cylindrically symmetric about the bond axis, and rotates at any angle along the bond axis direction, and the shape and symbol of the orbital do not change. For σ bond, the bond energy is large and the bond stability is high because the bonding atomic orbitals overlap along the bond axis resulting in maximum overlap when σ bond is formed. Taking the x-axis as the bond axis, *s*-*s*, s-p_x, p_x-p_x can overlap to form σ bond, as shown in Figure 8-4.

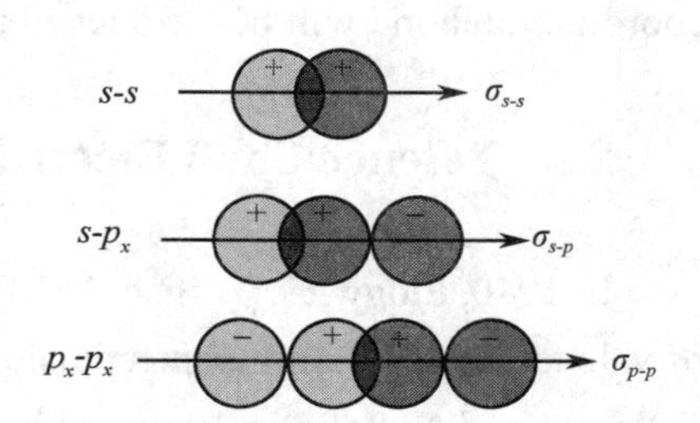

Figure 8-4 Schematic diagram of σ bond

(2) π bond Atomic orbitals overlap by side-on ("side-to-side", also visualized as "shoulder to shoulder", 肩并肩), such as p_y-p_y and p_z-p_z. The overlapping part of the orbital has mirror antisymmetry to the plane passing through a bond axis, which is called π bond (π键). The electrons that form the π bond are called π electrons, as shown in Figure 8-5.

Generally speaking, the overlapping degree of π bond is less than that of σ bond, so the bond energy of π bond is lower than that of σ bond, and the stability of π bond is also weaker than that of σ bond. π bond electrons have high energy and are easy to move, so they are active participants in chemical reactions. When a covalent single bond is formed between two atoms, it is σ bond usually; when a covalent double bond or triple bond is formed, there is a σ bond and the rest are π bonds. For example, N atom has three single electrons ($2p_x^1 2p_y^1 2p_z^1$). When two N atoms form a N_2 molecule, p_x-p_x forms a σ bond, while p_y-p_y and p_z-p_z form two mutually perpendicular π bonds, as shown in Figure 8-6.

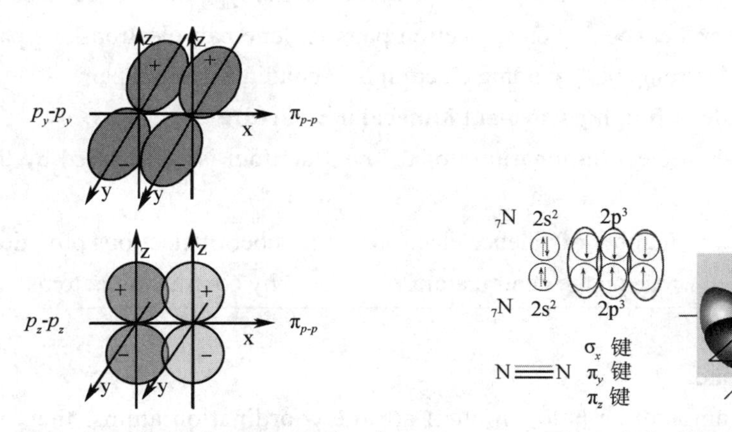

Figure 8-5 Schematic diagram of π bond

Figure 8-6 Chemical bond in N_2

2.1.5 Coordinate Covalent Bonds (Coordination Bond)

The common electron pairs of covalent bonds discussed above are composed of two electrons which are provided respectively by two atoms. In addition, there is another kind of covalent bond. The shared electron pair is not provided by two atoms, but provided unilaterally by one of the atoms. The covalent bond formed by the sharing of an electron pair provided by one atom for the sharing of two atoms is called coordinate covalent bond or coordination bond (配位共价键，或配位键). The formation condition of the coordination bond is that the valence electron shell of one atom has a lone electron pair (i.e. an unshared electron pair), and the valence electron shell of the other atom has an empty orbital that can accept the lone electron pair. As long as the conditions are met, coordination bonds may be formed within molecule, or between molecules, or between ions and or between molecule and ion (the theory of

coordination bond will be introduced in Chapter 9).

2.2 Valence Shell Electron Pair Repulsion Theory

In 1940, Sidgwick (British, 1873—1932) and Powell (England, 1903—1969), put forward the theory of valence shell electron pair repulsion (VSEPR, 价层电子对互斥理论), which was further developed by Gillespie and Nyholm in 1960s. This theory does not need the concept of atomic orbital, and it is simple and intuitive in explaining, judging and predicting the accuracy of polyatomic molecular geometry.

2.2.1 Basic Points of VSEPR

(1) The geometry of molecule or ion depends on the number of valence shell electrons around the central atom. The valence shell electron pair on central atom refers to the σ bond electron pair and the lone pair electron which are not involved in bonding.

(2) The valence pairs of central atoms should be as far away from each other as possible to minimize the repulsion force, which determines the geometry of the molecule. Table 8-2 shows the arrangement of the least electrostatic repulsion between the electron pairs in the valence electron shell.

Table 8-2 The arrangement of valence electron pairs with the least electrostatic repulsion

Number of valence electron pairs	2	3	4	5	6
Arrangement of electron pairs	linear	trigonal planar	tetrahedral	trigonal bipyramidal	octahedral

(3) The repulsion between valence electron pairs is related to the type of valence electron pairs. The order of electrostatic repulsion between valence electron pairs is: lone pair electron-lone pair electron > lone pair electron-bonding electron pair > bonding electron pair-bonding electron pair.

2.2.2 The General Rule of Judging Covalent Molecular Structure

(1) First, the valence shell electron logarithm of the central atom is determined by the following formula:

$$\text{Number of valence electron pairs} = \frac{\text{number of valence electrons on the central atom} + \text{number of electrons provided by coordination atoms}}{2} \tag{8-3}$$

In a normal covalent bond:

When the hydrogen atom and the halogen atom are the coordination atoms, they each provide one electron; when the halogen atom is the central atom of the molecule, it provides seven valence electrons.

When oxygen and sulfur atoms act as coordination atoms, they can be considered not to provide shared electrons, and when they act as central atoms, they can be considered to provide all 6 valence electrons.

If the species in question is an ion, the number of electrons corresponding to the charge should be added or subtracted. For example, the number of the valence electron pair on central atom N in NH_4^+ ion is (5+4-1)/2=4; the number of the valence electron pair on central atom S in SO_4^{2-} ion is (6+2)/2=4.

(2) According to the number of valence electron pair on the central atom, find out the corresponding electronic pair arrangement from table 8-2, in which can minimize the electrostatic repulsion between the

electronic pairs.

(3) Draw the structure diagram, arrange the coordination atoms around the central atom, each pair of electrons connects one coordination atom, and the remaining unconjugated pair of electrons is the lone pair of electrons. According to the repulsion force between lone pair electrons, bonding pairs, the stable structure with the minimum repulsion force is determined.

2.2.3 Examples of the Application of Valence Shell Electron Pair Repulsion Theory

(1) In CH_4 molecule, the central atom C has four valence electrons, and the four H atoms provide four electrons. Therefore, the total number of electrons in the valence shell on central atom C is 8, that is, there are 4 pairs of electrons. It can be seen from Table 8-2 that the valence electron pairs on C atom are arranged as tetrahedron. Because all the valence electron pairs are bonding electron pairs, the geometry of CH_4 molecule is tetrahedral.

(2) There are 7 valence electrons of the central atom Br in the BrO_3^- ion, and the O atom does not provide electrons. Considering one electron from the ion, the total number of valence electrons is 8, and the valence electron pair is 4. It can be seen from Table 8-2 that the arrangement of valence shell electron pairs of Br atom is tetrahedral, three top angles of tetrahedral are occupied by three O atoms, and the remaining top angles are occupied by lone pair electrons. This arrangement corresponds to only one type, so the BrO_3^- ion is trigonal pyramidal.

(3) In PCl_5 molecule, the central atom P has 5 valence electrons, 5 Cl atoms each provide 1 electron, the central atom has 5 valence shell electron pairs, and the spatial arrangement is trigonal bipyramidal.

(4) In ClF_3 molecule, the number of valence shell electron pair on Cl is 5, and the geometry of electron pairs is trigonal bipyramidal. However, there may be three different arrangements for the relative positions of bonding electron pair and lone electron pair, as shown in Figure 8-7:

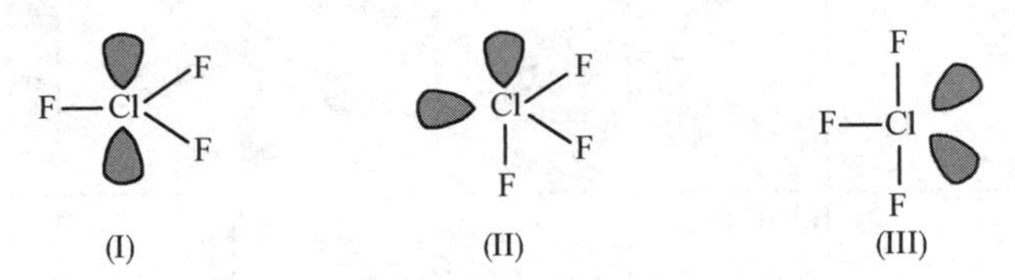

Figure 8-7 three possible geometries of ClF_3

There are three different angles between the electron pair and the central atom, 90°, 120°, 180°, of which the angle of 90° has the largest repulsive force. See Table 8-3.

表 8-3 Possible geometries analysis of ClF_3

Possible geometries of ClF_3	Opportunity to be at 90° angle		
	Lone pair electron-lone pair electron	Lone pair electron-bonding electron pair	Bonding electron pair-bonding electron pair
(I)	0	6	0
(II)	1	3	2
(III)	0	4	2

According to the comprehensive analysis, the (Ⅲ) structure is the most stable, so ClF_3 molecule is similar to T-shaped structure.

Using the VSEPR, we can predict the geometries of covalent compounds or ions formed by the atoms of most main group elements. The relationship between the arrangement of valence electron pairs on central atoms and the geometries of molecules is shown in Table 8-4.

Table 8-4 Relationship between central atom arrangement and geometry

Number of valence electron pairs	Arrangement of electronic pairs	Number of bonding electron pairs	Number of lone pair electrons	Molecular type	Arrangement of electronic pairs	Molecular geometry	Example
2	linear	2	0	AB_2	B═A═B	linear	$HgCl_2$
3	trigonal planar	3	0	AB_3		trigonal planar	BF_3
		2	1	AB_2		angle (V-shaped)	$PbCl_2$
4	tetrahedral	4	0	AB_4		tetrahedral	CH_4
		3	1	AB_3		trigonal pyramidal	NH_3
		2	2	AB_2		angle (V-shaped)	H_2O
5	trigonal bipyramidal	5	0	AB_5		trigonal bipyramidal	PCl_5
		4	1	AB_4		distorted tetrahedral (seesaw)	SF_4
		3	2	AB_3		T-shaped	ClF_3

(continued)

Number of valence electron pairs	Arrangement of electronic pairs	Number of bonding electron pairs	Number of lone pair electrons	Molecular type	Arrangement of electronic pairs	Molecular geometry	Example
5	trigonal bipyramidal	2	3	AB_2		linear	I_3^-
6	octahedral	6	0	AB_6		octahedral	SF_6
		5	1	AB_5		square pyramidal	IF_5
		4	2	AB_4		square planar	ICl_4^-

2.3 Hybrid Orbital Theory

We have described covalent bonding as "electron pair sharing" that results from the overlap of orbitals from two atoms. This is the basic idea of the valence bond (VB) theory--it describes how bonding occurs. In many examples throughout this chapter, we first use the VSEPR theory to describe the orientations of the electron groups. Then we use the VB theory to describe the atomic orbitals that overlap to produce the bonding with that geometry. We also assume that each lone pair occupies a separate orbital. Thus, the two theories work together to give a fuller description of the bonding. Valence shell electron pair repulsion theory can predict the geometries of polyatomic molecules, and the theory is intuitive and simple, but it cannot well explain the formation of bonds. The valence bond theory has successfully elucidated the nature and characteristics of covalent bond, but it has encountered some difficulties in explaining the geometries of polyatomic molecules. Pauling proposed hybrid orbital theory based on the valence bond theory, further complemented and developed the valence bond theory, and successfully explained the geometries of polyatomic molecules.

2.3.1 Basic Points of Hybrid Orbital Theory

We have learned that an isolated atom has its electrons arranged in orbitals in the way that leads to the lowest total energy for the atom. Usually, however, these "pure atomic" orbitals do not have the correct energies or orientations to describe where the electrons are when an atom is bonded to other atoms. When other atoms are nearby as in a molecule or ion, an atom can combine its valence shell orbitals to form a new set of orbitals that is at a lower total energy in the presence of the other atoms than

the pure atomic orbitals would be. This process is called hybridization (杂化), and the new orbitals that are formed are called hybrid orbitals (杂化轨道). These hybrid orbitals can overlap with orbitals on other atoms to share electrons and form bonds. Such hybrid orbitals usually give an improved description of the experimentally observed geometry of the molecule or ion. In fact, the concept of hybrid orbitals was developed specifically to explain the geometries of polyatomic ions and molecules.

The basic points are as follows.

(1) Only atomic orbitals with similar energy can be hybridized. Hybridization only occurs in the process of forming molecules, but it is impossible for isolated atoms to hybridize. The common hybrid types are *ns-np* hybridization, *ns-np-nd* hybridization, and $(n\text{-}1)d\text{-}ns\text{-}np$ hybridization.

(2) When hybrid orbitals form bonds, the principle of minimum repulsion between chemical bonds should be satisfied. The repulsive force between bond and bond determines the directionality of bond, that is, the angle between the hybrid orbitals, so the type of hybrid orbitals is related to the geometries of molecules.

(3) The ability of forming bond of hybrid orbitals is stronger than that of the original atomic orbitals, and the bond energy of the chemical bond is larger. Because the shape of atomic orbitals change after hybridization, the distribution of electron cloud is concentrated in a certain direction, which is more concentrated than that of *s*, *p*, and *d* orbital without hybridization, and the degree of overlap is increased, and the bonding ability is enhanced, as shown in Figure 8-8.

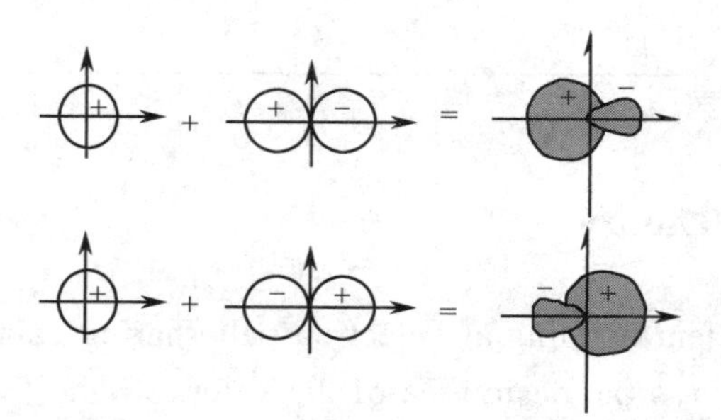

Figure 8-8 Schematic diagram of formation of hybrid orbitals

(4) The number of hybrid orbitals equals the total number of atomic orbitals participating in the hybridization.

2.3.2 Types of Hybridization

According to the different kinds and numbers of atomic orbitals, different types of hybrid orbitals can be formed, usually divided into equivalent hybridization and nonequivalent hybridization.

(1) Equivalent Hybridization The process of degenerate hybrid orbitals, which are formed by the hybridization of a group of unpaired atomic orbitals or empty orbitals, is called equivalent hybridization (完全由一组具有未成对电子的原子轨道或者是空轨道参与杂化而形成的简并杂化轨道的过程称为等性杂化), For example, sp^3, sp^2 and sp hybridization mentioned below.

***sp^3* hybridization** The hybridization of one *ns* orbital and three *np* orbitals in the same atom is called sp^3 hybridization (sp^3杂化), and the four hybrid orbitals newly formed are called sp^3 hybrid orbitals (sp^3杂化轨道). The characteristic of sp^3 hybrid orbital is that each hybrid orbital contains 1/4 *s* component and 3/4 *p* component. The angle between hybrid orbits is 109°28′, and the geometry is tetrahedral.

In gaseous CH_4, the outermost electron configuration of ground state C atom is $2s^22p^2$. Under the

influence of H atom, a 2*s* electron of C atom is excited into a $2p_z$ orbital, which makes C atom obtain the electronic shell configuration of $2s^1 2p_x^1 2p_y^1 2p_z^1$. One 2*s* orbital and three 2*p* orbitals are sp^3 hybrid to form four sp^3 hybrid orbitals with equal energy, as shown in Figure 8-9.

Each sp^3 hybrid orbital has an unpaired electron, which overlaps with the 1*s* orbital containing unpaired electron of four H atoms to form four σ bonds. Since the angle between the four sp^3 hybrid orbitals of C atom is 109°28′, the geometry of CH_4molecule is regular tetrahedron, as shown in Figure 8-10.

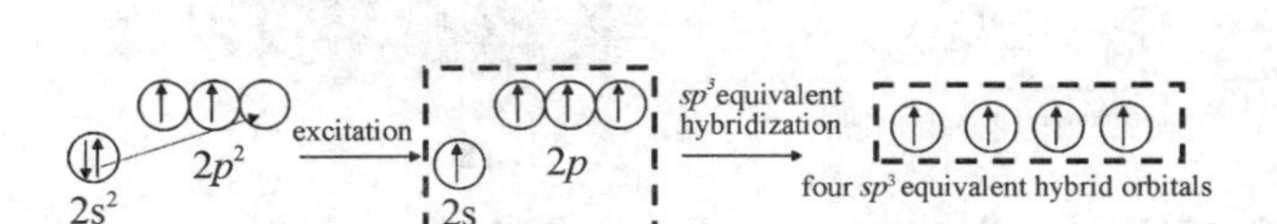

Figure 8-9 Schematic diagram of sp^3 hybrid of C atom

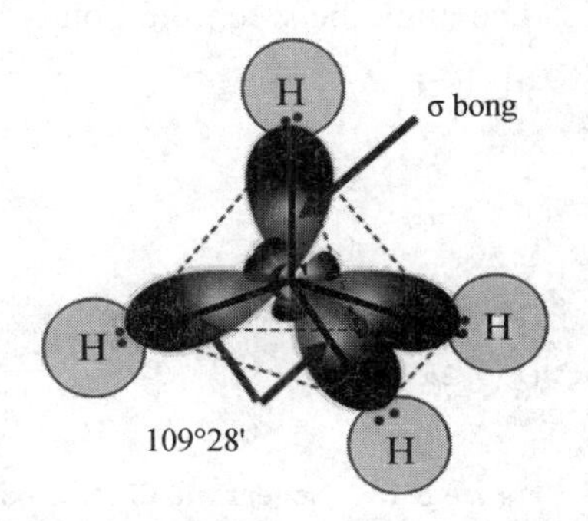

Figure 8-10 Schematic diagram of CH_4 molecular formation

***sp*² hybridization** The hybridization of one *ns* orbital and two *np* orbitals in the same atom is called sp^2 hybridization, and the three hybrid orbitals formed are called sp^2 hybrid orbitals. Each sp^2 hybrid orbital contains 1/3 *s* orbital component and 2/3 *p* orbital component. The angle between the hybrid orbitals is 120°, and the geometry is a trigonal planar.

For example, in the gaseous BF_3 molecule, the outermost electronic configuration of the ground state B atom is $2s^2 2p^1$. Under the influence of the F atom, a 2*s* electron of the B atom is excited into an empty 2*p* orbital, making the B atom obtain the electronic shell configuration of $2s^1 2p_x^1 2p_y^1$. One 2*s* orbital and two 2*p* orbitals are sp^2 hybrid to form three sp^2 hybrid orbitals with equal energy, as shown in Figure 8-11. Each sp^2 hybrid orbital contains an unpaired electron, which overlaps with 3*p* orbital with unpaired electron of three F atoms to form three σ bonds with an angle of 120°.The molecular geometry of BF_3 is a trigonal planar, as shown in Figure 8-12.

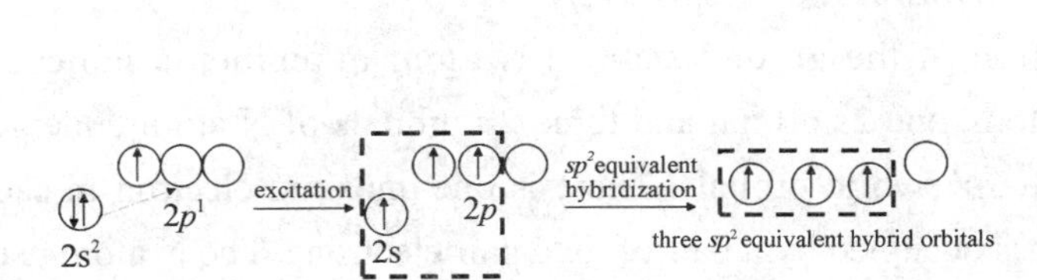

Figure 8-11 Schematic diagram of sp^2 hybrid of B atom

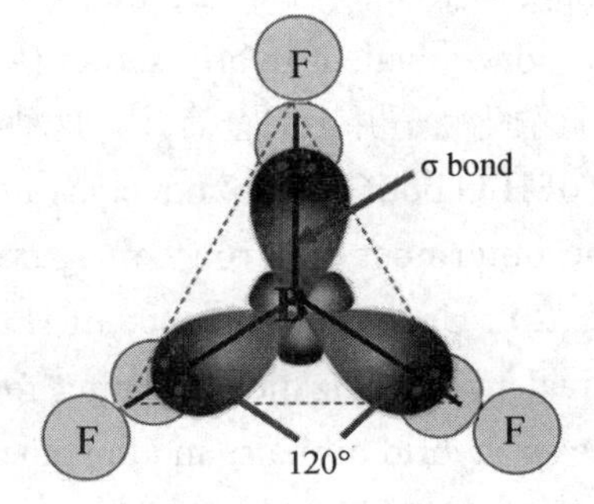

Figure 8-12 Schematic diagram of BF_3 molecular formation

***sp* hybridization** The hybridization of one *ns* orbital and one *np* orbitals in the same atom is called *sp* hybridization, and the two hybrid orbitals formed are called *sp* hybrid orbitals. Each *sp* hybrid orbital contains 1/2 of the *s* orbital component and 1/2 of the *p* orbital component. The angle between the two hybrid orbitals is 180°, and the geometry is linear.

For example, BeH_2 molecule, the electronic configuration of Be atom is $1s^2 2s^2$. Under the influence of H atom, a 2*s* electron of Be atom can be excited into an empty 2*p* orbital, making Be atom obtain the valence electronic configuration of $2s^1 2p_x^1$. One 2*s* orbital and one 2*p* orbital are *sp* hybrid, forming two *sp* hybrid orbitals, and each hybrid orbital has an unpaired electron, as shown in Figure 8-13. Two *sp* hybrid orbitals of Be atom overlap with 1*s* orbital of two H atoms containing unpaired electrons to form two σ bonds. The angle between the bonds is 180°, and the geometry of the molecule is linear, as shown in Figure 8-14.

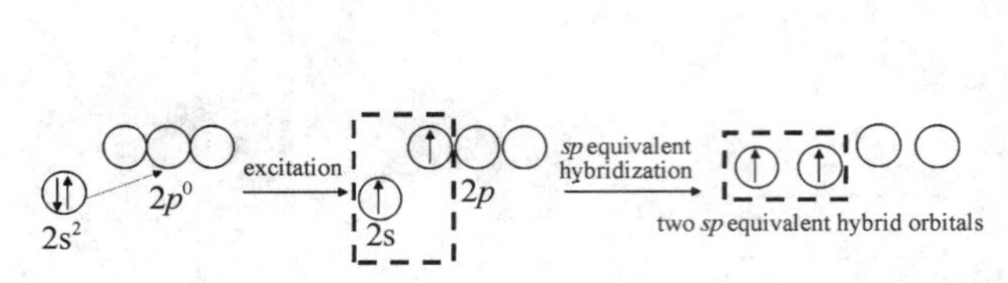

Figure 8-13 Schematic diagram of *sp* hybrid of Be atom

Figure 8-14 Schematic diagram of BeH_2 molecular formation

Besides *sp*, sp^2, and sp^3 hybridization, there are other types of hybridization as shown in Table 8-5.

Table 8-5 Other types of hybridization

Types	Number of orbitals	Molecular geometries	Example
sp	2	linear	$HgCl_2$
sp^2	3	trigonal planar	BF_3
sp^3	4	tetrahedral	CCl_4
dsp^2	4	square planar	$[CuCl_4]^{2-}$
sp^3d (or dsp^3)	5	trigonal bipyramidal	PCl_5
sp^3d^2 (or d^2sp^3)	6	octahedral	SF_6

微课

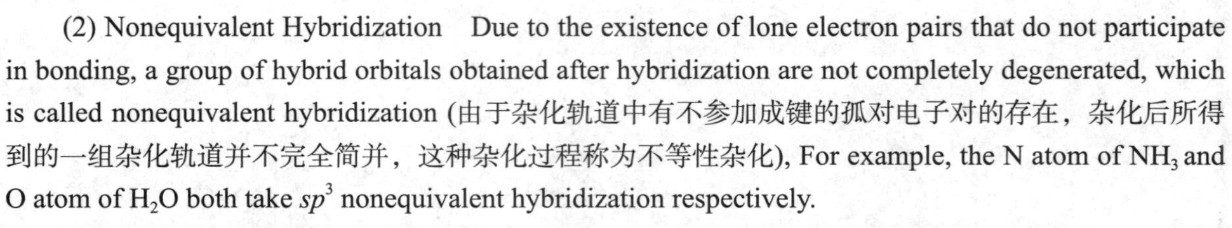

(2) Nonequivalent Hybridization Due to the existence of lone electron pairs that do not participate in bonding, a group of hybrid orbitals obtained after hybridization are not completely degenerated, which is called nonequivalent hybridization (由于杂化轨道中有不参加成键的孤对电子对的存在，杂化后所得到的一组杂化轨道并不完全简并，这种杂化过程称为不等性杂化), For example, the N atom of NH_3 and O atom of H_2O both take sp^3 nonequivalent hybridization respectively.

The outermost electron configuration of the ground state of N atom in ammonia molecule is $2s^2 2p_x^1 2p_y^1 2p_z^1$. Under the influence of H atom, one 2*s* orbital and three 2*p* orbitals of N atom undergo sp^3 nonequivalent hybridization, forming four sp^3 hybrid orbitals. There is one unpaired electron in each of the three sp^3 hybrid orbitals, and the other is occupied by a pair of lone pair electrons. The N atom uses its three sp^3 hybrid orbitals that each contains one unpaired electron overlapping with the 1*s* orbital of three H atoms and forming three N-H bonds. Because the electron cloud of lone pair electrons is concentrated around the N atom, which has a large repulsion effect on the electron cloud of three N-H bonds, the angle between N-H bonds is compressed to 107°18′, so the geometry of NH_3 molecule is trigonal pyramidal, as shown in Figure 8-15.

The outermost electron configuration of the ground state O atom in H_2O molecule is $2s^22p^4$. Under the influence of H atom, the O atom adopts sp^3 hybridization to form four sp^3 hybrid orbitals. There is an unpaired electron in each of the two hybrid orbitals, and the other two hybrid orbitals are occupied by lone pair electrons. The O atom uses its two sp^3 hybrid orbitals that each contains one unpaired electron overlapping with the 1*s* orbital of two H atoms and forming two O—H bonds. Because two pairs of lone pair electrons have greater repulsion to the bonding electrons of two O—H bonds, the angle between the O—H bonds is compressed to 104°45′, so the spatial geometry of water molecules is angular, as shown in Figure 8-16.

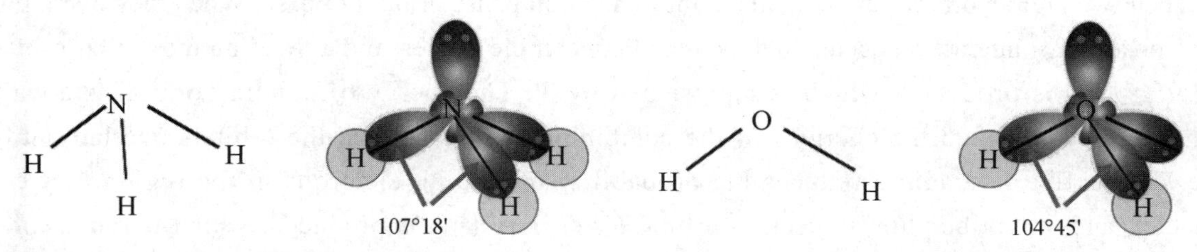

Figure 8-15 NH_3 molecular structure **Figure 8-16 H_2O molecular structure**

2.4 Molecular Orbital Theory

The valence bond theory successfully explains the nature and formation of covalent bond, and can predict and explain the geometries of molecules, making the geometry and chemical bond of molecules more intuitive and easier to understand. However, the valence bond theory has its limitations. It cannot explain ① the existence of single electron bond such as H_2^+, ② the paramagnetism of O_2, ③ the structure of some molecules with large π bonds. In 1932, American chemist Mulliken and German chemist Hund proposed the molecular orbital (MO) theory (分子轨道理论) in order to make up for the shortcomings of the valence bond theory.

Molecular orbital theory treats molecules as a whole, and regards the electrons in molecules as moving in the potential field formed by all atomic nuclei and electrons in molecules, which comprehensively reflects the movement states of electrons in molecules. The properties of the whole molecule are well described. In recent years, with the development and application of computer technology, the theory develops rapidly and plays a very important role in covalent bond theory.

2.4.1 The Concept of Molecular Orbital

Through learning the chapter of atomic structure, we know that the electrons in the atom move in the comprehensive potential field formed by the nucleus and other electrons. We can use ψ_{1s}, ψ_{2s}, ψ_{2p}…… wave function to represent the space motion state and energy of these electrons. These space motion states are commonly known as atomic orbitals.

Molecular orbital theory points out that in polyatomic molecules, electrons no longer belong to one atom, nor are they limited to two adjacent atoms, but move in the whole range of molecules. The spatial motion state of electrons in molecules can also be described by wave function ψ, that is to say, the spatial motion state of electrons in molecules is called molecular orbital (MO) (描述分子中的电子运动状态的概率波函数称为分子轨道). Each wave function ψ has its corresponding energy and shape. And the wave function of ψ^2 is called the probability density of electrons in the molecule in space (ψ^2为分子中的电子在空间出现的概率密度). The main difference between the molecular orbital and the atomic orbital is that the molecular orbital is polycentric (polyatomic), while the atomic orbital is one central (mononuclear).

2.4.2 Basic Points of Molecular Orbital Theory

(1) As an approximate treatment, it can be considered that molecular orbitals are formed by linear combination of atomic orbitals (LCAO, 原子轨道线性组合). The number of molecular orbitals is equal to the number of atomic orbitals combined.

When two atomic orbitals are combined, they can be in phase or out of phase. When they overlap in phase, constructive interaction occurs in the region between the nuclei, and a bonding molecular orbital (成键分子轨道) is produced, which is represented by Ψ. The energy of bonding orbital is always lower (more stable) than the energies of the combining orbitals. When the orbitals overlap out of phase, destructive interaction reduces the probability of finding electrons in the region between the nuclei, and an antibonding molecular orbital (反键分子轨道) is produced, which is represented by Ψ^*. This is higher in energy (less stable) than the original atomic orbitals, leading to repulsion between the two atoms. The overlap of two atomic orbitals always produces two MOs: one bonding and one antibonding.

For example, the linear combination of 1*s* atomic orbitals ψ_A and ψ_B of two hydrogen atoms can produce two molecular orbitals:

$$\Psi_{\sigma 1s} = C_1(\psi_A + \psi_B)$$
$$\Psi^*_{\sigma 1s} = C_2(\psi_A - \psi_B)$$

In the formula, C_1 and C_2 are constants. $\Psi_{\sigma 1s}$ is bonding molecular orbital, which is formed by overlapping two atomic orbitals with the same sign (wave function addition). Wave function with the same sign means that the electronic waves they represent are in the same phase. When they are combined, the two waves are superposed with each other to get the stronger waves, and the probability density of electron between two nuclei increases and the energy of the system decreases, so the energy of bonding molecular orbital is lower than that of atomic orbital. $\Psi^*_{\sigma 1s}$ is an antibonding molecular orbital, which is formed by the superposition of different atomic orbital (wave function subtraction). The different sign of wave function indicates that they represent different electronic wave phases. When they are combined with each other, some of them cancel each other, the probability density between nuclei decreases, and the increase of system energy is not conducive to bonding, so the energy of antibonding molecular orbital is higher than that of atomic orbital.

(2) In order to form molecular orbitals effectively, the atomic orbitals must meet the following three principles: symmetry matching principle (对称性匹配原则), energy approximation principle (能量相近原则) and maximum overlap principle (轨道最大重叠原则).

Symmetry Matching Principle: Only atomic orbitals with the same symmetry can be combined into molecular orbitals. Atomic orbitals have certain symmetry. For example, *s* orbitals are spherical symmetry, *p* orbitals are antisymmetric to the center (i.e., half of the wave function symbols of the angular distribution of atomic orbitals are positive and half are negative), *d* orbitals are central symmetric to the coordinate axis or a plane. In order to effectively combine into molecular orbital, it is necessary to require the same symmetry (matching) of the atomic orbitals participating in the combination, and the atomic orbitals with different symmetry cannot be combined into molecular orbitals. The so-called symmetry being the same, it can be understood that when two atomic orbitals rotate 180° with two atomic nuclei as their axes (designated as the X axis), the positive and negative signs of the angular distribution of the atomic orbitals both change or neither change, that is, the symmetry of the atomic orbitals is the same (matching); If one of the positive and negative signs changes, the other does not change, that is, the symmetry is not the same (mismatching). Figure 8-17 (a), (c) shows the symmetry mismatch of two atomic orbitals, while figure 8-17 (b), (d), and (e) shows the symmetry match of two atomic orbitals.

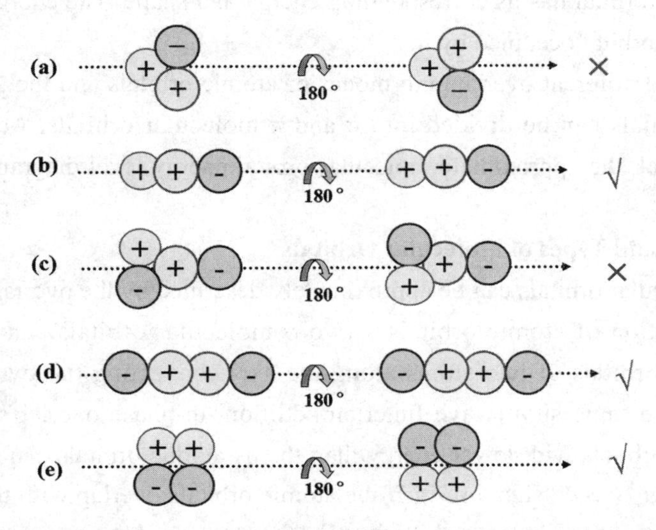

Figure 8-17 Schematic diagram of atomic orbital symmetry

Principle of Energy Approximation: Whether two atomic orbitals with the same symmetry can be combined into molecular orbitals depends on whether the energies of the two atomic orbitals are close. Only two symmetrically matched atomic orbitals with close energy can be effectively combined into molecular orbitals, and the closer the energies of the atomic orbitals are, the more efficient the combination is. This is called the principle of energy approximation. In homonuclear diatomic molecules, 1*s*-1*s*, 2*s*-2*s* and 2*p*-2*p* can be effectively combined into molecular orbital, and the principle of energy approximation is particularly important for heteronuclear diatomic or polyatomic molecules to be effectively combined into molecular orbitals.

For example, the energies of atomic orbitals of H and F atoms are as follows:

$$E_{1s}(\mathrm{H}) = -1.8\times10^{-18}\,(\mathrm{J})$$
$$E_{1s}(\mathrm{F}) = -1.12\times10^{-16}\,(\mathrm{J})$$
$$E_{2s}(\mathrm{F}) = -6.43\times10^{-18}\,(\mathrm{J})$$
$$E_{2p}(\mathrm{F}) = -2.98\times10^{-18}\,(\mathrm{J})$$

Because the 1*s* orbital energy of the H atom is close to the 2*p* orbital energy of the F atom, these two orbitals can be combined into molecular orbitals when the HF molecule is formed.

The Maximum Overlapping Principle: When two atomic orbitals with the same symmetry and close energies are combined into molecular orbitals, the overlapping degree of atomic orbitals should be maximized as much as possible, so as to decrease the energy of bonding molecular orbitals as much as possible and stabilize them, this is called the maximum overlap principle.

It can be seen that the principle of symmetry is the most important among the three principles of molecular orbital theory, which determines whether atomic orbitals can form molecular orbitals; while the principle of energy approximation and the maximum overlapping principle only determine the efficiency of combination.

(3) The arrangement of electrons in the molecular orbitals will follow the three principles of the lowest energy principle, Pauli exclusion principle, and Hund's rule, that is, to obtain the ground state electronic configuration of the molecule.

(4) Each molecular orbital has its corresponding energy and shape. The energy of the electron is the energy of the molecular orbital occupied by it.

(5) According to the different overlapping modes of atomic orbitals and the symmetry of molecular orbitals, molecular orbitals can be divided into σ and π molecular orbitals. According to the energy level of molecular orbital, the approximate molecular orbital energy level diagram (MO diagram) can be obtained.

2.4.3 Formation and Types of Molecular Orbitals

The shape of molecular orbitals can be approximately described by the overlap of atomic orbitals.

(1) *ns-ns* combination of atomic orbitals Two σ molecular orbitals can be obtained by linear combination of the *ns* orbitals of two atoms along the axis connecting the two nuclei. If the atomic orbitals overlap with the same sign (wave function addition, in-phase overlap, constructive overlap), the bonding molecular orbitals with lower energy than the *ns* atomic orbitals can be obtained, which are represented by the sign σ_{ns} (read "sigma-n*s*"); If the atomic orbitals overlap with the different sign (wave function subtraction, out-of-phase overlap, destructive overlap) the antibonding molecular orbital with higher energy than the *ns* atomic orbitals can be obtained, which is represented by σ^*_{ns} (read "sigma-n*s* -star"), as shown in Figure 8-18.

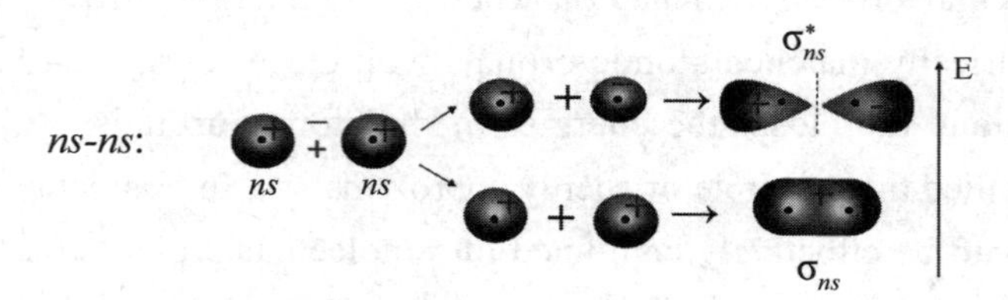

Figure 8-18 *s* atomic orbitals combined into molecular orbitals

(2) *ns-np$_x$* combination of atomic orbitals When the *ns* orbital of one atom overlaps the *np* orbital of another atom along the bond axis, since the *s* orbital only matches the symmetry of the p_x orbital, the n*s* and np_x orbitals can be combined into two molecular orbitals. The bonding molecular orbital of σ_{sp} and the antibonding molecular orbital of σ^*_{sp}, as shown in Figure 8-19.

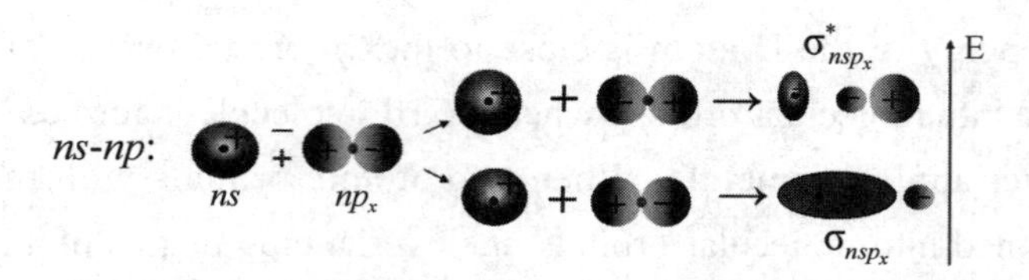

Figure 8-19 *s-p$_x$* atomic orbitals combined into molecular orbitals

(3) *np-np* combination of atomic orbitals When two n*p* atomic orbitals are combined into molecular orbitals, there are two ways of combination: head-on ("head to head") and side-on ("shoulder to shoulder"). Each atom has three *np* orbitals, namely np_x, np_y, np_z. Their spatial distribution is perpendicular to each other. The np_x orbitals of the two atoms are combined linearly along the axis (X axis) connecting the two nuclei in the way of "head to head", and two σ molecular orbitals can be obtained, i.e., the bonding molecular orbital of σ_{p_x} and the antibonding molecular orbital of $\sigma^*_{p_x}$, as shown in Figure 8-20.

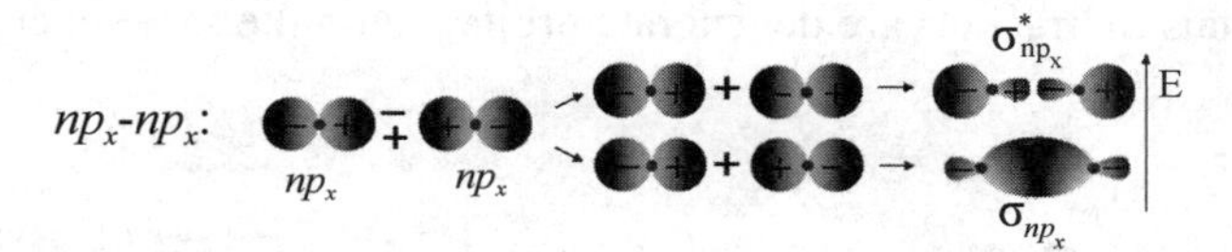

Figure 8-20 p_x-p_x atomic orbitals combined into molecular orbitals

When the bond axis is selected as X-axis, the np_y or np_z orbitals of two atoms overlap perpendicular to the bond axis, and are combined into two π molecular orbitals in a "shoulder to shoulder" way. The bonding molecular orbital π_{np_y} or π_{np_z}, as shown in Figure 8-21, and the antibonding molecular orbital $\pi^*_{np_y}$ or $\pi^*_{np_z}$ are combined in the same way, and they are degenerate orbitals mutually. Their shapes are same, and their energy are equal, but the spatial orientation is different, with 90° angle.

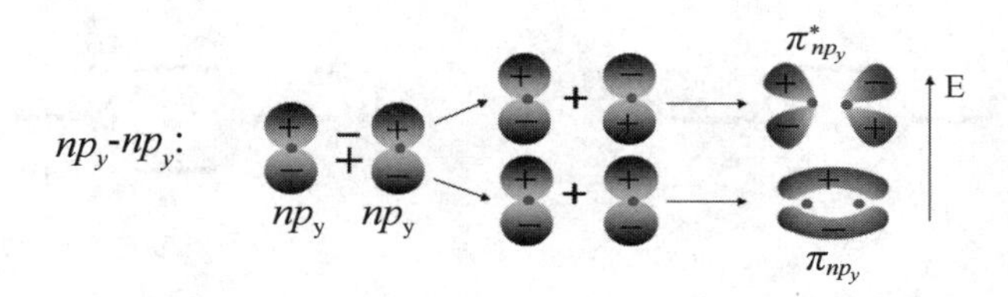

Figure 8-21 p_y-p_y atomic orbitals combined into molecular orbitals

It can be seen from the above discussion that the electrons distributed in the bonding σ molecular orbital are called bonding σ electrons, the electrons distributed in the antibonding σ molecular orbital are called antibonding σ electrons, the electrons distributed in the bonding π molecular orbital are called bonding π electrons, and the electrons distributed in the antibonding π molecular orbital are called antibonding π electrons. Bonding electrons increase the stability of molecules, while antibonding electrons decrease the stability of molecules.

There are also *p-d*, *d-d*, and other kinks of atomic orbitals overlapping in the molecular orbital theory, which are generally found in transition metal compounds and some oxyacids.

2.4.4 Molecular Orbital Energy Levels of Homonuclear Diatomic Molecules in the Second Period

Each molecular orbital has certain energy, and the molecular orbital energy of different molecule is different. Because the theoretical calculation of molecular orbital energy is very complex, it is mainly determined by molecular spectroscopy experiment at present.

Figure 8-22 shows the molecular orbital energy levels of homonuclear diatomic molecules in the second period. There are two kinds of order of the molecular orbital energy levels of the homonuclear diatomic molecules in the second period, here are the two cases.

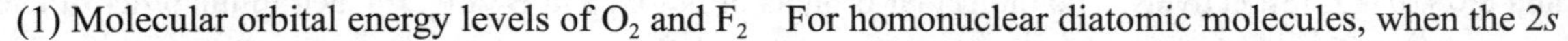

(1) Molecular orbital energy levels of O_2 and F_2 For homonuclear diatomic molecules, when the 2*s*

and $2p$ orbital energies of the constituent atoms differ greatly (generally more than 15eV or 2.4×10^{-19} J), for example, the energy difference between the $2p$ orbital and the $2s$ orbital of O atom is 2.64×10^{-18} J, and the energy difference between the $2p$ orbital and the $2s$ orbital of F atom is 3.45×10^{-18} J, when two atomic orbitals are combined into molecular orbitals, only *s-s* and *p-p* overlaps, and no interaction between $2s$ and $2p_x$ orbitals occurs. The order of energy levels of molecular orbitals is shown in Figure 8-22 (a). The molecular orbitals of O_2 and F_2 are arranged according to the order of energy levels, namely:

$$\sigma_{1s}\sigma_{1s}^{*}\sigma_{2s}\sigma_{2s}^{*}\sigma_{2p_x}(\pi_{2p_y}\pi_{2p_z})(\pi_{2p_y}^{*}\pi_{2p_z}^{*})\sigma_{2p_x}^{*}$$

The molecular orbitals in brackets are degenerate orbitals with the same energy.

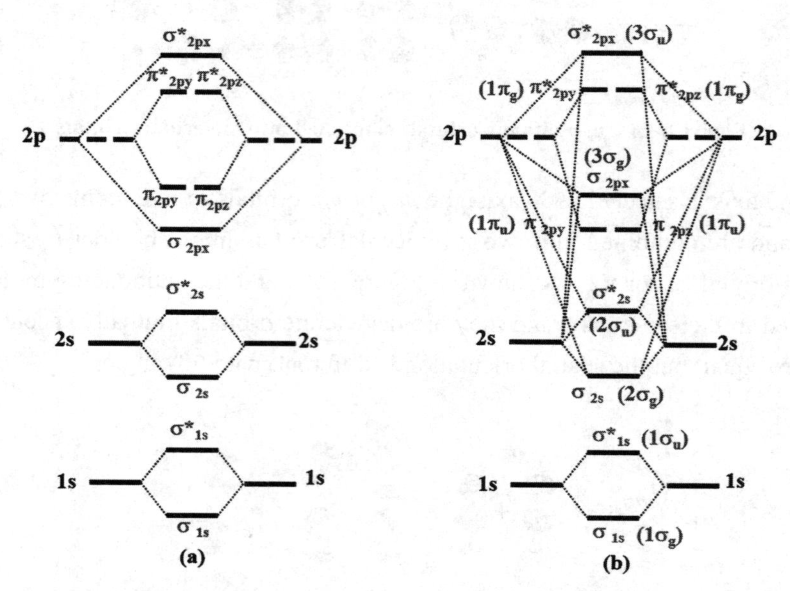

Figure 8-22 Molecular orbital energy levels of homonuclear diatomic molecules

(2) Molecular orbital energy diagram of H_2-N_2 For homonuclear diatomic molecules, if the energy levels of the $2s$ and $2p$ orbitals differs slightly (it is generally believed that about 10 eV or 1.6×10^{-19} J), not only the *s-s* and *p-p* overlaps, but also the interaction between the $2s$ and $2p_x$ orbitals. At this time, the molecular orbitals σ_{2s} and σ_{2p_x}, σ_{2s}^{*} and $\sigma_{2p_x}^{*}$ with the same symmetry, which are composed of two $2s$ and $2p_x$ orbitals, are further combined to form new molecular orbitals. At this time, the molecular orbitals are not simply the combination of $2s$-$2s$ and $2p_x$-$2p_x$ atomic orbitals. Therefore, they cannot be represented by symbols such as σ_{2s}, σ_{2p_x}, etc., but by symbols such as $1\sigma_g$, $1\sigma_u$, etc., where g represents central symmetry while u represents central anti-symmetry. (For homonuclear diatomic molecular orbitals, the center of the line between the nuclei can be taken as the center of symmetry. When the ψ symbols on both sides of the center of symmetry are the same, it is called centro-symmetry or g-symmetry; when the ψ symbols on both sides of the center of symmetry are opposite, it is called centro-anti-symmetry or u-symmetry. From the molecular orbital diagram, we can see that σ bonding orbitals are g-symmetric, σ antibonding orbitals are u-symmetric, π bonding orbitals are u-symmetric, and π antibonding orbitals are g-symmetric.) The order of the energy levels of the molecular orbitals after recombination is as follows:

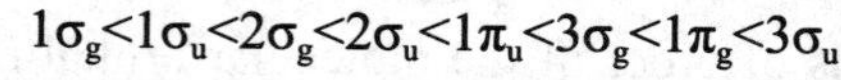

$$1\sigma_g<1\sigma_u<2\sigma_g<2\sigma_u<1\pi_u<3\sigma_g<1\pi_g<3\sigma_u$$

For example, the energy difference between $2p$ and $2s$ orbitals of N, C and B is 2.03×10^{-18} J, 8.4×10^{-19} J, and 8.0×10^{-19} J, respectively. The molecular orbitals of N_2, C_2 and B_2 are arranged in the order of energy

levels in Figure 8-22 (b).

Figure 8-22 the order of orbital energy levels is drawn according to the results of spectroscopy experiments. The energy levels of each orbital are listed in Table 8-6.

Table 8-6 Molecular orbital data of nitrogen and oxygen

orbital / energy ($\times 10^{-18}$J) / molecule	atomic orbitals		molecular orbital					
	$2s$	$2p$	σ_{2s}	σ_{2s}^*	σ_{2px}	π_{2p}	π_{2p}^*	σ_{2px}^*
N_2	-4.10	-2.07	-5.69	-3.00	-2.50	-2.74	1.12	4.91
O_2	-5.19	-2.55	-7.19	-4.79	-3.21	-3.07	-2.32	---

The application of molecular orbital theory is illustrated by several specific examples.

H_2 H_2 is the simplest homonuclear diatomic molecule. When two hydrogen atoms are close to each other, two $1s$ atomic orbitals (AOs) can overlap to form σ bonding and antibonding molecular orbitals (MOs). According to the principle of the lowest energy, two electrons with different spin modes will preferentially occupy the lowest energy of σ_{1s} bonding molecular orbital. And the energy of the system will be reduced to form a σ bond. The electronic arrangement of H_2 molecular orbital can be expressed as σ^2_{1s}, as shown in Figure 8-23.

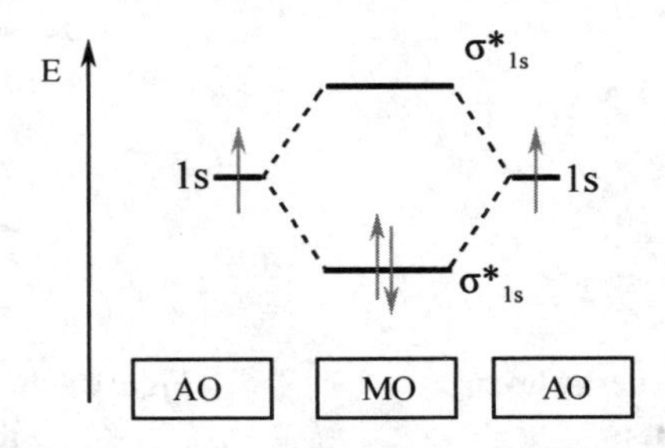

Figure 8-23 Atomic and molecular orbitals of hydrogen molecule

H_2^+ According to the principle of the lowest energy, the only electron in H_2^+ molecular ion enters into the σ bonding molecular orbital, and the energy of the system is reduced. Therefore, H_2^+ can exist stably and form a single electron σ bond with the bond energy of 255.48 kJ/mol. The electronic arrangement of H_2^+ molecular ion can be expressed as σ^1_{1s}, as shown in Figure 8-24.

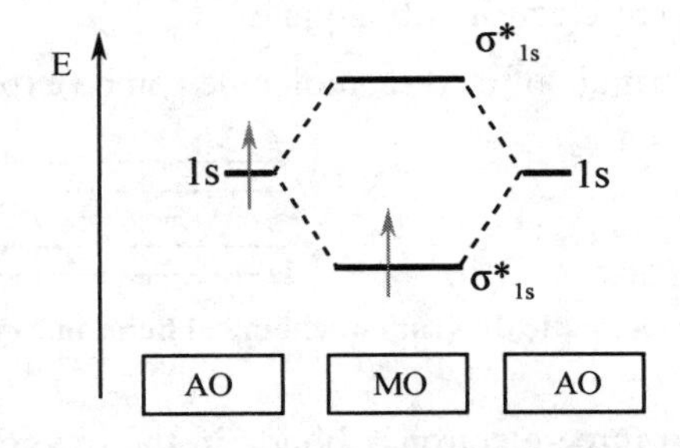

Figure 8-24 Atomic and molecular orbitals of hydrogen molecular ion

N_2 There are 7 electrons in each nitrogen atom and 14 electrons in N_2 molecule. According to the lowest energy principle, Pauli exclusion principle, and Hund's rule, the molecular orbital electronic arrangement of nitrogen molecule is as follows:

$$N_2: (1\sigma_g)^2\ (1\sigma_u)^2\ (2\sigma_g)^2\ (2\sigma_u)^2\ (1\pi_u)^2\ (1\pi_u)^2\ (3\sigma_g)^2$$

The main contribution to bonding are $(1\pi_u)^4(3\sigma_g)^2$ three pairs of electrons, which is equivalent to forming two π bonds and one σ bond. Because there are three bonds (N≡N) in nitrogen molecule, the bond order (键级) is 3, so N_2 molecule has special stability. The molecular orbital energy level diagram of the electronic arrangement of the N_2 molecule can also be expressed shown in Figure 8-25.

O_2 The oxygen molecule is composed of two oxygen atoms, each of which has eight electrons outside the nucleus, and there are 16 electrons in the oxygen molecule. According to the lowest energy principle, Pauli exclusion principle, and Hund's rule, the molecular orbital energy level diagram of the electronic arrangement of the oxygen molecule is shown in Figure 8-26.

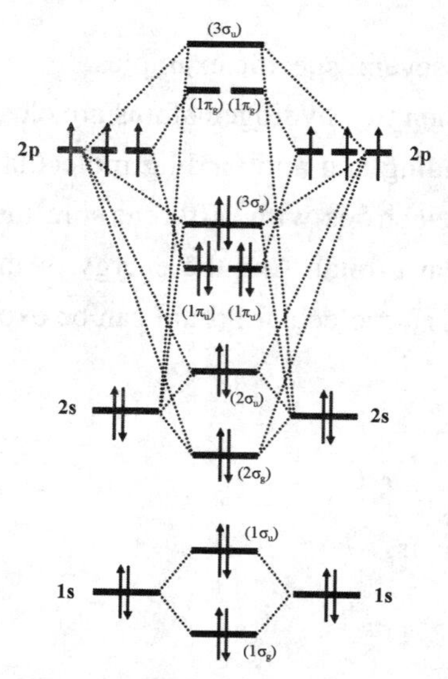

Figure 8-25 Molecular orbital energy level diagram of N_2 molecule

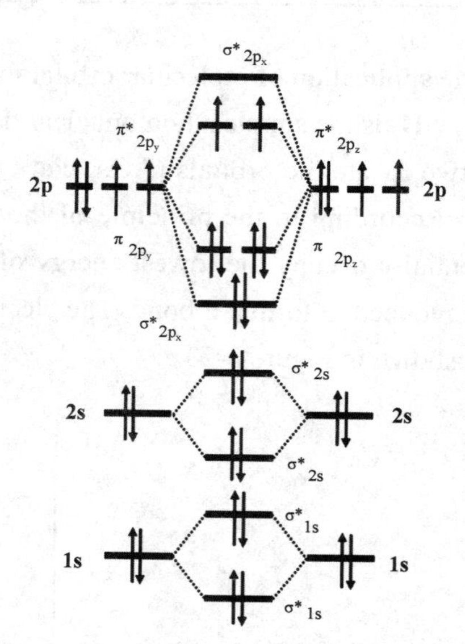

Figure 8-26 Molecular orbital energy level diagram of O_2 molecule

The molecular orbital electronic arrangement of O_2 molecule is:

$$O_2:\ (\sigma_{1s})^2(\sigma_{1s}^*)^2(\sigma_{2s})^2(\sigma_{2s}^*)^2(\sigma_{2p_x})^2(\pi_{2p_y})^2(\pi_{2p_z})^2(\pi_{2p_y}^*)^1(\pi_{2p_z}^*)^1$$

In the molecular orbital of O_2, the contribution of bonding $(\sigma_{2s})^2$ and antibonding $(\sigma_{2s}^*)^2$ to chemical bonding is mutually offset. In fact, what contributes to chemical bonding is $(\ _{2p})$ that forms σ bond, $(\pi_{2p_y})^2(\pi_{2p_y}^*)^1$ that forms a three-electron π bond and $(\pi_{2p_z})^2(\pi_{2p_z}^*)^1$ that forms another three-electron π bond. Therefore, the structural formula of oxygen molecule can be expressed as follows:

O —⋯— O or :O —— O:

Figure 8-27 Schematic diagram of chemical bond in oxygen molecule

There is one σ bond and two three-electron π bonds in the oxygen molecule. The bond energy of each three-electron π bond is only half of that of a single bond. The bond energy of two three-electron

π bonds is equivalent to that of a single bond, which is consistent with the fact that O_2 molecule has double bond energy. Therefore, it can be expected that the bond in O_2 molecule is not as strong as that in N_2 molecule. Experimental facts also prove that the energy required for breaking the chemical bond in O_2 molecule (i.e. the dissociation energy of O_2 molecule: 497.9 kJ/mol) is less than that required for breaking the chemical bond in N_2 molecule (the dissociation energy of N_2 molecule is 948.9 kJ/mol).

The molecular orbital energy diagram of O_2 clearly shows that there are two unpaired electrons with the same spin in O_2 molecule, so O_2 has paramagnetism, which has been proved by experiments. (The magnetism of matter is mainly caused by the spin of electrons. Usually there are unpaired electrons in paramagnetic matter, while the electrons in diamagnetic matter are all in pairs.) The paramagnetism of O_2 molecule cannot be explained by the Valence Bond Theory (electron pairing), but it is a natural conclusion when the molecular orbital theory is used to deal with the structure of O_2 molecule.

2.5 Bond Parameter

The properties of covalent bonds can be quantitatively discussed by quantum mechanical calculation, and can also be qualitatively or semi-quantitatively described by some physical quantities characterizing the properties of bonds. For example, bond order and bond energy are used to characterize the strength of bond. Bond length and bond angle are used to describe the geometries of molecules, etc. We call these physical quantities that characterize the properties of chemical bond collectively as bond parameter (表征化学键性质的物理量统称为键参数).

2.5.1 Bond Order

In the molecular orbital theory, the concept of bond order is put forward as bellow:

bond order = (number of electrons in bonding MOs— number of electrons in antibonding MOs)/2

键级 = (成键分子轨道中的电子数—反键分子轨道中的电子数)/2

The bond order is actually the net number of pair of bonding electrons. One pair of bonding electrons forms one covalent bond, so the bond order is generally equal to the number of bonds. For example, the bond order of He_2, H_2, O_2 and N_2 is 0, 1, 2, and 3, respectively. Compared with the atomic system, the more the number of electrons in the bonding molecular orbital, the more the energy of the molecular system is reduced, the more stable the molecule is. So the larger the bond order is, the stronger the bond is. On the contrary, the increase of the number of electrons in the antibonding molecular orbital weakens the stability of the molecule. The zero bond order means that stable molecules cannot be formed between atoms.

Therefore, it can be seen from the above analysis that the order of the stability of the following molecules is:

$$N_2 > O_2 > H_2 > He_2$$

2.5.2 Bond Energy

The strength of the covalent bond formed between atoms can be measured by the amount of energy needed to break the bond. In diatomic molecules, at 100kPa and 298.15K, the energy required to break 1mol of gaseous molecules into ideal gaseous atoms is called the bond dissociation energy (kJ/mol，在双原子分子中，于 100kPa 和 298.15K下，将 1mol 气态分子断裂成理想气态原子所需要的能量称做

键的离解能), that is:

$$AB(g) = A(g) + B(g) \qquad D\,(A\text{-}B)$$

In polyatomic molecules, to break the bond into a single atom, requires multiple dissociations, so the dissociation energy is not equal to the bond energy, but the average value of the multiple dissociation energy is equal to the bond energy (在多原子分子中，键能等于多次离解能的平均值).

For example:

$$CH_4(g) = CH_3(g) + H(g) \qquad D_1 = 435.34\ \text{kJ/mol}$$
$$CH_3(g) = CH_2(g) + H(g) \qquad D_2 = 460.46\ \text{kJ/mol}$$
$$CH_2(g) = CH(g) + H(g) \qquad D_3 = 426.97\ \text{kJ/mol}$$
$$CH(g) = C(g) + H(g) \qquad D_4 = 439.07\ \text{kJ/mol}$$

Although there are four equivalent C-H bonds in CH_4 molecule, the energy needed to separate them is different. The so-called bond energy refers to the average value of the energy required for each bond when 1mol gas molecule is disassembled into gas atoms at 100kPa and 298.15K (键能通常是指在100kPa 和 298.15K 下将 1mol 气态分子拆开成气态原子时，每个键所需能量的平均值). The sign of bond energy is expressed by E and the sign of bond dissociation energy is expressed by D. It is obvious that the bond energy is equal to the bond dissociation energy for diatomic molecules. For example, at 298.15K, the bond energy $E_{(H\text{-}H)}=D_{(H\text{-}H)}=436$ kJ/mol for H_2; for polyatomic molecules, the bond energy and dissociation energy are different. For example, the bond energy of C-H bond in CH_4 molecule should be the average value of the dissociation energy of four C—H bonds:

$$E_{(C\text{-}H)}=(D_1+D_2+D_3+D_4)/4= D_{totle}/4 = 1661.84/4 = 415.46\ \text{kJ/mol}$$

The bond energy is usually determined by thermochemical method or spectroscopy experiment. Table 8-7 lists the average bond energy of some bonds.

Table 8-7 The average bond energy of some bonds

Covalent bond	Bond length l (pm)	Bond energy E (kJ/mol)	Covalent bond	Bond length l (pm)	Bond energy E (kJ/mol)
H—H	74	436	C—C	154	346
H—F	92	570	C═C	134	602
H—Cl	127	432	C≡C	120	835
H—Br	141	366	N—N	145	159
H—I	161	298	N≡N	110	946
F—F	141	159	C—H	109	414
Cl—Cl	199	243	N—H	101	389
Br—Br	228	193	O—H	96	464
I—I	267	151	S—H	134	368

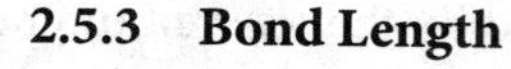

2.5.3 Bond Length

The equilibrium distance between two bonded nuclei in a molecule is called bond length (分子中两个成键的原子核之间的平衡距离称为键长). In theory, the bond length can be calculated by quantum mechanics approximation. But because of the complexity of molecular structure, the bond length is often measured by spectroscopy, X-ray diffraction, and other experimental methods. For example, the bond length of C—C is 154 pm, and the bond length between H-H is 74 pm.

The bond length of the same bond in different molecules is almost constant. See the bond length data of some chemical bonds listed in Table 8-7 above.

Generally speaking, the longer the bond, the smaller the bond energy and the weaker the bond strength. The bond length of single bond, double bond and triple bond is shortened in turn, and the bond energy is increased in turn, but the bond length of double bond and triple bond is not double or triple compared with that of single bond.

2.5.4 Bond Angle

The angle between two adjacent bonds in a molecule is called bond angle (分子中相邻两个键之间的夹角叫键角). Figure 8-28 shows the bond lengths, bond angles, and geometries of some molecules.

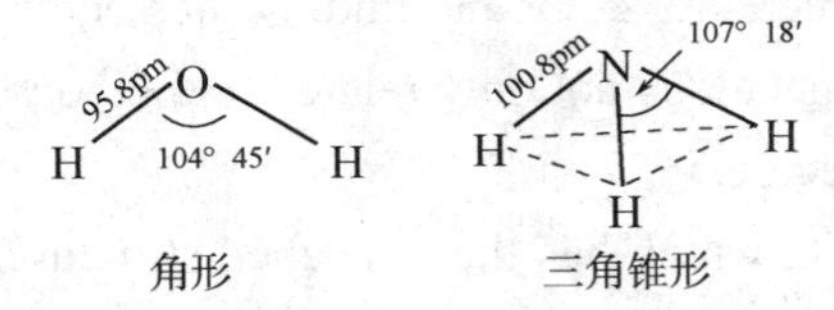

Figure 8-28 Bond lengths, bond angles and geometries of H_2O and NH_3

Bond angle is usually measured by spectroscopy and other experimental techniques. Bond angle and bond length are important parameters that determine the molecular geometries.

3. Polarity of Bond and Molecule

3.1 Polarity of Bond

Covalent bonds may be either polar or nonpolar. Due to the different electronegativity of atoms, the pairs of electrons shared in covalent bonds are not necessarily shared equally. In a nonpolar covalent bond (非极性共价键) such as that in the hydrogen molecule, H_2, the electron pair is shared equally between the two hydrogen nuclei. This is true for all homonuclear diatomic molecules because the two identical atoms have identical electronegativities. Covalent bonds, such as the one in HF, in which the electron pairs are shared unequally are called polar covalent bond (极性共价键).

For a covalent bond between two atoms, the bond polarity can be measured by bond moment (键矩) expressed as μ:

$$\vec{\mu} = q \times r \tag{8-4}$$

Bond moment is a vector whose direction is from positive to negative. Where q (C, Coulomb) is the quantity of charge, and r (m, Meter) is the distance between the poles. The unit of bond moment is C·m or Debye (D) that is always used in molecular physics.

The relationship between the two units is 1Debye = 3.336×10^{-30} C · m.

For polar covalent bonds, the greater difference the electronegativity between two atoms, the greater the bond moment, the greater the polarity of the bonds. When the electronegativity difference between

them reaches a certain degree, the shared electron pair completely deviates to one end of the atom with larger electronegativity, which makes it an anion, and the other part becomes a cation. At the same time, the covalent bond is converted into an ionic bond.

3.2 Polarity of Molecule

For diatomic molecules, since there is only one bond, the polarity of the bond determines the polarity of the molecule.

For homonuclear diatomic molecules, the distance between the positive and negative poles is 0, and the bond moment of covalent bond is 0, which makes a nonpolar covalent bond.

For polyatomic molecules that have several polar bonds, the overall dipole moments (偶极矩) can be made by considering the bond moment as vector and finding the vector sum.

The dipole moment of the molecule is not only related to the bond moment in the molecule, but also related to the geometry of the molecule.

Considering the H_2O molecule, which has the V-shaped structure, O—H bond is a polar bond, and the vectors of the bond moments of the two O—H bonds point to the angle between them, which makes H_2O a polar molecule. The C—H bond in CH_4 molecule is also a polar bond, but the tetrahedral structure of CH_4 molecule belongs to the central symmetric structure, which can offset the bond moments of the four C—H in different directions, making the positive and negative charge centers of the whole molecule completely coincide. So CH_4 molecule is a nonpolar molecule.

If the molecule is in a trigonal planar, tetrahedral and other central symmetric structure, the polarity of each bond can offset each other, and the vector sum of the bond moment is finally 0, thus the molecule has no polarity. For example, CO_2, BF_3, and CCl_4 with central symmetric structure are non-polar.

Table 8-8 Molecular dipole moment and geometry

Molecule	$\mu(\times10^{-30})$/C·m	Geometry	Molecule	$\mu(\times10^{-30})$/C·m	Geometry
H_2	0	Linear	H_2O	6.17	V-shaped
Cl_2	0	Linear	BF_3	0	Trigonal planar
CO_2	0	Linear	NH_3	4.90	Trigonal pyramidal
HI	1.27	Linear	CH_4	0	Tetrahedral
H_2S	3.67	V-shaped	CCl_4	0	Tetrahedral

4. Intermolecular Forces and Hydrogen Bonds

PPT

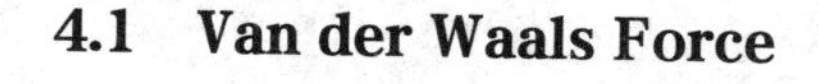

4.1 Van der Waals Force

Matter is composed of charged particles, so there is a good reason to speculate that there is an electrostatic force between molecules. In fact, the electrostatic attraction between the dipoles makes the intermolecular force (分子间作用力). These intermolecular forces affect chemical and physical properties. Due to the great contribution of the physicist Van der Waals (Dutch, 1837~1923) to the intermolecular force, the intermolecular force is also called Van der Waals force (范德华力). There are three types of Van der Waals forces: orientation force (dipole-dipole force), induction force and dispersion force.

4.1.1 Orientation Force

The positive and negative charge centers of polar molecules do not overlap, which leads to dipoles. This kind of dipole is caused by the structure of the molecule itself, which is called the inherent dipole (固有偶极，或永久偶极) of molecule. When two polar molecules get close to each other, the electrostatic attraction generated by their inherent dipoles causes the positive pole of one molecule to attract the negative pole of the other one. This electrostatic force, which is generated by the inherent dipoles of two polar molecules, is generally called the orientation force (取向力). The larger the dipole moment is, the greater the orientation force is, as shown in Figure 8-29.

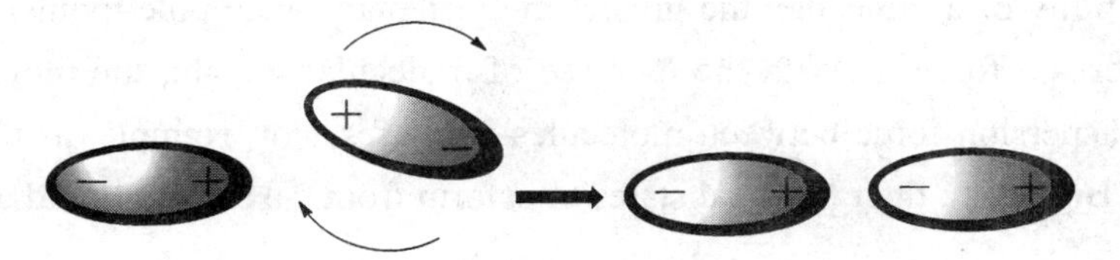

Figure 8-29 Formation of orientation forces

4.1.2 Induction Force

The electrons in molecules and atoms can be moved somewhat under the influence of electric field. The positive and negative charge centers of non-polar molecules overlap. Under the electric field of neighbor polar molecules, the centers of non-polar molecules are separated. This makes the positive and negative charge centers of non-polar molecules no longer overlap, resulting in a dipole moment. This dipole moment is induced by the inherent dipole of neighbor polar molecules, which is called induction dipole (诱导偶极). The electrostatic attraction between the induced dipole and the inherent dipole is defined as induction force (诱导力), as shown in the Figure 8-30.

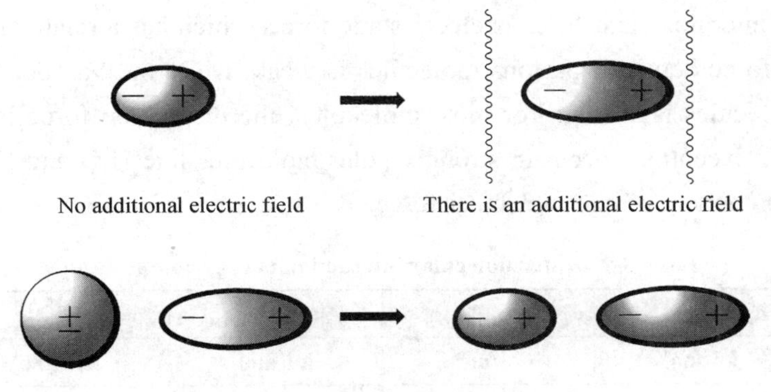

Figure 8-30 Formation of induction force

Similarly, the inherent dipole of polar molecules can also be induced by the external electric field of neighboring polar molecules, so that the inherent dipole moment increases. The electrostatic attraction between the induced dipole moment and inherent dipole moment also belongs to the induction force.

In summary, the induction force can occur not only between polar and non-polar molecules, but also between polar and polar molecules.

4.1.3 Dispersion Force

Many non-polar molecules exist in the form of solid or liquid at room temperature, such as liquid bromine and elemental iodine. This indicates that there must be a third intermolecular force besides orientation force and induction force. Statistically the positive and negative charge centers of non-polar molecules overlap. But if we look at a single molecule, its nucleus and electrons are in constant motion, which leads to the instantaneous separation of the charge centers. The resulting dipole is defined as instantaneous dipole (瞬时偶极). The force produced by the instantaneous dipole is defined as the dispersion force (色散力), as shown in the Figure 8-31. The dispersion force explains the existence of non-polar molecules in the form of condensed matter.

The motion of nucleus and electron exists in all molecules, there will be instantaneous dipoles no matter whether there are dipoles in molecules or not. That is to say, the dispersion force exists in all molecules. The strength of the dispersion force is related to the deformability of molecules. The greater the deformability of a molecule, the greater the instantaneous dipole moment, and the larger the corresponding dispersion force is. With the increase of molecular weight, the molecular deformability increases, and the dispersion force between molecules increases. For example, as the molecular weight increases of F_2, Cl_2, Br_2 and I_2, their physical states transform from gaseous to liquid until solid.

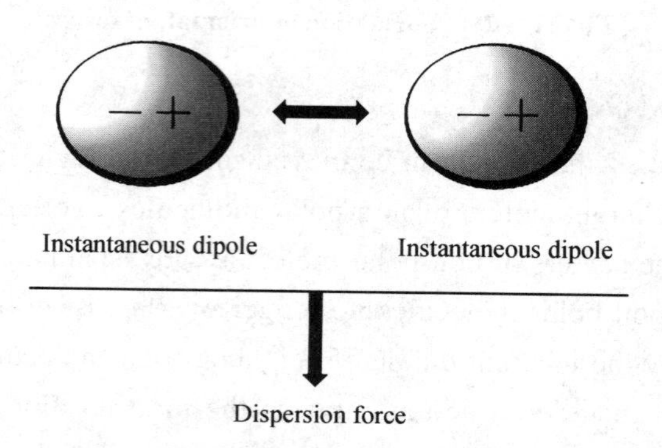

Figure 8-31 Formation of Dispersion Force

The essence of intermolecular force is electrostatic force, which has a relatively small action range. The Van der Waals force between gaseous molecules is weak, while the Van der Waals force between liquid and solid molecules is strong. For most molecules, the dispersion force is the most important intermolecular force. Exceptions occur in strongly polar molecules like H_2O and NH_3. Orientation force becomes the main component of intermolecular force.

Table 8-9 Intermolecular interaction energy composition

Molecule	$E_{(Orientation\ force)}$ kJ/mol	$E_{(Induction\ force)}$ kJ/mol	$E_{(Dispersion\ force)}$ kJ/mol	$E_{(sum)}$ kJ/mol	$E_{(Dispersion\ force)}/E_{(sum)}$
Ar	0	0	8.49	8.49	100%
HCl	3.31	1.00	16.83	21.14	79.61%
HBr	0.686	0.502	21.94	23.128	94.88%
HI	0.025	0.113	25.87	26.008	99.47%
NH_3	13.31	1.55	14.73	29.59	49.78%
H_2O	36.39	1.93	9	47.32	19%

4.2 Hydrogen Bonds

Hydrogen bond (氢键) is a phenomenon that pertains to many areas of the chemical sciences, and it is an important type of molecular interaction. Of all the atoms, only hydrogen leaves a completely bare nucleus when it forms a single covalent bond to another atom. The exposed nucleus has great electrostatic attraction to the negative center nearby, which is defined as hydrogen bond (氢键).

Generally, the higher the molecular weight is, the higher the boiling point. However, the boiling point of H_2O is much higher than those of other homologous compounds with larger molecular weight, as shown in the figure below (Figure 8-32). This abnormal phenomenon occurs in the hydride of halogen group elements and the hydride of nitrogen group elements as well. This could not be explained by the Van der Waals forces discussed above, but the hydrogen bond theory.

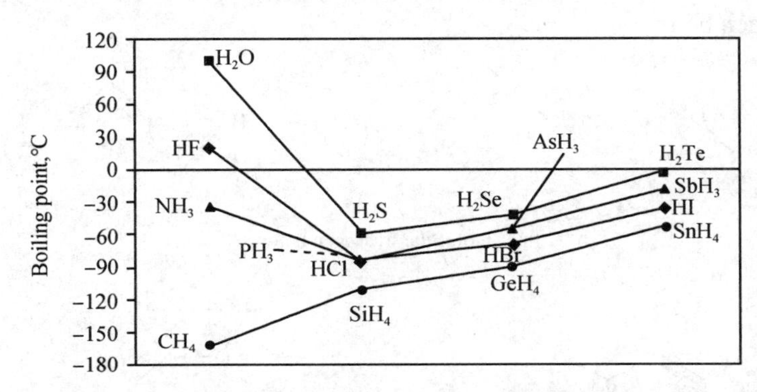

Figure 8-32 Hydrogen bond results in abnormal boiling points of compounds

4.2.1 The Formation of Hydrogen Bond

When H atom forms a covalent bond with atom X (such as N, O, and F) which has a large electronegativity and a small radius, atom X exerts a strong electronegativity in a small distance, attracting the shared electron pair to X, depriving the electrons on the H atom at the other end of the covalent bond, which makes H almost a naked proton. When the naked proton gets close to another atom Y (such as N, O, and F) with a great electronegativity and a lone pair of electrons, the proton uses its positive charge to simultaneously attract the negatively charged X and Y, forming the "X—H……Y" structure.

$$X—H\cdots\cdots Y \tag{8-5}$$

Hydrogen bond occurs in many matters such as proteins, cellulose, starch, leather and so on. Even solid materials such as ice have strong hydrogen bond between units. Water and other liquids that have —OH groups on the molecules have extensive hydrogen bond.

4.2.2 The Types of Hydrogen Bond

There are two types of hydrogen bonds: Intramolecular hydrogen bond and intermolecular hydrogen bond.

For intermolecular hydrogen bond (分子间氢键), X—H and Y belong to two different molecules, as shown in the figure below (Figure 8-33). The intermolecular hydrogen bond increases the force between different molecules and polymerizes them together. Intermolecular hydrogen bonds significantly increase the boiling point of matter. As mentioned above, the boiling point of water is 100°C while the boiling point of H_2S is 61°C. The abnormally high boiling points of H_2O, HF and NH_3 compared with the

congeners are caused by hydrogen bonds. Materials with intermolecular hydrogen bond need to overcome extra intermolecular force——hydrogen bond when they are melting or gasifying.

If hydrogen bonds occur between solvent molecules and solute molecules, the enhanced force between them will increase solute solubility. For example, H_2O and ethanol can be mutually soluble in any proportion.

For intramolecular hydrogen bonds (分子内氢键), X-H and Y belong to one molecule in the structure, as shown in the figure below (Figure 8-33). For example, in HNO_3, the common electron pair of O—H bond is attracted by O, and H becomes a nearly naked proton. The proton attracts O belongs to N—O through electrostatic attraction, which forming intramolecular hydrogen bond. In O-nitrophenol, the common electron pair on the O-H deviates from the attraction of O, H is positive, and O of the nitro group is negative. Both of them form a stable six membered ring by electrostatic attraction, thus forming intramolecular hydrogen bond.

Inter-molecular hydrogen bond

intra-molecular hydrogen bond

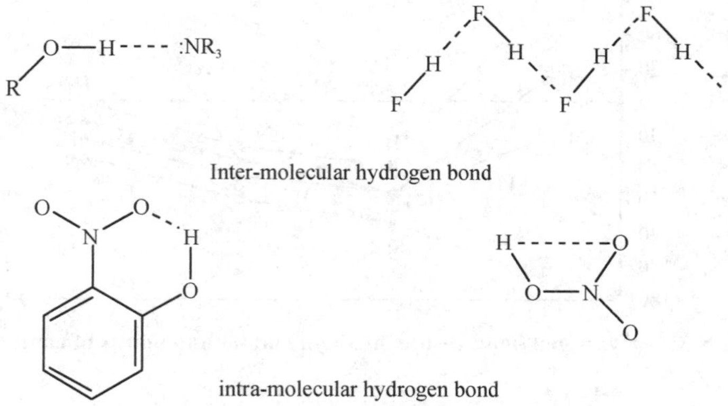

Figure 8-33 Different types of hydrogen bonds

图 8-33 不同类型的氢键

The ring structure formed by intramolecular hydrogen bond makes the polarity of the compound decrease, so the melting point and boiling point of the compound decrease. Therefore, the boiling point of nitric acid (83°C) producing intramolecular hydrogen bond is much lower than that of sulfuric acid (338°C), which produces intermolecular hydrogen bond.

4.2.3 The Characteristics of Hydrogen Bond

In a covalent bond, a pair of shared electrons charged negatively attracts the positive centers simultaneously at both ends, thus forming a strong covalent bond. Similar to covalent bonds, H in the structure of hydrogen bonds acts as a positive charge center and also acts as an adhesive to connect X and Y. The volume of H atom is small. When the intermolecular hydrogen bond is formed, if another electronegative ion Z approaches this H, it will be strongly repelled by the strong electronegative X and Y in the hydrogen bond. So in hydrogen bond, the coordination number of hydrogen is generally 2, which is called saturation of hydrogen bond (氢键的饱和性), as shown in Figure 8-34 (a).

Because of the small size of the hydrogen atom, X and Y should be as far away as possible in order to reduce the repulsion between them. The included angle of X…H—Y is very close to 180º, which is called directionality of the hydrogen bond (氢键的方向性), as shown in the Figure 8-34 (b). But the bond angle of intramolecular hydrogen bond is not 180º. Saturability and directionality of hydrogen bond are similar to those of covalent bonds.

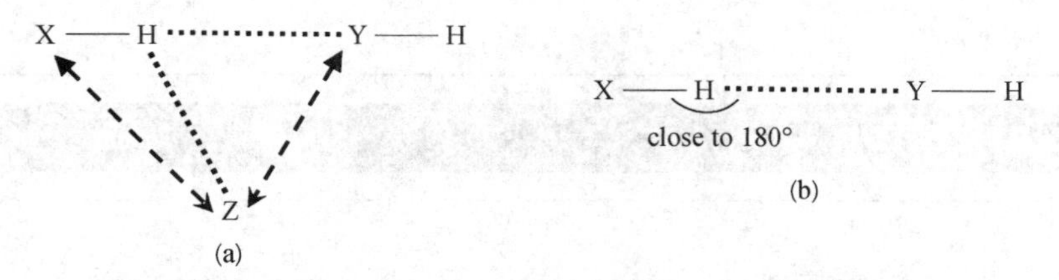

Figure 8-34 Saturability and directionality of hydrogen bonds

(a) saturability of hydrogen bonds (b) directionality of hydrogen bonds

4.2.4 The Effects of Hydrogen Bond on the Properties of Compounds

The bond energy of hydrogen bond is between covalent bond and Van der Waals force, which mainly affects the physical properties of matter. For example, intermolecular hydrogen bonds increase the melting and boiling point of compounds, while intramolecular hydrogen bond decrease the melting and boiling point of compounds. The moderate strength of the hydrogen bond makes the substance relatively stable. The hydrogen bond can be disassembled under appropriate conditions, which is great significance for the biological process.

The hydrogen bond between C═O and N—H in the amino acid residues of peptide chains is the main force to form the secondary and tertiary structure of the protein. In DNA, a large number of hydrogen bonds between the two main chains form a spiral three-dimensional structure. Under suitable conditions, the hydrogen bond in the double helix structure can break off again, and a new DNA molecule can be formed by base pair replication. Figure 8-35 shows the hydrogen bonds in the protein.

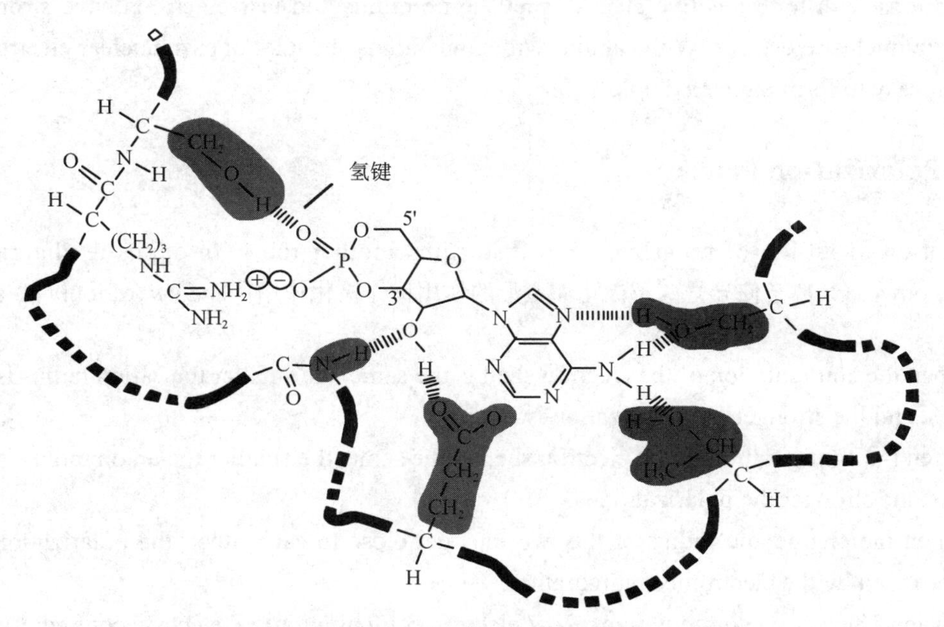

Figure 8-35 Hydrogen bonds in protein molecules

5. Ionic Polarization

5.1 Definition of Ionic Polarization

The theory of ionic polarization (离子极化) is the extension and supplement of the ionic bond theory. According to the ionic bond theory, the positive and negative ions in ionic compounds are bonded by electrostatic attraction, which inevitably affects the structure of two kinds of ions. When the positive and negative ions get close to each other, under the action of opposite electric field, the electron cloud deforms, making the positive and negative charge centers no longer coincide. The ability of inducing a dipole deformed is called ionic polarization; the ability of deforming under the polarization of an ion is called ionic deformation (离子变形).

Ionic polarization is a mutual process. Firstly, the charge of an ion will generate an electric field around it, which will polarize the opposite ion in its effective range, resulting in the deformation of the electron cloud of opposite ion; on the other hand, the electric field of the opposite ion will also cause the polarization deformation of the ion. In this process, each ion has dual properties of deformation and polarization.

The cations with high positive charge, small ionic radius and less electrons have strong binding ability of extranuclear electrons. While anions with large radius and a lot of extranuclear electrons deform easily, being easy to form induced dipole.

5.2 Polarization Power

The action of an ion of polarizing and distorting another ion with opposite charge is called polarization power (某离子使异号离子极化而变形的作用称作极化作用). The law of ionic polarization is as follow.

(1) When the configuration of the electron shell is the same, the smaller the cation radius is, the more the charge is, and the stronger the polarization is.

(2) When the configuration of the electron shell is the same, the smaller the anion radius is, the more the charge is, the stronger the polarization is.

(3) When the charge and radius of the two ions get close to each other, the polarization ability is mainly determined by the electronic configuration:

(18+2) and 18 electron configuration> 9-17 electron configuration > 8 electron configuration

In a word, the more electrons in the outer *d* shell are, the stronger the polarization ability is.

5.3 Ionic Deformation

Under the influence of external magnetic field, the action of ions taking place of deformation and generating induced dipole moment is called ionic deformation (ionic distortion) (离子受外磁场影响发生变形，产生诱导偶极矩的现象称作离子的变形). The main factors affecting ionic deformation are:

(1) The cation with less charge and larger radius will increase the deformability of cations.

(2) The anion carries more charges and larger radius will increase the deformability of anions.

(3) When the charges and radii of the two ions get close to each other, the configuration of the electron shell determines the deformability.

(18 + 2) and 18 electron configuration > 9-17 electron configuration > 8 electron configuration

5.4 Additional Polarization of Ions

Positive ions with 18 or (18+2) electron configuration have great polarization and deformation. At this time, the positive ion should consider both polarization and deformation. For instance, AgCl is a compound composed of cation and anion, but it shows typical characteristics of covalent compound. Ag^+ is of 18-electron configuration, with great polarization and great deformation; Cl^- is of 18-electron configuration with great deformation and great polarization as well. When two ions get closed, the Ag^+ with great polarization ability of deforms the electron cloud of Cl^-. The polarization ability of the deformed Cl^- is further strengthened. In turn, the electron cloud of Ag^+ deformed by Cl^- enhancing the polarization ability of Ag^+ and polarizing Cl^-. The interaction promotes the polarization ability of opposite ions and enhances its own deformability. Finally it makes the electron cloud of the cation and anion change greatly, and results in the overlap of the electron cloud. A large number of overlapping electron cloud between the cation and the anion decreases the ionic bond composition and increases the covalent bond composition. The type of bond changes from the ionic bond to the covalent bond. Figure 8-36 shows a schematic diagram of the gradual transition from ionic bond to covalent bond due to ionic polarization.

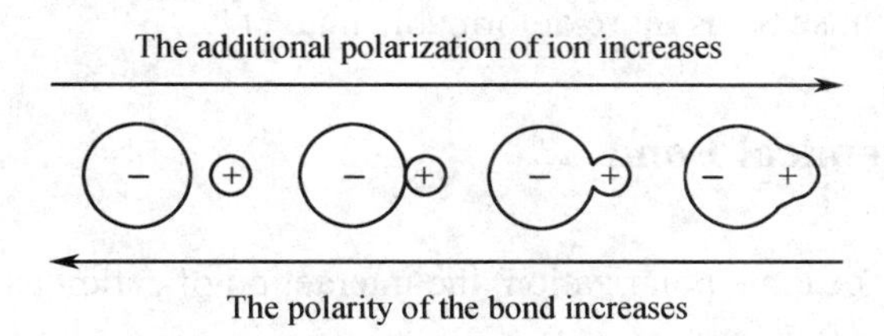

Figure 8-36 The effect of ionic polarization on bond type

5.5 The Effect of Ionic Polarization on the Bond Type and Properties of Compounds

5.5.1 Effect on the Boiling Point of Matter

In general, ionic compounds have higher melting point and boiling point than covalent compounds. For instance, $AlCl_3$ and NaCl have the same anion, but the melting point of $AlCl_3$ (463K) is much lower than that of NaCl (1074K). Al^{3+} and Na^+ both belong to the 8 electron configuration, but Al^{3+} has more charges and smaller radius, so it has stronger polarization ability to cause more significant deformation to Cl^-. The composition of covalent bond in $AlCl_3$ increased. $BeCl_2$ and $MgCl_2$ have the same anions and their cations have the same charge, but the radius of Be^{2+} is smaller and the polarization ability is stronger. Therefore, the covalent bond composition of $BeCl_2$ is higher and the melting point is lower.

5.5.2 Effect on Solubility

In general, ionic compounds have higher solubility than covalent compounds. For example, AgF is easy to dissolve in water, but AgCl, AgBr and AgI are difficult to dissolve in water and the solubility decreases in turn. With the increase of atomic number, the electron shells number, the radius of ions, the deformability and the polarization ability of the halide anions are gradually increased. As a result, the silver compounds show more covalent features with the increase of atomic number. Comparing the solubility of Na_2S with ZnS, Na^+ ion has a large radius and a small charge, which results in a small ability of polarize to S^{2-}, so Na_2S is easy to dissolve in water; Zn^{2+} ion with small radius and large charge has great ability to polarize S^{2-} ions, so ZnS is difficult to dissolve in water.

5.5.3 Effect on Material Color

When substance selectively absorbs light in some band of visible light, it will present the complementary color. Generally speaking, if the cation and anion of the compound are colorless, the compound is colorless, just like K^+ and Cl^- forming a colorless KCl. Ag^+ and I^- are colorless, but AgI is yellow. This can be explained by ionic polarization theory. Substances absorb electromagnetic wave of specific wavelength to gain energy, making the electrons in ground state jump to excited state. If the absorbed electromagnetic wave length is in the visible light, the substance reflects other electromagnetic waves that cannot be utilized, making the substance has color. Ionic polarization reduces the energy difference between ground state and excited state. After absorbing part of the visible light, the color of the compound becomes darker.

The colors of AgCl, AgBr, and AgI are white, light yellow, and yellow, respectively. The reason is that the anion polarization ability of the compounds is enhanced, leading to mutual polarization and resulting in the deepening of the substance color. Another example is that the color of sulfide is usually darker than that of oxide, because S^{2-} is more deformable than O^{2-}.

5.6 Ionicity of Chemical Bond

According to the theory of ionic polarization, the interaction of cation and anion leads to the overlap of electron cloud, and the overlap of electron cloud between two atoms is the reason for the formation of covalent bond, so there is a certain covalent bond component in the ionic bond. In fact, although there is essential difference between ionic bond and covalent bond, there is no strict limit between them. Even in the bond consisting of the most electronegative F and the least electronegative Cs, Cs-F bond has 92% ionic bond composition. In this most pure ionic bond, there are still overlaps of atomic orbitals. Polar molecules with incomplete overlap of positive and negative charge centers indicate that polar molecules have certain ionic bonding components.

In order to quantify the relationship between ionic bond and covalent bond, Pauling proposed the concept of ionic bond percentage: ΔX represents the electronegativity difference between bonding atoms. When $\Delta X = 1.7$, the proportion of ionic type in a chemical bond is 50%. When $\Delta X > 1.7$, the bond is considered as an ionic bond; when $0 < \Delta X < 1.7$ the bond is considered as a polar covalent bond; when $\Delta X = 0$, the bond is a nonpolar covalent bond.

重点小结

1. 化学键 化学键指的是分子和晶体中相邻原子之间强烈的相互吸引作用。根据成键原因，化学键可分为：离子键、共价键和金属键三类。

2. 共价键 按键的极性分成：极性共价键和非极性共价键。

按原子轨道重叠部分的对称性分成：σ 键和 π 键。

3. 价层电子对互斥理论 可以解释、判断和预见多原子的分子的空间构型，该理论比较简单和直观。中心原子价层电子数与配位原子的价电子数之和的一半就是中心原子的价层电子对数，中心原子的价层电子对之间尽可能远离，以使斥力最小，并由此决定了分子的空间构型。

4. 杂化轨道理论 中心原子轨道的杂化理论用于解释多原子分子的空间几何构型，常见原子轨道杂化类型有等性 sp、sp^2、sp^3 杂化和不等性杂化，杂化轨道的成键能力比原来未杂化的轨道的成键能力强，这是因为杂化后原子轨道的形状发生变化，电子云分布集中在某一方向上，原子轨道的重叠程度大，形成的化学键稳定。杂化轨道类型决定分子的几何构型。

5. 分子轨道理论 分子轨道是由原子轨道的线性组合形成的，有几条原子轨道参加组合就生成几条分子轨道，其中一半成键，一半反键。原子轨道有效组合成分子轨道的三原则是对称性匹配原则、能量近似原则、轨道最大重叠原则。

电子填入分子轨道时，遵循的三原则是能量最低原理和泡利不相容原理。

6. 键参数 键长、键角、键级、键能的概念是衡量化学键性质重要的参数。

7. 分子间作用力与氢键 范德华力包括取向力、诱导力和色散力，其作用能小于化学键，没有饱和性和方向性。氢键有饱和性和方向性主要影响物质的物理性质，如溶解度、存在状态、熔点、沸点、颜色等。

Exercises

1. Indicate the electron shell configurations of the following ions.

Be^{2+}, Fe^{2+}, Ca^{2+}, F^-, Ag^+, Pb^{2+}, S^{2-}, Na^+, Cl^-, Zn^{2+}, Sn^{2+}

2. Indicate which of the following combinations can form a molecular orbital and the type of orbital formed.

p_x-s, p_x-p_x, p_y-p_y, p_y-p_z, s-s, p_y-s

3. Compare the sizes of the following ionic radii.

(1) Na^+, Mg^{2+}, Al^{3+}

(2) F^-, Cl^-, Br^-, I^-

(3) Fe^{3+}, Fe^{2+}

4. Determine the spatial configuration of the following molecules by VSEPR theory.

BF_3, SiF_4, CCl_4, XeF_4, ClF_3, PCl_5

5. Determine the type of hybridization of BF_3 and $HgCl_2$.

6. Write the molecular orbital electron arrangements for the following molecules and ions and indicate the stability of the compound.

H_2^+, O_2, N_2, O_2^+

7. Which of the following molecules have hydrogen bonds in their structures?

Toluene, HF, CH_3COOH, HNO_3, O-nitrophenol

8. Indicate the type of van der Waals force between the following substances.

I_2 and H_2O, C_6H_6 and CCl_4, CH_3OH and H_2O, HI and HBr

9. Compare the melting points of the following substances.

NaF and MgO, CaO and MgO, $AlCl_3$ and NaCl, NaCl and NaI

10. Explain the following phenomena by ionic polarization theory.

(1) The colors of AgCl, AgBr, AgI are white, light yellow and yellow

(2) The solubility of AgCl, AgBr and AgI decreased in turn

11. Are the following statements true or false? Explain why?

(1) When NH_3 and BF_3molecules are formed, the central atom adopts the same hybrid type.

(2) Linear molecules are chemical polarity, non linear molecules are polar molecules.

(3) dsp^2 hybrid orbitals are four new hybrid orbitals which are composed of $3s$, $3p$ and $3d$ orbitals.

(4) There is one σ bond and two three-electron π bonds in the O_2 molecule.

(5) The formation of hydrogen bonds always raises the melting and boiling points of substances.

12. Compare the cationic polarizability of the following compounds.

(1) $ZnCl_2$, $FeCl_2$, $CaCl_2$, KCl

(2) $SiCl_4$, $AlCl_3$, $MgCl_2$, NaCl

13. Compare the anionic deformability of the following compounds.

(1) KF, KCl, KBr, KI

(2) Na_2O, Na_2S, NaF

14. Indicate whether the following molecules are polar or not.

CCl_4, BCl_3, NH_3, SO_2, CO_2, $HgCl_2$

15. Which of the following compounds do not have lone electron pairs?

H_2O, NH_4^+, NH_3, H_2S, ClF_3

16. Please illustrate the difference between ionic bond and covalent bond.

17. Why covalent bond has both saturation and directionality?

18. Explain the following phenomena with molecular orbital theory.

(1) He_2 cannot exist.

(2) O_2 is a paramagnetic molecule.

目标检测

1. 指出下列离子的电子层构型。

Be^{2+}，Fe^{2+}，Ca^{2+}，F^-，Ag^+，Pb^{2+}，S^{2-}，Na^+，Cl^-，Zn^{2+}，Sn^{2+}

2. 指出下列哪些可以形成分子轨道以及形成的轨道类型。

p_x-s，p_x-p_x，p_y-p_y，p_y-p_z，s-s，p_y-s

3. 比较下列离子半径的大小。

（1）Na^+，Mg^{2+}，Al^{3+}

（2）F^-，Cl^-，Br^-，I^-

（3）Fe^{3+}，Fe^{2+}

4. 利用价层电子对互斥理论，判断下列分子的空间构型。
BF_3，SiF_4，CCl_4，XeF_4，ClF_3，PCl_5
5. 分析 BF_3 和 $HgCl_2$ 的杂化类型。
6. 写出下列分子及离子的分子轨道电子排布式，并指出分子或离子的稳定性。
H_2^+，O_2，N_2，O_2^+
7. 下列分子中存在氢键的是?
甲苯，HF，CH_3COOH，HNO_3，邻硝基苯酚
8. 指出下列物质间的范德华力类型。
I_2 和 H_2O，C_6H_6 和 CCl_4，CH_3OH 和 H_2O，HI 和 HBr
9. 比较下列物质的熔点。
NaF 和 MgO，CaO 和 MgO，$AlCl_3$ 和 NaCl，NaCl 和 NaI
10. 利用离子极化理论解释下列现象。
(1) AgCl，AgBr，AgI 颜色分别为白色、淡黄色、黄色
(2) AgCl，AgBr，AgI 溶解度依次减小
11. 下列说法是否正确，说明原因。
(1) 形成 NH_3 和 BF_3 分子时，中心原子采用相同的杂化方式。
(2) 线型分子都是非极性分子，非线型分子都是极性分子。
(3) dsp^2 杂化轨道是由 3s、3p 和 3d 轨道混合起来形成的 4 条新杂化轨道。
(4) 在 O_2 分子中有一个 σ 键和 2 个三电子 π 键。
(5) 氢键的形成总是使物质的熔点和沸点升高。
12. 比较下列化合物阳离子极化能力的大小。
(1) $ZnCl_2$，$FeCl_2$，$CaCl_2$，KCl
(2) $SiCl_4$，$AlCl_3$，$MgCl_2$，NaCl
13. 比较下列化合物阴离子变形性的大小。
(1) KF，KCl，KBr，KI
(2) Na_2O，Na_2S，NaF
14. 指出下列分子是极性分子还是非极性分子。
CCl_4，BCl_3，NH_3，SO_2，CO_2，$HgCl_2$
15. 指出下列化合物中哪些不具有孤电子对。
H_2O，NH_4^+，NH_3，H_2S，ClF_3
16. 举例说明离子键和共价键的区别。
17. 说明为什么共价键既有饱和性，又有方向性。
18. 用分子轨道理论解释。
(1) He_2 不能存在。
(2) O_2 为顺磁性分子。

PPT

Chapter 9 Coordination Compounds
第九章 配位化合物

学习目标

知识要求

1. **掌握** 配位化合物的定义、分类；配合平衡的基本概念；配合平衡和稳定常数及不稳定常数的概念。

2. **熟悉** 配位化合物价键理论的基本要点、内轨型和外轨型配合物、配离子的空间结构、配合物的性质；配合平衡与酸碱平衡、沉淀溶解平衡、氧化还原反应平衡的关系及有关计算。

3. **了解** 配位化合物的类型（简单配合物，螯合物）；影响配位化合物稳定性的因素、配位化合物的平衡移动、配位化合物在生命科学中的应用。

能力要求

通过学习能熟练运用配合物的基本知识，进行有关配合物的命名、结构解释和计算。

1. Brief Introduction

Coordination compound (配合物), also called complex compound (络合物), is a class of compounds with coordination compositions and wide application. For example, the familiar blue vitriol and Prussiam blue are both cooridination compounds.

The study of modern coordination chemistry can be traced back to 1798, when the French chemist B. M. Tassaert first prepared $CoCl_3 \cdot 6NH_3$ with bivalent cobalt salt, ammonium chloride and ammonia, and found that metals such as chromium, nickel, copper and so on reacting with Cl^-, CN^-, CO, H_2O and C_2H_4 can also produce similar compounds.

However, the chemical theory at that time could not explain the bonding and properties of these compounds. Until 1893, the Swiss chemist Alfred Werner (1866—1919) summarized the previous theories, and first proposed a series of basic concepts such as modern coordination bond, coordination number and coordination compound structure, which successfully explained the conductivity, isomerism and magnetism of many coordination compounds, and lay the foundation of modern coordination chemistry. Werner was also known as the "father of coordination chemistry" and won the Nobel Prize in chemistry in 1913.

In 1923, the British chemist Sidgwick proposed the "effective atomic number" rule (EAN), which revealed the relationship between the number of electrons in the central atom and its coordination number. Many coordination compounds (especially carbonyl coordination compounds) conform to this rule, but there are exceptions. Although the principle can only partly reflect the essence of the formation of coordination compounds, its idea greatly promotes the development of coordination chemistry. The development of modern coordination chemistry depends largely on the development of molecular orbital theory. At present, valence bond theory, crystal field theory and coordination field theory have been widely used to explain the structure and properties of coordination compounds.

Driven by modern structural chemistry theory and modern physical experimental methods, coordination chemistry has developed into a discipline with rich research contents and fruitful achievements in inorganic chemistry, and has been widely used in daily life, industry, agriculture, biology, medicine and other fields. The research results of coordination chemistry promote the development of separation technology, coordination catalysis, functional coordination compounds, bioinorganic chemistry, atomic cluster chemistry, molecular biology, atomic energy, rocket and other fields. Chemical simulation of nitrogen fixation, photosynthesis, artificial simulation and solar energy utilization are closely related to coordination chemistry. In a word, coordination chemistry has become one of the most important fields in Modern Inorganic Chemistry, which has extremely important theoretical and practical significance in the whole field of chemistry. In this chapter, the basic concepts of coordination compounds, valence bond theory (价键理论) and crystal field theory (晶体场理论), coordination equilibrium and its influencing factors, and the application of coordination compounds are introduced.

2. Basic Concepts of Coordination Compounds

2.1 Definition and Composition of Coordination Compounds

2.1.1 Definition of Coordination Compounds

A small amount of ammonia is added to $CuSO_4$ solution to form sky blue $Cu_2(OH)_2SO_4$ precipitate. The dark blue transparent solution can be obtained by adding ammonia continuously. The results show that the concentration of Cu^{2+} in the dark blue solution is very low. The solution is evaporated by alcohol, and the dark blue crystal $[Cu(NH_3)_4]SO_4$ is obtained.

Another typical example is that cyanide. NaCN and KCN are highly toxic cyanide salts. However, although both potassium ferrocyanide ($K_4[Fe(CN)_6]$) and potassium ferrocyanide ($K_3[Fe(CN)_6]$) have no toxicity though contain cyanide. The reason is that the free CN^- is very low, so the compounds are almost non-toxic.

Among the above three substances $[Cu(NH_3)_4]SO_4$, $K_4[Fe(CN)_6]$ and $K_3[Fe(CN)_6]$, there is a relatively stable coordination compound structural unit formed by coordination bond combination in square brackets, which is called coordination unit (inner sphere). The coordination unit can be a cation, such as $[Cu(NH_3)_4]^{2+}$, or an anion, such as $[Fe(CN)_6]^{2-}$, which is called coordination ion. Coordination ion form coordination compounds with ions having opposite charges, and their properties are slightly like

inorganic salts, also known as inorganic salts. Coordination units without charge are also coordination compounds, also known as coordination molecules, such as $[Ni(CO)_4]$.

To sum up, coordination units with specific composition and spatial configuration, which are composed of simple cation or atoms and a certain number of neutral molecules or anions, are called coordination ions or coordination molecules. Compounds composed of coordination ions or coordination molecules are called coordination compounds.

2.1.2 Composition of Coordination Compounds

微课

The coordination compound is composed of inner sphere (内界) and outer sphere (外界). Inner sphere is coordination unit, and the outer sphere is simple ion. The inner sphere of the coordination compound is composed of central atom and ligands with coordination bond, which is difficult to dissociate. When writing chemical formula, it is usually enclosed by square brackets. The coordination ions are far away from other ions, and the coordination ions and other ions are bound by ionic bonds. This part other than the coordination ions is often called outer sphere. The composition of the coordination compound can be illustrated as follows:

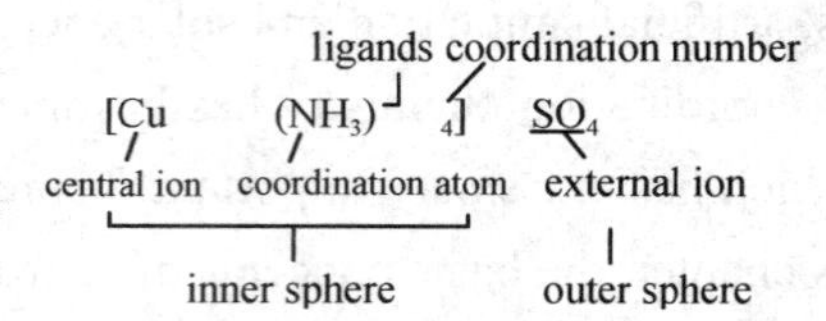

Figure 9-1 Structure of coordination compounds

If there are two or more kinds of ligands in the inner sphere of the coordination compound, the coordination compound is called mixed-ligand coordination compound (混合配体配合物). Most of the coordination compounds in life body are mixed-ligand coordination compound.

(1) Central atom In the inner sphere of the coordination compound, the positively charged ion or atom located at the center of coordination unit is called the center ion or atom of the coordination compound. The central atom is the acceptor of pairs of electron, most of which are *d*-block or *ds*-block elements, such as Cu^{2+} of $[Cu(NH_3)_4]^{2+}$ and Ni of $[Ni(CO)_4]$. Some *p*-block nonmetals can also be used as central atoms, such as Si^{4+} of $[SiF_6]^{2-}$.

(2) Ligand and coordination atom In coordination compounds, anion or molecule that binds to central ion or atom by coordination bonds is called ligand. For example, NH_3 of $[Cu(NH_3)_4]^{2+}$, CN^- of $[Fe(CN)_6]^{2-}$. Ligand combines with the central ion to form the inner sphere of the coordination compound.

In principle, any molecule or ion that has an unshared pair of electrons (lone pair electrons) and can form a coordination bond with metal ions can be used as a ligand, such as NH_3, H_2O, OH^-, X^-, CN^-, etc. In fact, any ligand usually contains high electronegative elements (C, N, O,S, F, Cl, Br, I, etc.). The formed ligand can be molecule (NH_3, H_2O, etc.) or ion (OH^-, X^-, CN^-, etc.). The atom giving the lone pair electron in the ligand is called coordination atom (配位原子), such as N of NH_3, O in OH^-, C in CN^-, etc.

Ligands can be divided into monodentate (single-toothed) ligands and polydentate (multi-toothed) ligands according to the number of coordination atoms.

The monodentate ligands contain only one coordination atom and form only one coordination bond with the central ion, such as NH_3, H_2O, OH^-, X^-, CN^-, etc. The polydentate ligands contain two or more coordination atoms. For example, the ethylenediamine ($H_2NCH_2CH_2NH_2$, commonly abbreviated to en)

has two coordination atoms N in its molecule, which is called a bidentate ligand. Another example is oxalic ion ($^-O_2CCO_2^-$, ox), contains two coordination atoms O. Coordination compound with annular structure formed by polydentate ligand and the same central ions is called chelate. So polydentate ligand is also called chelator. In the six-dentate ligand ethylenediaminetetraacetic acid (EDTA), two electron pairs bonded to the central ion are provided by nitrogen atoms, and the other four are provided by four oxygen atoms, as shown below:

$$\begin{matrix} ^-\ddot{O}OCCH_2 \diagdown & & \diagup CH_2CO\ddot{O}^- \\ & \ddot{N}CH_2CH_2\ddot{N} & \\ ^-\ddot{O}OCCH_2 \diagup & & \diagdown CH_2CO\ddot{O}^- \end{matrix}$$

EDTA can form stable coordination compounds with most metal ions, which are widely used in medicine and biology. It is also used to remove calcium and magnesium ions from hard water. Formulas and names of some ligands some ligands are shown in Table 9-1.

Table 9-1 Some common ligands

Classification	Anion	Anion name	Ligand	Ligand name
Anionic ligands	F^-	Fluoride	$:F^-$	Fluoro
	Cl^-	Chloride	$:Cl^-$	Chloro
	Br^-	Bromide	$:Br^-$	Bromo
	I^-	Iodide	$:I^-$	Iodo
	OH^-	Hydroxide	$:OH^-$	Hydroxo
	NO_2^-	Nitrite	$:ONO^-$	Nitrito
	$S_2O_3^{2-}$	Thiosulfate	$:SO_3^{2-}$	Sulfido
	$C_2O_4^{2-}$	Oxalate	$^-:O(CO)_2O:^-$	Oxalato
	NH_2^-	Imide	$:NH_2^-$	Nitro
Neutral ligands	CO	Carbon monoxide	:CO	Carbonyl
	H_2O	Water	$H_2O:$	Aqua
	NH_3	Ammonia	$:NH_3$	Ammine
	$H_2NCH_2CH_2NH_2$	Ethylenediamine	$:NH_2CH_2CH_2H_2N:$	Ethylenediamine

(3) Coordination number The coordination number (配位数) is the number of coordination bonds directly connected to the central ion or atom. The coordination number is also the total number of coordination atoms directly combined with the central ion or the central atom. If the structure of coordination ion is known, the coordination number can be determined.

Coordination compounds with monodentate ligands:

coordination number = the number of ligands.

Coordination compounds with polydentate ligands:

coordination number = the number of coordination atoms × coordination number

For example, the coordination number of Zn^{2+} in $[Zn(NH_3)_4]SO_4$ is 4, the coordination number of Cr^{3+} in $[Cr(H_2O)_4Cl_2]^+$ is 6; the coordination number of Cu^{2+} in $[Cu(en)_2]^{2+}$ is 4, the coordination number of Ca^{2+} in $[Ca(EDTA)]^{2-}$ is 6.

Many metal ions have an optimal coordination number. The coordination numbers usually are 2, 4, 6, and the most common ones are 4 and 6. The coordination numbers of some common ions are shown in Table 9-2.

Table 9-2 Coordination number of common ions

Coordination number	Central atom	Example
2	Au^+, Ag^+, Cu^+	$[Ag(NH_3)_2]^+$, $[Au(CN)_2]^-$
4	Cu^{2+}, Pt^{2+}, Zn^{2+}, Sn^{2+}, Cd^{2+}, Hg^{2+}, Al^{3+}	$[Zn(NH_3)_4]$, $[Cu(NH_3)_4]$, $[HgI_4]^{2-}$, $[Pt(NH_3)_2Cl_2]$
6	Fe^{2+}, Fe^{3+}, Al^{3+}, Cr^{3+}, Co^{3+}, Co^{2+}, Pt^{2+}, Ni^{2+}	$[FeF_6]^{3-}$, $[Cr(NH_3)_6]^{3+}$, $[PtCl_6]^{4-}$

The coordination number is often influenced by the charge and radius of the central atom, the charge and radius of the ligand, and the external conditions of the formation of the coordination compound.

The coordination number increases with the increasing the central atom size, such as $[AlF_6]^{3-}$ and $[BF_4]^-$. It should be noted that the increase of the radius of the central atom is not conducive to the formation of coordination compounds with high coordination number. For example, the coordination number of Hg^{2+} in $[HgCl_4]^{2-}$ is smaller than that of Cd^{2+} in $[CdCl_6]^{4-}$, although the radius of Hg^{2+} ion is larger than that of cadmium ion. The coordination number decreases with the increase of ligand size, e.g. $[FeF_6]^{3-}$ and $[FeCl_4]^-$.

The external conditions, especially the temperature and concentration, also affect the coordination number. Generally, the coordination compounds with high coordination number are easy to form under the condition of lower temperature and higher concentration of ligands.

In a word, the factors affecting the coordination number are very complex. In addition to the above factors, the valence shell electron configuration of the central atom and the spatial geometries of the ligand will affect the coordination number. However, a central ion or atom often has a specific coordination number binding with different ligands.

(4) The Charge number of coordination ion The charge number of the coordination ion is equal to the algebraic sum of the charge number of the central atom or ion and each ligand. For example, $[Fe(CN)_6]^{3-}$ with 3 unit negative charges is composed of one Fe^{3+} and six CN^-, and the algebraic sum of positive and negative charges is –3. For $[PtCl_2(en)]$, because it is composed of one Pt^{2+}, two $C1^-$, and one neutral en, the algebraic sun of positive and negative charge is 0.

2.2 Nomenclature of Coordination Compounds

Coordination compounds are named systematically through a set of rules by the International Union of Pure and Applied Chemistry (IUPAC). The naming principles are as follows.

(1) For coordination compound, like inorganic salts, cations are first named and separated from anions by space, regardless of whether the coordination ions are cations or anions, as is the case for all ionic compounds. However, the names of neutral coordination compounds are given without spaces. For exmple,

$K_3[Fe(CN)_6]$ potassium ferricyanide (Ⅲ)

$[Co(NH_3)_6]Cl$ hexaamminecobalt (Ⅲ) chloride

(2) In naming the coordination compound (neutral, cationic or anionic), the ligands begins first. The

metal is listed next, following in parentheses by the oxidation state of the metal as a Roman numeral, for example,

$[Cu(NH_3)_4]SO_4$ Tetraammoniated copper(Ⅱ) sulfate

The complete ligand name consists of a Greek prefix, as shown in Table 9-1.

(3) When more than one ligand in a given ligand binds to the same metal atom or ion, the number of these ligands is specified by the following prefixes (see Table 9-3).

Table 9-3 The prefixes for the number of ligands

number	2	3	4	5	6	7	8	9	10	11	12
prefixes	di	tri	tetra	penta	hexa	hepta	octa	nona	deca	undeca	dodeca

However, when the name of the ligand already contains numeric prefixes, which is usually enclosed in parenthesis, and the numerical prefixes are changed to bis, tris, tetrakis, pentakis, hexakis, heptakis, octakis, ennea for 2, 3, 4, 5, 6, 7, 8 and 9 ligands, respectively.

(4) Supposing ligands contain more than one atom, that can connect isomers have specific names for each connection mode. For example, CN^- is termed C-cyano, NC^- termed N-isocyano, SCN^- termed S-thiocyanato, NCS^- termed N-thiocyanto, NO_2^- termed nitro, ONO^- termed nitrito.

(5) If there are many kinds of ligands to bond with the central ion, the negative ligand is written first, and the neutral ligand followed. The categories ligands are listed in alphabetical order. For example, $[PtNH_2NO_2(NH_3)_2]$ Amino nitro diamine platinum (Ⅱ)

2.3 The Common Types of Coordination Compounds

Common coordination compounds can be divided into simple coordination compounds, chelates and polynuclear coordination compounds.

2.3.1 Simple Coordination Compounds

A coordination compound formed by the direct coordination of monodentate ligands with only one central atom is called a simple coordination compound. For example, $K_4[Fe(CN)_6]$, $[Cu(NH_3)_4]SO_4$, and $K[Ag(SCN)_2]$ are all simple coordination compounds. There is no ring structure in the molecule or ion of simple coordination compound, and the phenomenon of gradual formation and dissociation will occur in the solution.

2.3.2 Chelate

Chelate, also known as internal coordination compound, is a coordination compound with ring structure formed by central atom and polydentate ligand. Polydentate ligand has two or more coordination atoms that combine with a central ion at the same time. Two to three other atoms are separated between the two coordination atoms in the ligand to form a stable five membered or six membered rings with the central ion.

For example, diethylenediamine cadmium (Ⅱ) ion, en provides two nitrogen atoms to cooperate with Cd^{2+} to form a ring structure.

$$\left[\begin{matrix} CH_2-NH_2 & \searrow & & \swarrow & NH_2-CH_2 \\ | & & Cd & & | \\ CH_2-NH_2 & \nearrow & & \nwarrow & NH_2-CH_2 \end{matrix}\right]^{2+}$$

In the chelate, the ratio of the number of central atoms to the number of ligands is called the mixture ratio. Because of the ring structure, the chelates have high stability and seldom dissociate step by step.

Polydentate ligands are also called chelating agents. The chelating agent must have two conditions: ① The ligand must contain two or more coordination atoms, and mainly O, N, S and other atoms can give electron pairs. ② Two coordination atoms must be separated by two or more other atoms. Only in this way can stable ring structure be formed. For example, hydrazine H_2N-NH_2, which is a single dentate ligand, cannot form chelates.

The chelating agent ethylenediaminetetraacetic acid, commonly expressed as H_4Y, is called EDTA for short. Because its solubility in water and acid is very small, the commonly used salt is disodium salt: $Na_2H_2Y \cdot 2H_2O$, also abbreviated as EDTA. There are six coordination atoms (four oxygen atoms and two nitrogen atoms), which form a special stable chelate with five-member rings with metal ions.

Most of the chelates have five or six membered rings, which are more stable than the general coordination compound. Many chelating reactions can be used for quantitative titration and masking metal ions. Some vital substances such as chlorophyll and hemoglobin are also chelates.

2.3.3 Polynuclear Coordination Compounds

Coordination compounds with two or more central ions in the inner sphere are called polynuclear coordination compounds. Many central ions in the polynuclear coordination compounds are connected by the "bridge group" of the ligand. As a bridge group, the ligand must have two or more pairs of lone pair electrons in the coordination atom, which can form coordination covalent bond with two or more metal ions at the same time. The common bridge groups are O^{2-}, OH^-, NH_2^-, X^-, etc. as shown below:

```
C2H4 \      / Cl \      / Cl
       Pd          Pd
Cl   /      \ Cl /      \ C2H4
```

The interaction between metal ions in the polynuclear coordination compounds through electron transfer and their coordination and influence with the bridging and terminal ligands make them exhibit many physical functions, chemical properties, and biological activities different from mononuclear coordination compounds. In recent years, some coordination clusters with paramagnetic ions (including transition metal ions and rare earth metal ions) have been synthesized.

3. The Chemical Bond Theories of Coordination Compounds

Some physical and chemical properties of the coordination compound depend on the inner structure of the coordination compound, especially the binding force between the ligand and the central atom in the inner sphere. The chemical bond theory of coordination compounds is to clarify the nature of the binding force, and use it to explain some properties of coordination compounds, such as coordination number, geometry, magnetism, etc. The chemical bond theory of coordination compounds mainly includes valence bond (VB) theory, crystal field theory and coordination field theory. This chapter focuses on the valence bond theory and crystal field theory.

3.1 Valence Bond Theory

3.1.1 Basic Content of VB Theory

In 1931, L. Pauling, an American chemist, applied the theory of hybrid orbital to coordination compounds and proposed the VB theory of coordination compounds. The basic points are as follows:

(1) In the forming of coordination compounds, central atoms (or ions) must provide empty orbitals to accept the lone-pair electrons of ligands, forming coordination covalent bonds (the ligand acts as Lewis base to donate a pair of electrons while the central metal acts as Lewis acid to provide several equivalent hybrid orbitals). That is to say, the hybrid orbital of the central atom overlaps with the orbital of the coordination atom,and is filled with the lone electron pair to form the coordination bond.

(2) In order to enhance the bonding ability and form a well-balanced and low-energy coordination compound, the empty orbital provided by the central atom is first hybridized to form a hybrid orbital with the same number, energy and certain spatial extension direction. The hybrid orbital of the central atom and the lone pair electron orbital of the coordination atom overlap to form a bond in the bond axis direction.

(3) The number of hybrid orbitals provided by the central atom is equal to the coordination number, and the hybrid type determines the spatial configuration, stability and magnetism of the coordination compounds. The hybrid types for various coordination numbers and geometries are shown in Table 9-4.

3.1.2 Inner-orbital and Outer-orbital Coordination Compound

When the transition element is the central atom, its valence electron orbital often includes the *d* orbital of the sub-outer shell. According to the electronegativity of the coordination atom providing the lone pair electrons and the filling of the outer electrons of the central atom, there are two types of empty orbits provided by the hybridization of the central atom.

Table 9-4 Hybrid types and spatial configuration of coordination compounds

Coordination number	Hybrid type	Geometric	Examples
2	sp	Linear	$[Ag(NH_3)_2]^+$, $[Cu(CN)_2]^-$
4	sp^3	Tetrahedral	$[Zn(CN)_4]^{2-}$, $[FeCl_4]^-$
4	dsp^2	Square planar	$[Ni(CN)_4]^{2-}$, $[Cu(NH_3)_4]^{2+}$
5	d^2sp^2	Square pyramidal	$[CoCl_4]^{2-}$
5	dsp^3	Trigonal bipyramidal	$[Fe(CO)_5]$
6	d^2sp^3	octahedral	$[Fe(CN)_6]^{3-}$, $[Fe(CN)_6]^{4-}$
6	sp^3d^2	octahedral	$[FeF_6]^{3-}$, $[Co(NH_3)_6]^{2+}$

(1) Outer-orbital coordination　For outer-orbital coordination compound (外轨型配合物), all the hybrid orbitals involved in bonding are provided by the empty outermost orbitals (*ns*, *np*, *nd*) of the central atom or ion. When the coordination atom or ion is a high electronegativity element with weak electron-donating ability (such as halogen or oxygen), the outer electrons of the central atom cannot be affected. The intrinsic configuration of the electronic shell of the central atom remains unchanged (in accordance with the Hund's rule). Only the outer *ns, np* and *nd* space orbits hybridize, forming a group of hybrid orbital with equal energy. They can be used for bonding to ligands. The outer-orbital coordination compounds are formed. For example, $[Ni(NH_3)_4]^{2+}$, the electron configuration of Ni^{2+} is $3d^8$, which can

be hybridized with one 4*s* orbital and three 4*p* orbital. The four sp^3 hybrid orbits and four N atoms in NH_3 form four coordination bonds, thus forming the outer-orbital coordination ion $[Ni(NH_3)_4]^{2+}$ with tetrahedron spatial geometry, which belongs to the outer-orbital coordination. The orbital diagram for the coordination compound is

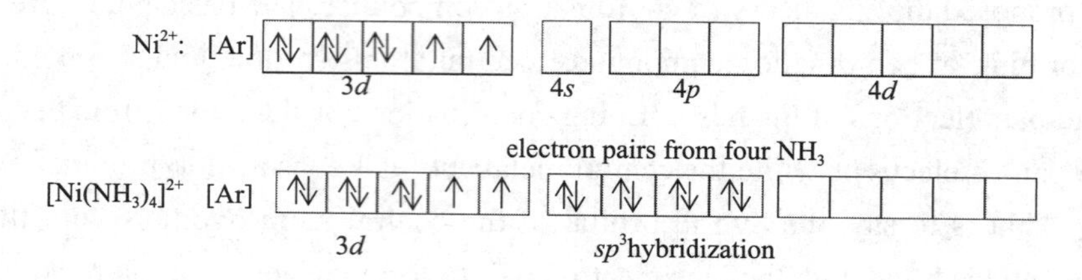

For $[FeF_6]^{3-}$, the eletron configuration of Fe^{3+} is $[Ar]3d^5$. According to Hund's rule, the five electrons of $3d^5$go into degenerate orbitals. When Fe^{3+} forms $[FeF_6]^{3-}$ with F^-, due to the strong electronegativity and weak electron-donating ability of F^-, it can only be hybridized with one 4*s* orbital, three 4*p* orbital and two 4*d* in the outermost shell. The six sp^3d^2 hybrid orbits with equal energy and six F^- form six coordination bonds form $[FeF_6]^{3-}$ with octahedral spatial configuration, which belongs to the outer-orbital coordination. The schematic diagram is

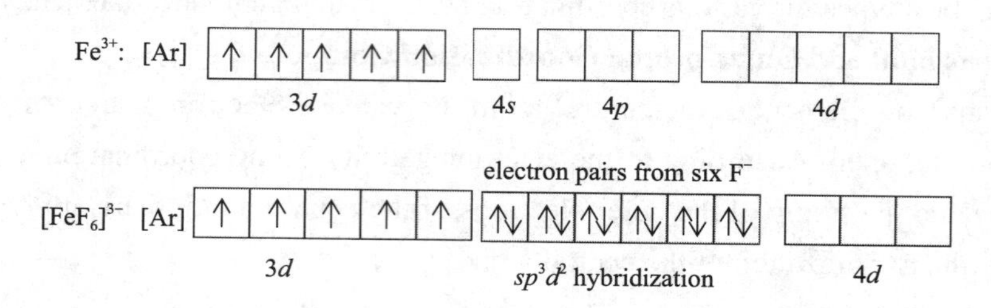

(2) Inner-orbital coordination For inner-orbital coordination (内轨型配合物), the (*n*-1)*d* electrons of the central atom or ion are rearranged, and some empty (*n*-1)*d* is used for hybridization, the inner-orbital coordination compound is formed. If the coordination atom is a low electronegativity element with strong electron-donating ability (such as C, N, etc.), (*n*-1)*d* electrons of central atom or ion are forced to pair. The empty (*n*-1)*d* with lower energy, n*s* and n*p* hybridize, forming a group of hybridized orbits with equal energy. They can be used for bonding to ligands. For coordination compound with ligands equaling to six, typical example is $[Fe(CN)_6]^{3-}$. In forming $[Fe(CN)_6]^{3-}$, 3*d* single electrons of Fe^{3+} are forced to pair, since the electronegativity of C is low and the ability of offering electron is strong. Five single electrons are combined in three 3*d* orbitals, the single electron is reduced from five to one, and two 3*d* orbits are vacated. Then the empty two 3*d*, one 4*s*, two 4*p* of Fe^{3+} undergoes d^2sp^3 hybridization, which combine with six CN^- to form a positive octahedral geometry $[Fe(CN)_6]^{3-}$. Schematic diagram of the formation process of coordination ions is

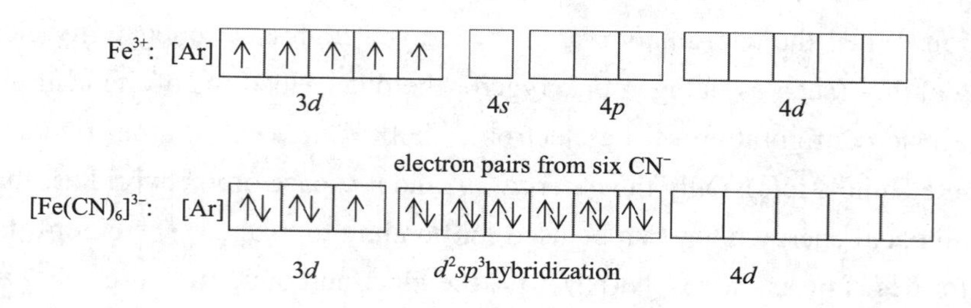

For coordination compounds with CN^- equaling to 4, $[Ni(CN)_4]^{2-}$, the electron configuration of Ni^{2+} is $3d^8$ (three double occupied 3*d* orbitals and two singly occupied 3*d* orbital). In forming $[Ni(CN)_4]^{2-}$, the two unpaired 3*d* electrons are forced to pair, giving an empty orbital. Then Ni^{2+} uses an unoccupied 3*d* orbital, one 4*s* and two 4*p* orbitals, undergoes dsp^2 hybridization to form four dsp^2 hybrid orbitals, combines with four CN^- to form square planar coordination compound. The process is shown as below.

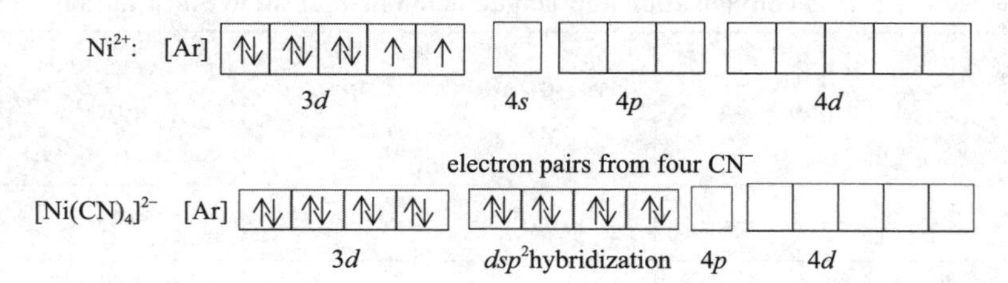

The metal ions with d^1, d^2 and d^3, electron configurations can only undergo d^2sp^3 hybridization to form inner-orbital coordination compounds (its coordination number is six) with any ligands. Contrarily, the electron number of $(n\text{-}1)d$ is between 7-10 such as Ni^{2+}, Co^{2+}, and so on, can only undergo sp^3d^2 hybridization to form outer-orbital coordination compounds (its coordination number is six) with any ligands, because there is not enough inner orbital being available, such as $[CoF_6]^{4-}$, $[Co(CN)_6]^{2+}$.

However, there are two cases for transition mental ions with d^4, d^5, and d^6 configurations to bond with ligands. One is bond to ligands by sp^3d^2 hybridization, forming outer-orbital coordination compounds; the other is bond to ligands by d^2sp^3 hybridization, forming inner-orbital coordination compounds. In general, the coordination atom with low electronegativity is easy to form inner-orbital coordination compounds, and the central mental uses sp^3d^2 hybridization. On the contrary, the coordination atom with high electronegativity in the ligand is easy to form outer-orbital coordination compounds, and the central mental uses d^2sp^3 hybridization. Because the energy of $(n\text{-}1)d$ orbital is lower than that of n*d* orbital, the inner-orbital coordination compounds of the same central atom are more stable than the outer-orbital coordination compounds.

3.1.3 Magnetic Moment of Coordination Compounds

In general, the magnetic moment of coordination compound (配合物的磁矩) is used to determine whether the coordination compound is an inner-orbital coordination compounds or outer-orbital coordination compounds. According to the magnetic moment μ of the coordination compounds, the number of unpaired electrons in the coordination compound can be calculated and the type of hybrid orbital can be determined. The approximate calculation:

$$\mu=\sqrt{n(n+2)}\text{B.M.} \tag{9-1}$$

Where μ is magnetic moment of the coordination compounds, n is number of unpaired electrons, B.M. is the unit of Bohr magnetion (B. M.). The magnetic moment and corresponding n are listed in Table 9-5. If all the electrons in a molecule and ion appear in pairs, the spin direction is opposite, and the spin magnetic moment is cancelled. This kind of material is diamagnetic coordination compound (反磁性配合物). Substance with one or more unpaired electrons is paramagnetic coordination compound (顺磁性配合物). The electronic structure and magnetic moment of some coordination compounds are shown in table 9-6.

Table 9-5 The number of unpaired electrons and the calculated magnetic moment

n	0	1	2	3	4	5
μ/B. M.	0.00	1.73	2.83	3.87	4.90	5.92

Table 9-6 The electron configuration and magnetic moment of some coordination compounds

Coordination ion	(*n*-1)*d* configuration	Hybridization type	Unpaired electrons	Theoretical magnetic moment (B.M.)	Measured magnetic moment (B.M.)
$[FeF_6]^{3-}$	Fe^{3+} ↑↓ ↑↓ ↑ — —	sp^3d^2	5	5.92	5.88
$[Fe(H_2O)_6]^{2+}$	Fe^{2+} ↑↓ ↑ ↑ ↑ ↑	sp^3d^2	4	4.90	5.30
$[Fe(CN)_6]^{3-}$	Fe^{3+} ↑↓ ↑↓ ↑ — —	d^2sp^3	1	1.73	2.3
$[CoF_6]^{3-}$	Co^{3+} ↑↓ ↑ ↑ ↑ ↑	sp^3d^2	4	4.90	-
$[Co(NH_3)_6]^{2+}$	Co^{2+} ↑↓ ↑↓ ↑ ↑ ↑	sp^3d^2	3	3.87	3.88
$[Co(NH_3)_6]^{3+}$	Co^{3+} ↑↓ ↑↓ ↑↓ — —	d^2sp^3	0	0	0
$[Ni(CN)_4]^{2-}$	Ni^{2+} ↑↓ ↑↓ ↑↓ ↑↓ —	dsp^2	0	0	0
$[Mn(CN)_6]^{4-}$	Mn^{2+} ↑↓ ↑↓ ↑ — —	d^2sp^3	1	1.73	1.70

3.2 Crystal Field Theory

Valence bond theory has successfully explained the coordination number, geometry, magnetism and stability of coordination compounds, but there are two limitations: one is that it cannot explain the color of coordination compounds, the other is that it cannot explain the properties of coordination compounds quantitatively and semi quantitatively. These problems can be satisfactorily explained by crystal field theory (CFT) and coordination field theory.

3.2.1 Basic Content of CFT

Crystal field theory (晶体场理论) is a model that considers that the interaction between the central ion and the ligands in the coordination compound is electrostatic interaction, which is similar to the electrostatic interaction between the positive and negative ions in ionic crystals, so it is called crystal field theory. The influence of ligands on the energy of *d* orbitals of central ion is mainly considered. The basic points are as follows.

(1) Crystal field theory assumes that the central atom and ligand form coordination compounds through electrostatic force, which makes the energy of the system decrease. The ligand is a point charge with negative charge in a certain space around the central ion, and the electrostatic field formed around the central ion is called crystal field (晶体场).

(2) Under the action of crystal field, the energy levels of five degenerate *d* orbitals split, and the energy is no longer equal. The energy of *d*-orbitals directed toward ligands increases, while energy of the

d-orbitals directed between ligands decreases.

(3) Due to the splitting of d orbitals energy level, the electrons in the d orbital of the central atom rearrange and occupy the orbit with lower energy preferentially, which makes the total energy of the system lower and the coordination compound more stable.

In this paper, the crystal field theory is introduced with octahedral coordination compound as an example.

3.2.2 Splitting of *d* Orbitals in an Octahedral Field of Ligands

Most of the central atoms of the coordination compounds are transition metal ions. The spatial orientations of the five degenerate d orbitals in the valence shell of the transition metal ions are different, so they will be affected by different electrostatic fields of the ligands with different symmetries. It is assumed that the negative charges of the six ligands are uniformly distributed on the spherical surface with the central atom as the center (i.e. spherical symmetry). The repulsion force of the negative electric field on the five d orbitals is the same, and the energy levels are still the same although they are all increased. In fact, the six ligands approach the central ion in the positive and negative directions ($\pm x$, $\pm y$, $\pm z$) of three coordinate axes respectively, as shown in Figure 9-2. The maxima of the electron cloud of the orbitals just collide with the ligands head-on, so the energy of the two orbitals $d_{x^2-y^2}$ and d_{z^2} is increased (compared with the spherical field). However, the maximum values of d_{xy}, d_{yz} and d_{xz} of the other three orbitals are inserted between the ligands, so the repulsion effect is small and the energy is lower than that of the spherical field and higher than that of the free ion.

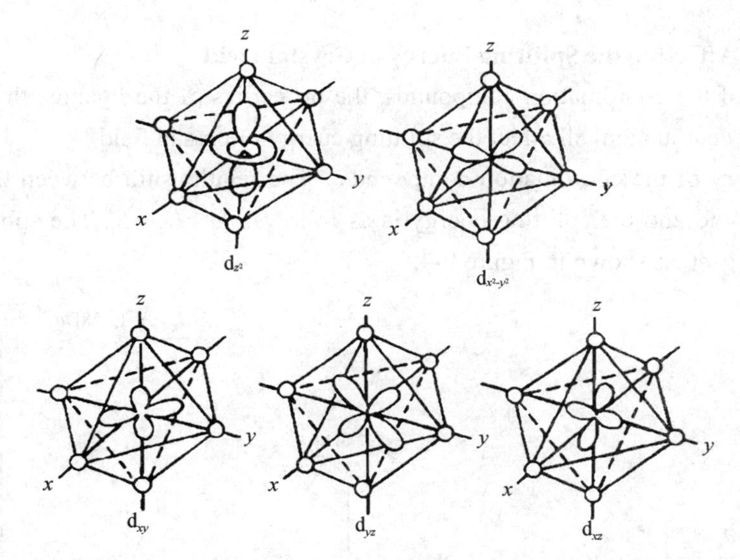

Figure 9-2 The relative space relationship between the ligands and the *d* orbitals of the central ion in the octahedral field

As a result, the energy levels of the d orbital of the central atom or ion in the octahedral coordination compounds are divided into two groups: one is the high-energy double degenerate orbital which is designated e_g (d_γ) orbital, the other is the low-energy d_{xy}, d_{xz} and d_{yz} triple degenerate orbital, which is called the t_{2g} (d_ε) orbital. The splitting of orbital energies is called the crystal field effect, and the energy difference between e_g and t_{2g} sets of orbitals is the crystal field splitting energy, designated as E_s, as shown in Figure 9-3. In the figure, E_0 is the energy of d orbital of free mental ion, and E_1 is the energy

of *d* orbital of metal ion in spherical field. The splitting of five *d* orbitals into two energy levels does not change the total energy. Assuming that E_s = 0 Dq for the spherical field and Δ_O = 10 Dq for the octahedral field (八面体场), Δ_O is equivalent to the energy required for an electron to be excited from t_{2g} orbital to e_g orbital.

$$\Delta_O = E(e_g) - E(t_{2g}) = 10\ \text{Dq} \quad (9\text{-}2)$$

$$2\,E(e_g) + 3\,E(t_{2g}) = 0 \quad (9\text{-}3)$$

The results that are solved by equations (9-2) and (9-3) are

$$E(e_g) = +\frac{3}{5}\Delta_o = +6\text{Dq}$$

$$E(t_{2g}) = -\frac{2}{5}\Delta_o = -4\text{Dq}$$

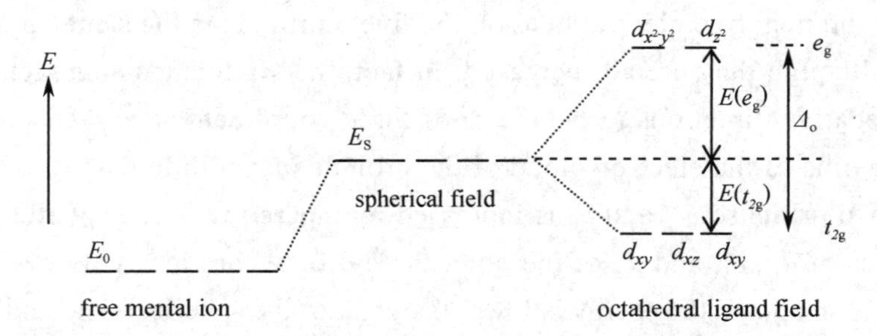

Figure 9-3 Splitting of *d* orbital energies by an octahedral field of ligands

3.2.3 Factors Affecting the Splitting Energy of Crystal Field

The geometry of the coordination compounds, the properties of the ligands, the oxidation number and the radius of the central atom all affect the splitting energy of crystal field.

(1) The geometry of the coordination compounds The relationship between the geometry of the coordination compound and the splitting energy is as follows: $\Delta_s > \Delta_o > \Delta_t$. The splitting energy of the square field is the largest, as shown in Figure 9-4.

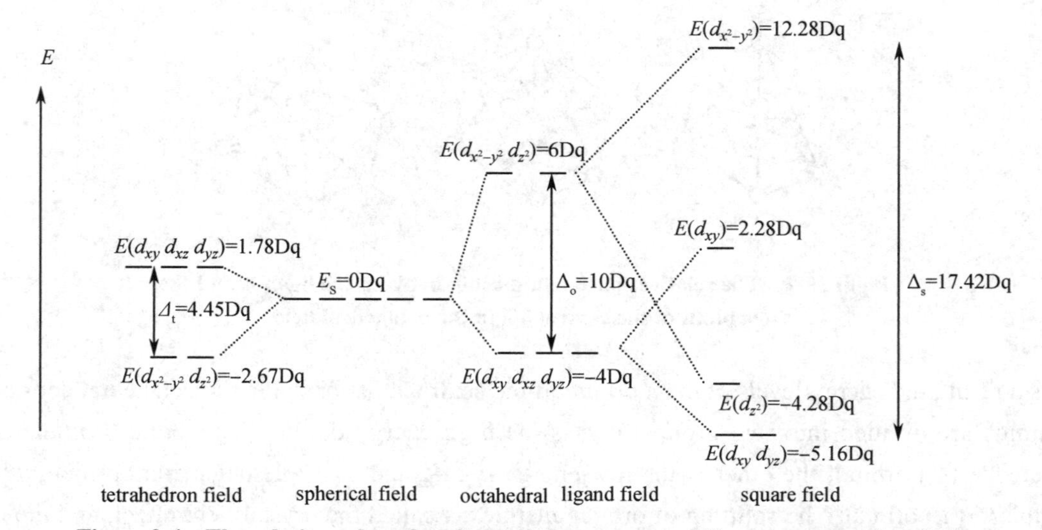

Figure 9-4 The relative value of orbital splitting energy in different coordination fields

(2) Field strength of ligands For a given central atom, the splitting energy depends on the field strength of the ligands. The larger the field strength is, the larger the splitting energy is. Strong-field ligands lead to a larger splitting energy, weak-field ligands lead to a smaller splitting energy. According to the spectral experiments of octahedral coordination compounds, the order of field strength of ligands from weak to strong is as follows:

$$I^- < Br^- < Cl^- < SCN^- < F^- \approx \text{urea} < S_2O_3^{2-} < OH^- \approx ONO^- < C_2O_4^{2-} < H_2O < NCS^- \approx$$
$$\text{EDTA} < \text{py} \approx NH_3 < \text{en} < \text{bpy} < \text{phen} < SO_3^{2-} < NO_2^- < CN^- < CO$$

This sequence is called spectro-chemical series. For example, F^- is a weak-field ligand; CN^- is a strong-field ligand. The magnitude of splitting energy is directly related to the color and magnetism of the coordination compounds.

(3) Central metal

① oxidation value of central metal For coordination compounds with the same ligands, the splitting energy depends on the oxidation value of the central atom. The higher the oxidation value of the central metal is, the greater the splitting energy is. For example, the Δ_o of the following coordination compounds are respectively:

$[Co(H_2O)_6]^{2+}$ 111.3kJ/mol
$[Co(H_2O)_6]^{3+}$ 222.5kJ/mol
$[Fe(H_2O)_6]^{2+}$ 124.4kJ/mol
$[Fe(H_2O)_6]^{3+}$ 163.9kJ/mol

Obviously, for the same ligand, the Δ_o with high central ion charge is much larger. The higher the oxidation number of the central metal is, the greater the electrostatic attraction to the ligand is. The closer the distance between the central metal and the ligand is, the greater the repulsion between the *d*-electrons of the central metal and the ligand is, and so the greater the splitting energy is.

② Period of metal ions (number of electron shells) For ligands and the charge of metal ion are the same, the Δ_o of metal ions in the same group increases from top to bottom. The Δ_o of transition system elements increases in the following order:

First transition system < second transition system < third transition system

For example, the Δ_o of the following coordination compounds are respectively:

$[Co(NH_3)_6]^{3+}$ $\Delta_o = 275.1$ kJ/mol
$[Rh(NH_3)_6]^{3+}$ $\Delta_o = 405.4$ kJ/mol
$[Ir(NH_3)_6]^{3+}$ $\Delta_o = 478.4$ kJ/mol

3.2.4 Electron Configurations in octahedral Coordination Compounds

In octahedral coordination compounds, the *d* electron configuration for the central atom tends to reduce the energy of the system. For $d^{1\sim3}$ and $d^{8\sim10}$, according to the lowest energy principle and Hund's rule, there is only one arrangement for both strong and weak field ligands. For example, d^3, the single electron is arranged in the same spin direction on each t_{2g} (d_ε) orbitals. The distributions of *d* electrons of octahedral coordination compounds are shown in Table 9-7.

Table 9-7 CFSE of octahedral and tetrahedron field coordination compounds

d^n	octahedral field						Tetrahedron field		
	Weak field (high-spin coordination compound)			Strong field (low-spin coordination compound)					
	Electron configuration	Unpaired electron	CFSE/Dq	Electron configuration	Unpaired electron	CFSE/Dq	Electron configuration	Unpaired electron	CFSE/Dq
d^0	$t_{2g}^0 e_g^0$	0	0	$t_{2g}^0 e_g^0$	0	0	$e_g^0 t_{2g}^0$	0	0
d^1	$t_{2g}^1 e_g^0$	1	−4	$t_{2g}^1 e_g^0$	1	−4	$e_g^1 t_{2g}^0$	1	−2.67
d^2	$t_{2g}^2 e_g^0$	2	−8	$t_{2g}^2 e_g^0$	2	−8	$e_g^2 t_{2g}^0$	2	−5.34
d^3	$t_{2g}^3 e_g^0$	3	−12	$t_{2g}^3 e_g^0$	3	−12	$e_g^2 t_{2g}^1$	3	−3.56
d^4	$t_{2g}^3 e_g^1$	4	−6	$t_{2g}^4 e_g^0$	2	−16+P	$e_g^2 t_{2g}^2$	4	−1.78
d^5	$t_{2g}^3 e_g^2$	5	0	$t_{2g}^5 e_g^0$	1	−20+2P	$e_g^2 t_{2g}^3$	5	0
d^6	$t_{2g}^4 e_g^2$	4	−4	$t_{2g}^6 e_g^0$	0	−24+2P	$e_g^3 t_{2g}^3$	4	−2.67
d^7	$t_{2g}^5 e_g^2$	3	−8	$t_{2g}^6 e_g^1$	1	−18+P	$e_g^4 t_{2g}^3$	3	−5.34
d^8	$t_{2g}^6 e_g^2$	2	−12	$t_{2g}^6 e_g^2$	2	−12	$e_g^4 t_{2g}^4$	2	−3.56
d^9	$t_{2g}^6 e_g^3$	1	−6	$t_{2g}^6 e_g^3$	3	−6	$e_g^4 t_{2g}^5$	1	−1.78
d^{10}	$t_{2g}^6 e_g^4$	0	0	$t_{2g}^6 e_g^4$	0	0	$e_g^4 t_{2g}^6$	0	0

For $d^{4\sim7}$ configuration, d electrons can be arranged in two ways: one is according to the principle of the lowest energy, the d electrons are arranged in the d orbitals with the lowest energy; the other is according to the Hund's rule, the d electrons occupy the d-orbitals as much as possible and have the same spin direction, so the energy is the lowest. The electron arrangement depends on the relative size of the splitting energy (分裂能), Δ_o and the electron pairing energy (成对能), P. When an electron pairs up with another electron in an orbital, energy must be given to overcome the mutual repulsion between the electrons, this energy is called electron pairing energy, P.

If the ligand is a strong field, $\Delta_o > P$, the d-electron fill the lower energy orbitals t_{2g}, so the number of unpaired electrons in the coordination compound ion is less than that in the free ion. This distribution is called low-spin distribution, which forms low-spin coordination compounds (低自旋配合物).

If the ligand is a weak field, $\Delta_o < P$, the energy required for electron pairing is higher, and the d-electrons are filled with more d-orbitals (t_{2g} and e_g) as much as possible. The number of unpaired electrons in the coordination compound ion is the same as in the free ion. This distribution is called high-spin distribution, which forms high-spin coordination compounds (高自旋配合物). For example the d^4 distribution of Cr^{2+} in $[Cr(H_2O)_6]^{2+}$ and $[Cr(CN)_6]^{4-}$ is shown in Figure 9-5.

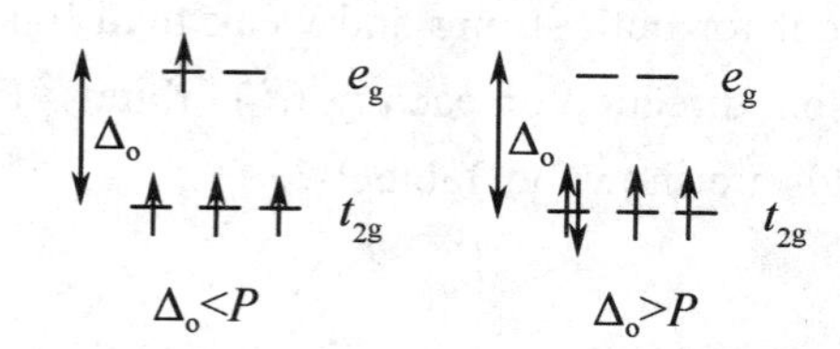

Figure 9-5 High-spin and low-spin coordination compound ions of Cr^{2+}

High-spin states have more unpaired electrons and larger magnetic moment, while low spin states have less unpaired electrons and smaller magnetic moment. In terms of stability, the high-spin state is equivalent to the outer-orbital coordination compound, with poor stability; the low spin state is equivalent to the inner-orbital coordination compound, with good stability. But there are differences between them. The high and low spin states compare the stabilization energy, while the inner and outer orbit types compare the energy differences between the inner- orbitals and outer-orbitals.

3.2.5 Crystal Field Stabilization Energy (CFSE)

When the *d*-electrons enter the split *d*-orbital (in the crystal field), the total energy of the system is lower than that of the degenerate *d* orbitals in a spherical field, which is called the crystal field stability energy (CFSE, 晶体场稳定化能). The lower the total energy of a system is, the more stable it is. Orbital splitting brings additional stabilization energy to the coordination compounds. The greater the absolute value of CFSE is, the more the energy of the system is reduced and the better the stability of the coordination compound is.

CFSE is related to the number of *d* electrons of central metal, the strength of crystal field formed by ligands, and the spatial configuration of coordination compound, as shown in Table 9-8. The CFSE of octahedral coordination compounds can be calculated as follows:

$$\mathrm{CFSE} = x\,E\,(t_{2g}) + y\,E\,(e_g) + (n_2 - n_1)\,P \qquad (9\text{-}4)$$

where x is the number of e_g electrons, y is the number of t_{2g} electrons, n_1 is the number of the paired electrons in degenerate d orbitals in a spherical field, n_2 is the number of the paired electrons in d orbitals in a coordination compound.

Example 9-1 Calculate the CFSE of Cr^{2+} in weak field and strong field ligands, respectively, compare their stabilities (Fig.9-5).

Solution Let us start with the electron configurations of Cr^{2+} ($3d^4$)

In strong field: $\mathrm{CFSE} = 4\,E\,(t_{2g}) + 0\,E\,(e_g) + (1 - 0)\,P$

$= 4\times(-4\ \mathrm{Dq}) + P$

$= -16\ \mathrm{Dq} + P < -6\ \mathrm{Dq}\ \ (\Delta_o > P)$

In weak field: $\mathrm{CFSE} = 3\,E\,(t_{2g}) + 1\,E\,(e_g) + (0 - 0)\,P$

$= 3\times(-4\ \mathrm{Dq}) + 1\times(6\ \mathrm{Dq})$

$= -6\ \mathrm{Dq}$

The results indicate that the low-spin coordination compound is more stable than the high-spin coordination compound because the greater absolute value of CFSE (the more negative CFSE).

3.2.6 Explaining the Colors of Transition Metal Coordination Compound

Crystal field theory can explain not only the magnetism and stability of coordination compounds, but also the color of transition metal coordination compounds.

Visible light is a mixture of light of various wavelengths. The color of substance under visible light is caused by the selective absorption of substance to mixed light. If the substance absorbs the red light in the visible light, it will be blue-green; if it absorbs the blue-green light, it will be red. That is to say, the color presented by the substance and the color of the absorption light selected by the substance are complementary colors. Table 9-8 shows the complementary relationship between the color of the substance and the color of the absorption light.

Table 9-8 Relationship between material color and absorption light color

Material color	Color absorbed	Absorption wavelength range (nm)
Yellow green	Purple	400-425
Yellow	Dark blue	425-450
Orange yellow	Blue	450-480
Orange	Greenish-blue	480-490
Red	Blue-green	490-500
Purplish red	Green	500-530
Purple	Yellow green	530-560
Dark blue	Orange	560-600
Greenish-blue	Orange	600-640
Blue-green	Red	640-750

The experimental results show that the splitting energy Δ of the coordination compounds is equivalent of the energy of visible light. The transition metal ions undergo energy level splitting under the crystal field. If the *d* electrons selectively absorb visible light with energy equal to the splitting energy Δ, the electrons are exited from the lower-energy level *d* orbital to the higher-energy level d orbital. This transition is called the *d-d* transition.

Most of the transition element coordination compounds with $d^{1\sim9}$ configuration have color, which is because the *d* orbital at the high energy level is unfilled. The energy absorbed by the d-electron transition is generally 10000~30000 cm^{-1}, including all visible light 14286~25000 cm^{-1}, so it will selectively absorb the light of specific color and present its complementary color. The frequency of *d–d* transition is generally in the near ultraviolet and visible region, so the transition metal coordination compounds generally have color. For example, the coordination ion of $[Ti(H_2O)_6]^{3+}$ is purplish red in aqueous solution, and the electron configuration of Ti^{3+} is $3d^1$. In the octahedral field, the electron is arranged in one of the three lower-energy t_{2g} orbitals. When the white light is used to irradiate $[Ti(H_2O)_6]^{3+}$ aqueous, the electron in t_{2g} orbitals absorbs the light with the wavelength of 492.7 nm (blue-green light) in the visible light, and is excited to one of the e_g orbitals. The energy of 492.7 nm photons is 242.79 kJ/mol (equivalent to the wavelength of absorption peak in Figure 9-6). If wave number (= 1/λ) is used, it is 20400 cm^{-1} (1 cm^{-1} = 11.96 J/mol), which is exactly equal to the splitting energy Δ of the coordination ion. For the d^{10} or d^0 coordination compound, t_{2g} orbitals are full of electrons or empty, *d-d* transition cannot produce, so their coordination compounds have no color.

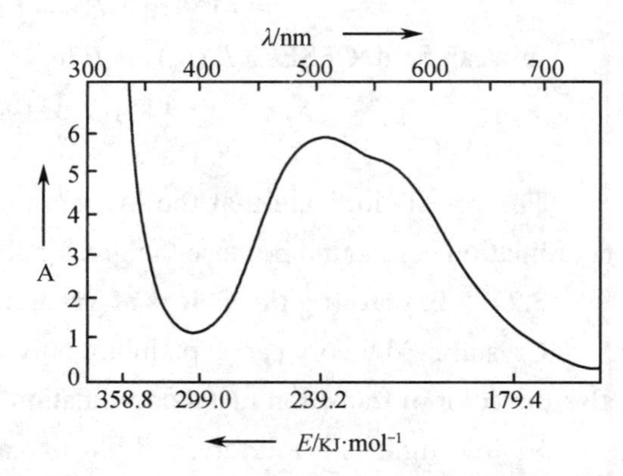

Figure 9-6 Absorption spectrum of $[Ti(H_2O)_6]^{3+}$

4. Coordination Equilibrium and Stabilization of Coordination Compounds

4.1 Equilibria Involving Coordination Ions

A coordination compound typically consists of a complex ion (配离子), a metal ion with its attached ligands, and counter ions, anions and cations as needed to produce a compound with no net charge. Metal ions exhibit marked preference for certain ligands. The substance $[Ag(NH_3)_2]Cl$ is a typical coordination compound. The brackets indicate the composition of the complex ion, in this case $[Ag(NH_3)_2]^+$, and the Cl^- counters ions is shown outside the brackets. In the solid state this compound consists of the large $[Ag(NH_3)_2]^+$ and the Cl^- counters ions, all packed together as efficiently as possible. When dissolved in water, the solid behaves like any ionic solid; the cations and anions are assumed to separate and move about independently.

Now let us consider the formation of $[Ag(NH_3)_2]^+$. When solutions containing Ag^+ ions and NH_3molecules are mixed, the following reactions take place:

$$Ag^+ + 2NH_3 \rightleftharpoons [Ag(NH_3)_2]^+$$

The equilibrium in reaction is shifted far to the right is a stable complex ion. At a certain temperature, the equilibrium constant for this reaction is

$$K_s^{\ominus} = \frac{[Ag(NH_3)_2{}^+]}{[Ag(NH_3)^+][NH_3]^2} \tag{9-5}$$

Where, $K_s^{\ominus}$ is called complex ion formation constants or stability constants (稳定常数). The higher the value of K_s is, the more stable the complex ion is.

For the reverse reaction, $K_{dis}^{\ominus}$ is used for the dissociation of $[Ag(NH_3)_2]^+$ into one Ag^+ and two NH_3. It is instability constant (不稳定常数). $K_{dis}^{\ominus}$ is the reciprocal of $K_s^{\ominus}$.

$$K_s^{\ominus} = \frac{1}{K_{dis}^{\ominus}} \tag{9-6}$$

In fact, complex ions are formed in solution by stepwise reaction, and the equilibrium constant can be written for each step. The successive equilibrium constants for the formation of $[Ag(NH_3)_2]^+$ is

$$Ag^+ + NH_3 \rightleftharpoons Ag(NH_3)^+ \qquad (1) \qquad K_1^{\ominus} = 2.1\times10^3$$

$$Ag(NH_3)^+ + NH_3 \rightleftharpoons Ag(NH_3)_2{}^+ \qquad (2) \qquad K_2^{\ominus} = 8.2\times10^3$$

where $K_1^{\ominus}$ and $K_2^{\ominus}$ are stepwise stability constants (also called formation constants) (逐级稳定常数).

When reaction (1) + (2), it gives

$$Ag^+ + 2NH_3 \rightleftharpoons [Ag(NH_3)_2]^+$$

$$K_s^{\ominus} = K_1^{\ominus} \cdot K_2^{\ominus} = 1.1\times10^7 \tag{9-7}$$

In a solution containing Ag^+ and NH_3, all the species NH_3, Ag^+, $Ag(NH_3)^+$, and $Ag(NH_3)_2{}^+$ exist at equilibrium. Calculating the concentrations of all these components can be complicated. However, when the total concentration of the ligands is much larger than the total concentration of the metal ion,and approximations can greatly simplify the problems.

For example, consider a solution prepared by mixing 100.0 ml of 2.0mol/L NH_3 with 100.0 ml of

1.0×10^{-3}mol/L $AgNO_3$. Before any reaction occurs, the mixed solution contains the major species Ag^+, NO_3^-, NH_3, and H_2O. What reaction or reactions will occur in this solution? From our discussions of acid-base chemistry, we know that one reaction is

$$NH_3\,(aq) + H_2O\,(l) \rightleftharpoons NH_4^+\,(aq) + OH^-\,(aq)$$

However, we are interested in the reaction between NH_3, and Ag^+ to form coordination compound ions, and since the position of the preceding equilibrium lies far to the left ($K_b^\ominus$, for NH_3 is 1.8×10^{-5}), we can neglect the amount of NH_3 used up in the reaction with water. Therefore, before any coordination compound ion formation, the concentrations in the mixed solution are

$$[Ag^+]_0 = \frac{(100.0\text{ml})(1.0\times10^{-3}\text{mol/L})}{(200.0\text{ml})} = 5.0\times10^{-4}\ \text{mol/L}$$

Total volume

$$[NH_3]_0 = \frac{(100.0\text{ml})(2.0\text{mol/L})}{(200.0\text{ml})} = 1.0\ \text{mol/L}$$

As mentioned already, the Ag^+ ion reacts with NH_3 in a stepwise way to form $AgNH_3^+$ and then $Ag(NH_3)_2^+$

$$Ag^+ + NH_3 \rightleftharpoons Ag(NH_3)^+ \qquad K_1^\ominus = 2.1\times10^3$$

$$Ag(NH_3)^+ + NH_3 \rightleftharpoons Ag(NH_3)_2^+ \qquad K_2^\ominus = 8.2\times10^3$$

Since both $K_1^\ominus$ and $K_2^\ominus$ are large, and since there is a large excress of NH_3, both reaction can be assumed to go essentially to complain. This is equivalent to writing the net reaction in the solution as follows.

$$Ag^+ + 2NH_3 \longrightarrow Ag(NH_3)_2^+$$

The relevant stoichiometric calculations are as follows.

	Ag^+	+	$2NH_3$	$\longrightarrow$	$Ag(NH_3)_2^+$
Before reaction	5.0×10^{-4}mol/L		1.0mol/L		0
After reaction	0		$1.0-2(5.0\times10^{-4})\approx1.0$mol/L		5.0×10^{-4}mol/L

Note that in this case we have used molarities when performing the stoichiometry calculations and assumed this reaction to be complete, using all the original Ag^+ to form $Ag(NH_3)_2^+$. In reality, *a* very small amount of the $Ag(NH_3)^+$ and Ag^+. However since the amount of $Ag(NH_3)_2^+$ dissociating will be so small, we can safely assume that $[Ag(NH_3)_2^+]$ is 5.0×10^{-4} mol/L at equilibrium. Also, we know that since so little NH_3 has been consumed, $[NH_3]$ is 1.0 mol/L at equilibrium. We can use these concentration to calculate $[Ag^+]$ and $[Ag(NH_3)^+]$ using the $K_1^\ominus$ and $K_2^\ominus$ expressions.

To calculate the equilibrium concentration of $Ag(NH_3)^+$, we use

$$K_2^\ominus = 8.2\times10^3 = \frac{[Ag(NH_3)_2^+]}{[Ag(NH_3)^+][NH_3]}$$

Since $[Ag(NH_3)_2^+]$ and $[NH_3]$ are known. Rearranging and solving for $[Ag(NH_3)^+]$ gives

$$[Ag(NH_3)] = \frac{[Ag(NH_3)^+]}{K_2[NH_3]} = \frac{5.0\times10^{-4}}{(8.2\times10^3)(1.0)} = 6.1\times10^{-8}\text{M}$$

Now the equilibrium concentration of Ag^+ can be calculated using $K_1^\ominus$

$$K_1^\ominus = 2.1\times10^3 = \frac{[Ag(NH_3)^+]}{[Ag^+][NH_3]} = \frac{6.1\times10^{-8}}{[Ag^+](1.0)}$$

$$[Ag^+] = \frac{6.1\times10^{-8}}{(2.1\times10^3)(1.0)} = 2.9\times10^{-11}\ \text{mol/L}$$

So far we have assumed that is the dominant silver-containing species in solution. Is this a valid assumption? The calculated concentrations are

$$[Ag(NH_3)_2^+] = 5.0\times10^{-4}\ \text{mol/L}$$
$$[Ag(NH_3)^+] = 6.1\times10^{-8}\ \text{mol/L}$$
$$[Ag^+] = 2.9\times10^{-11}\ \text{mol/L}$$

These values clearly support the conclusion that

$$[Ag(NH_3)_2^+] >> [Ag(NH_3)^+] >> [Ag^+]$$

Thus the assumption that $[Ag(NH_3)_2^+]$ is the dominant Ag^+-containing species is valid, and the calculated concentration are correct

This analysis shows that although coordination compound ion equilibria have many species present and look complicated, the calculation are actually quite straightforward, especially if the ligand is in large excess.

Example 9-2 Calculate the concentration of Ag^+, $Ag(S_2O_3)^-$, and $Ag(S_2O_3)_2^{3-}$ in a solution prepared by mixing 150.0mL of 1.00×10^{-3}mol/L $AgNO_3$ with 200.0mL of 5.00mol/L. The stepwise formation equilibria are

$$Ag^+ + S_2O_3^{2-} \rightleftharpoons Ag(S_2O_3)^- \qquad K_1^\ominus = 7.4\times10^8$$
$$Ag(S_2O_3)^- + S_2O_3^{2-} \rightleftharpoons Ag(S_2O_3)_2^{3-} \qquad K_2^\ominus = 3.9\times10^4$$

Solution The concentration of ligand and metal ion in the mixed solution before any reaction occur are

$$[Ag^+]_0 = \frac{(150\text{mL})(1.0\times10^{-3}\text{mol / L})}{(150.0\text{mL} + 200.0\text{mL})} = 4.29\times10^{-4}\ \text{mol/L}$$

$$[S_2O_3^{2-}]_0 = \frac{(200.0\ \text{ml})(5.0\ \text{mol/L})}{(150.0\ \text{ml} + 200.0\ \text{ml})} = 2.86\ \text{mol/L}$$

Since $[S_2O_3^{2-}] >> [Ag^+]_0$, and since $K_1^\ominus$ and $K_2^\ominus$ are large, both formation reactions can be assumed to go to completion, and the net reaction in the solution is as follows.

	Ag^+ +	$2S_2O_3^{2-}$ ⟶	$Ag(S_2O_3)_2^{3-}$
Before reaction:	4.29×10^{-4}mol/L	2.86 mol/L	0
After reaction:	≈ 0	$2.86-2(2.94\times10^{-4}) \approx 2.86$ mol/L	4.29×10^{-4} mol/L

Note that Ag^+ is limiting and that the amount of $S_2O_3^{2-}$ consumed is negligible. Also note that since all these species are in the same solution, the molarities can be used to do the stoichiometry problem.

Of course, the concentration of Ag is not zero at equilibrium, and there is some $Ag(S_2O_3)^-$ in the solution. To calculate the concentrations of these species, we must use the K_1 and K_2, expressions. We can calculate the concentration of

$Ag(S_2O_3)^-$ from $K_2^\ominus$:

$$3.94\times10^4 = K_2^{\ominus} = \frac{[Ag(S_2O_3)_2^{\ 3-}]}{[Ag(S_2O_3)^-][S_2O_3^{\ 2-}]} = \frac{4.29\times10^{-4}}{[Ag(S_2O_3)^-](2.86)}$$

$$[Ag(S_2O_3)^-]=3.8\times10^{-9}\text{mol/L}$$

We can calculate $[Ag^+]$ from $K_1^{\ominus}$:

$$7.4\times10^8 = K_1^{\ominus} = \frac{[Ag(S_2O_3)^-]}{[Ag^+][S_2O_3^{\ 2-}]} = \frac{3.8\times10^{-9}}{[Ag^+](2.86)}$$

$$[Ag^+]=1.8\times10^{-18}\text{mol/L}$$

These results show that

$$[Ag(S_2O_3)_2^{\ 3-}] >> [Ag(S_2O_3)^-] >> [Ag^+]$$

Thus the assumption is valid that essentially all the original Ag^+ is converted to $Ag(S_2O_3)_2^{\ 3-}$ at equilibrium.

4.2 Complex Ions and Solubility

Often ionic solids that are very nearly water-insoluble must be dissolved somehow in aqueous solutions. For example, when the various qualitative analysis groups are precipitated out, the precipitates must be redissolved to separate the ions within each group. Consider a solution of cations that contains Ag^+, Pb^{2+}, and $Hg_2^{\ 2+}$, among others. When dilute aqueous HCl is added to this solution, the Group I ions will form the insoluble chlorides AgCl, $PbCl_2$, and Hg_2Cl_2. Once this mixed precipitate is separated from the solution, it must be redissolved to identify the cations individually. How can this be done? We know that some solids are more soluble in acidic than in neutral solutions. What about chloride salts? For example, can AgCl be dissolved by using a strong acid? The answer is no, because Cl^- ions have virtually no affinity for H^+ ions in aqueous solution. The position of the dissolution equilibrium

$$AgCl(s) \rightleftharpoons Ag^+(aq) + Cl^-(aq)$$

is not affected by the presence of H^+

How can we pull the dissolution equilibrium to the right, even though Cl is an extremely weak base? The key is to lower the concentration of Ag^+ in solution by forming coordination compound ions. For example, Ag^+ reacts with excess NH_3, to form the stable coordination compound ion $Ag(NH_3)_2^{\ +}$. As a result, AgCl is quite soluble in concentrated ammonia solutions. The relevant reactions are

$$AgCl(s) \rightleftharpoons Ag^+ + Cl^- \qquad K_{sp}^{\ominus} = 1.6\times10^{-10}$$

$$Ag^+ + NH_3 \rightleftharpoons Ag(NH_3)^+ \qquad K_1^{\ominus}=2.1\times10^3$$

$$Ag(NH_3)^+ + NH_3 \rightleftharpoons Ag(NH_3)_2^{\ +} \qquad K_2^{\ominus}=8.2\times10^3$$

The Ag+ ion produced by dissolving solid AgCl combines with NH_3 to form $Ag(NH_3)_2^{\ +}$,which cause more AgCl to dissolve, until the point at which

$$[Ag^+][Cl^-] = K_{sp}^{\ominus} = 1.6\times10^{-10}$$

Here $[Ag^+]$ refers only to the Ag^+ ion that is present as a separete species in solution. It is not the total silver content of the solution,which is

$$[Ag]_{\text{total dissolved}} = [Ag^+] + [Ag(NH_3)^+] + [Ag(NH_3)_2^{\ +}]$$

For reasons discussed in the previous section, virtually all the Ag^+ from the dissolved AgCl ends up in the coordination compound ion $Ag(NH_3)_2^{\ +}$. Thus, we can represent the dissolving of solid AgCl in excess NH_3 by the equation

$$AgCl(s) + 2NH_3(aq) \rightleftharpoons Ag(NH_3)_2^+(aq) + Cl^-(aq)$$

Since this equation is the sum of the three stepwise reaction given above, the equilibrium constant for the reaction is the product of the constant for the three reactions. (Demonstrate this to yourself by multiplying together the three expressions for $K_{sp}^{\ominus}$, $K_1^{\ominus}$, and $K_2^{\ominus}$.) The equilibrium expression is

$$K^{\ominus} = \frac{[Ag(NH_3)_2^+][Cl^-]}{[NH_3]_2} = K_{sp}^{\ominus} \times K_1^{\ominus} \times K_2^{\ominus} = (1.6\times10^{-10})(2.1\times10^3)(8.2\times10^3) = 2.8\times10^3$$

Using this expression, we will now calculate the solubility of solid AgCl in a 10.0 mol/L NH_3 solution. If we let x be the solubility of solid AgCl in a 10.0mol/L NH_3 solution, we can then write following expressions for the equilibrium concentrations of the pertinent species:

$[Cl]^- = x$

$[Ag(NH_3)_2^+] = x$

— x mol/L of AgCl dissovle to produce x mol/L of Cl^- and x mol/L of $Ag(NH_3)_2^+$

$[NH_3] = 10.0-2x$ ← Formation of x mol/L $Ag(NH_3)_2^+$ requires $2x$ mol/L NH_3 since each complete ion contains two NH_3 ligands

Substituting these concentration into the equilibrium expressions gives

$$K^{\ominus} = 2.8\times10^{-3} = \frac{[Ag(NH_3)_2^+][Cl^-]}{[NH_3]^2} = \frac{(x)(x)}{(10.0-2x)^2} = \frac{x^2}{(10.0-2x)^2}$$

No approximations are necessary here. Taking the square root of both sides of the equation, it gives

$$\sqrt{2.8\times10^{-3}} = \frac{x}{10.0-2x}$$

$$x = 0.48\text{mol/L} = \text{solubility of AgCl(s) in 10.0mol/L } NH_3$$

Thus, the solubility of AgCl in 10.0mol/L NH_3 is much greater than its solubility in pure water, which is

$$\sqrt{K_{sp}^{\ominus}} = 1.3\times10^{-5}\text{mol/L}$$

In this chapter we have considered two strategies for dissolving a water-insoluble ionic solid. If the anion of the solid is a good base, the solubility is greatly increased

5. Application of Coordination Compounds

The ability of media ions to coordinate with and release ligands and to easily undergo oxidation and reduction makes them ideal for use in biologic systems. For example, metal ion coordination compounds are used in humans for the transport and storage of oxygen, as electron-transfer agents, as catalysts, and as drugs. Most of the first-row transition metals are essential for human health, as is summarized in Table 9-9. We will concentrate on iron's role in biologic systems, since several of its coordination coordination compounds have been studied extensively.

Iron palys a central role in almost all living cells. In mammals, the principal source of energy comes from the oxidation of carbohydrates, proteins, and fats.

Table 9-9 The first-row transition metals and their biologic significance

First-row transition meta	Biologic function(s)
Scandium	None known.
Titanium	None known.
Vanadium	None known in humans.
Chromium	Assists insulin in the control of blood sugar, may also be involved in the control of cholesterol.
Manganese	Necessary for a number of enzymatic reactions.
Iron	Component of hemoglobin and myoglobin; involved in the electron-transport chain.
Cobalt	Component of vitamin B_{12}, which is essential for the metabolism of carbohydrates, fats, and proteins.
Nickel	Component of the enzymes urease and hydrogenase.
Copper	Component of several enzymes; assists in iron storage; involved in the production of color pigments of hair, skin, and eyes.
Zinc	Component of insulin and many enzymes.

A protein is a large molecule assembled from amino acids, which have the general structure:

$$H_2N-\underset{H}{\overset{R}{C}}-COOH$$

in which R represents various groups.

Although oxygen is the oxidizing agent for these process, it does not react directly with these molecules. Instead, the electrons from the breakdown of these nutrients are passed along a coordination compound chain of molecules, called the respiratory chain, eventually reaching the O_2molecule. The principal electron-transfer molecules in the respiratory chain are iron-containing species called cytochromes, consisting of two main parts: an iron coordination compound called heme and a protein. The structure of the heme coordination compound is shown in Figure 9-7. Note that it contains an iron ion (it can be either Fe^{2+} or Fe^{3+}) coordinated to a rather complicated planar ligand called a porphyrin. As a class, porphyrins all contain the same central ring structure but have different substituent groups at the edges of the ring. The various porphyrin molecules act as tetradentate ligands for many metal ions, including iron, cobalt, and magnesium. In fact, *chlorophyll*, a substance essential to the process of photosynthesis, is a magnesium-porphyrin coordination compound of the type shown in Figure 9-8.

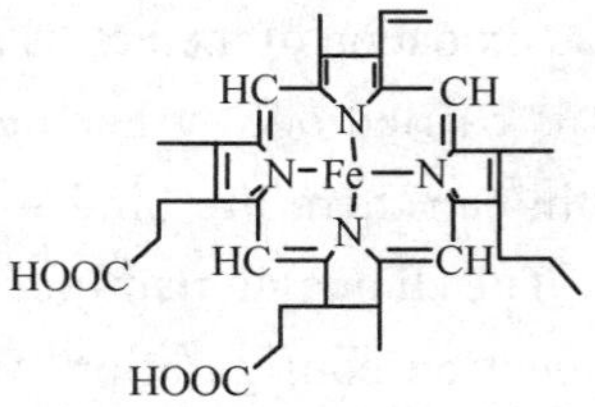

Figure 9-7 The heme coordination compound

(Fe^{2+} ion is coordinated to four nitrogen atoms of a planar porphyrin ligand)

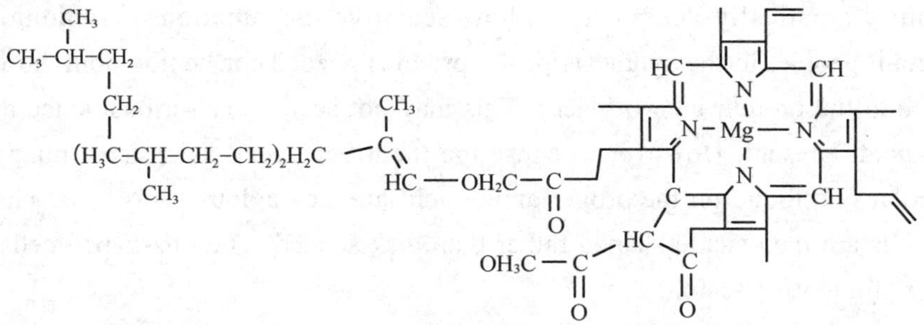

Figure 9-8 Chlorophyll is a porphyrin coordination compound of Fe^{2+}

(There are two similar forms of Chlorophyll, one of which is shown here)

In addition to participating in the transfer of electrons from nutrients to oxygen, iron also plays a principal role in the transport and storage of oxygen in mammalian blood and tissues. Oxygen is stored in a molecule called myoglobin, which consists of a heme coordination compound and a protein in a structure very similar to that of the cytochromes. In myoglobin, the Fe^{2+} ion is coordinated to four nitrogen atoms of the porphyrin ring and to a nitrogen atom of the protein chain. Since Fe^{2+} is normally six-coordinate, this leaves one position open for attachment of an O_2 molecule.

One especially interesting feature of myoglobin is that it involves an O_2 molecule attaching directly to Fe^{2+}. However, if gaseous O_2 is bubbled into an aqueous solution containing heme, the Fe^{2+} is immediately oxidized to Fe^{3+}. This oxidation of the Fe^{2+} in heme does not happen in myoglobin. This fact is of crucial importance because Fe^{3+} does not form a coordinate covalent bond with O_2, and myglobin would not function if the bound Fe^{2+} could be oxidized. Since the Fe^{2+} in the "bare" heme coordination compound can be oxidized, it must be the protein that somehow prevents the oxidation. How? Based on much research, the answer seems to be that the oxidation of Fe^{2+} to Fe^{3+} involves an oxygen bridge between two iron ions (the circles indicate the ligands):

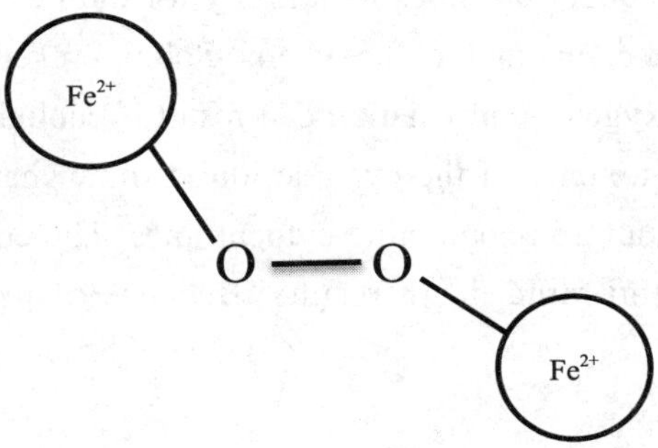

$$Fe^{2+}-O-O-Fe^{2+}$$

The bulky protein around the heme group in myoglobin prevents two molecules from getting close enough to from the oxygen bridge, and so oxidation of the Fe^{2+} is prevented.

The transport of O_2 in the blood is carried out by hemoglobin, a molecule consisting of four-myoglobin-like units. Each hemoglobin can therefore bind four O_2 molecules to form a bright red diamagnetic coordination compound. The diamagnetism occurs because oxygen is a strong-field ligand toward Fe^{2+}, which has a $3d^6$ electron configuration. When the oxygen molecule is release, water molecules occupy the sixth coordination position around each Fe^{2+}, giving a bluish paramagnetic coordination compound (H_2O is a weak-filed ligand toward Fe^{2+}) that gives venous blood its characteristic bluish tint.

Hemoglobin dramatically demonstrates how sensitive the function of a biomolecule is to its structure. In certain people, in the synthesis of the proteins needed for hemoglobin, an improper amino acid is inserted into the protein in two places. This may not seem very serious, since there are several hundred amino acids present. However, because the incorrect amino acids have nonpolar substituent instead of the polar one found on the proper amino acid, the hemoglobin drastically changes its shape. The red blood cells are then sickle-shaped rather than disk-shaped. The misshapen cells can aggregate, causing clogging of intense research.

Our knowledge of the working of hemoglobin allows us to understand the effects of high altitudes on humans. The reaction between hemoglobin and oxygen can be represented by the following equilibrium:

$$\underset{\text{Hemoglobin}}{Hb\,(aq)} + 4O_2\,(g) \rightleftharpoons \underset{\text{Oxyhemoglobin}}{Hb(O_2)_4\,(aq)}$$

At high altitudes, where the oxygen content of the air is lower, the position of his equilibrium will shift of the left, according to Le Châtelier's principle. Because less oxyhemoglobin is formed, fatigue, dizziness, and even a serious illness called high-altitude sickness can result. One way to combat this problem is to use supplemental oxygen, as most high-altitude mountain climbers do. However, this is impractical for people who live at high elevations. In fact, the human body adapts to the lower oxygen concentrations by making more hemoglobin, causing the equilibrium to shift back to the right. Some moving from Chicago to Boulder, Colorado (5300 feet) would notice the effects of the new altitude for a couple of weeks, but as the hemoglobin level increased, the effects would disappear. This change is called high-altitude acclimatization, which explains why athletes who want to compete at high elevations should practice there for several weeks prior to the event.

Our understanding of the biologic role of iron also allows us to explain the toxicities of substances such as carbon monoxide and the cyanide ion. Both CO and CN^- are very good ligands toward iron and so can interfere with the normal workings of the iron coordination compounds in the body. For example, carbon monoxide has about 200 times the affinity for the Fe^{2+} in hemoglobin as oxygen does. The resulting stable coordination compound, carboxyhemoglobin, prevents the normal uptake of O_2, thus depriving the body of needed oxygen. Asphyxiation can result if enough carbon monoxide is present in the air. The mechanism for the toxicity of the cyanide ion is somewhat different. Cyanide coordinates strongly to cytochrome oxidase, an iron-containing cytochromes. The coordinated cyanide thus prevents the electron-transfer process and rapid death results. Because of its behavior, cyanide is called a respiratory inhibitor.

Exercises

1. Which of the following compounds are double salts? What are chelates? What is simple salt? Which are coordination compounds?

(1) $[CoCl\,NH_3(en)_2]Cl_2$ (2) $[Cu(NH_3)_4]SO_4$

(3) $CuSO_4 \cdot 5H_2O$ (4) $(NH_3)_2SO_4 \cdot FeSO_4 \cdot 6H_2O$

(5) $(NH_4)_2[Fe(Br)_5(H_2O)]$ (6) $KAl(SO_4)_2 \cdot 12H_2O$

2. Name the following coordination compounds and point out the central ion and coordination number of the coordination compounds.

(1) $[Zn(NH_3)_4](OH)_2$ (2) $Na_3[Ag(S_2O_3)_2]$ (3) $[CoCl(NH_3)_5]Cl_2$

(4) $H_2[PtCl_6]$ (5) $[CoCl_2(NH_3)_3(H_2O)]Cl$

3. What is chelate and chelate effect? Which of the following compounds can be used as chelating agent?

$H_2N—NH_2$, NH_3, $(CH_3)_2N—NH_2$, $H_2NCH_2CH_2NH_2$

4. $AgNO_3$ solution can precipitate 1/3 of Cl^- from $Fe(NH_3)_4C1_3$ solution to AgCl, but only 1/4 of Cl^- can be precipitated in $Pt(NH_3)_3Cl_4$. Try to write the chemical formula of these two coordination compounds according to these facts.

5. According to the magnetic moment measured by the experiment, the following coordination compounds are determined to be the inner or outer orbit type, and the geometric configuration is determined.

(1) $[Fe(CN)_6]^{2-}$ $\mu = 0$ B.M.

(2) $[Fe(H_2O)_6]^{2+}$ $\mu = 5.30$ B.M.

(3) $[Co\,(NH_3)_6]^{3+}$ $\mu = 0$ B.M.

(4) $[Co\,(NH_3)_6]^{2+}$ $\mu = 3.88$ B.M.

(5) $[Ni\,(CN)_4]^{2-}$ $\mu = 0$ B.M.

(6) $[Ni\,(NH_3)_4]^{2+}$ $\mu = 3.0$ B.M.

6. According to the rules of soft and hard acid and alkali, compare the stability of the following coordination ions.

(1) F^-, SCN^- and Fe^{3+} form coordination ions

(2) F^-, I^- and Hg^{2+} form coordination ions

(3) SCN^-, ROH and Pd^{2+} form coordination ions

(4) CN^-, NH_3 and Cd^{2+} form coordination ions

7. Calculate the crystal field stabilization energy of the following coordination ions.

(1) $[Co(CN)_6]^{3-}$ (2) $[CoF_6]^{3-}$

8. The relationship between the splitting energy δ and the pairing energy P of the coordination compounds is known as follows: (1) $[Co(CN)_6]^{3-}$, $\Delta > p$ (2) $[Co(NH_3)_6]^{3-}$, $\Delta < p$. Please judge whether the coordination ion has high or low spin? The number of unpaired electrons for each center ion, and the number of electrons for the t_{2g} and e_g orbitals are indicated.

9. (1) It is known that $K_2[\,Ni\,(CN)_4]$ and $[Ni(CO)_4]$ are diamagnetic. Please judge the spatial configuration of these two coordination compounds and the type of central ion hybridization.

(2) Explain why most of the transition elements have colorless coordination ions, while most of the

Zn (Ⅱ) have colorless coordination ions.

10. The following statement is correct

(1) The coordination compound must have both internal and external boundaries.

(2) The metal ions A^{3+} and B^{2+} can form $[A(NH_3)_6]^{3+}$ and $[B(NH_3)_6]^{2+}$ espectively, and their stability constants are 4×10^5 and 2×10^{10} espectively. In the same concentration of $[A(NH_3)_6]^{3+}$ and $[B(NH_3)_6]^{2+}$ solution, the concentration relationship between A^{3+} and B^{2+} is $c(A^{3+}) > c(B^{2+})$.

11. According to the stability constant of the coordination compound, the direction of the following reactions is determined.

(1) $[HgCl_4]^{2-} + 4CN^- \rightleftharpoons [Hg(CN)_4]^{2-} + 4Cl^-$

(2) $[Cu(NH_3)_4]^{2+} + Cd^{2+} \rightleftharpoons [Cd(NH_3)_4]^{2+} + Cu^{2+}$

(3) $[FeF_6]^{3-} + 3C_2O_4^{2-} \rightleftharpoons [Fe(C_2O_4)_3]^{3-} + 6F^-$

12. To dissolve 0.1mol/L AgCl in 1 L, what is the initial concentration of ammonia water? The known conditions: $K_{sp}^{\ominus}(AgCl) = 1.77 \times 10^{-10}$, $K_S^{\ominus}([Ag(NH_3)_2]^+) = 1.12 \times 10^7$

13. The following data are known:

$Fe^{3+} + e^- = Fe^{2+}$ $\quad E^{\ominus} = 0.77V$

$Fe^{2+} + 6CN^- = [Fe(CN)_6]^{4-}$ $\quad K_S^{\ominus}([Fe(CN)_6]^{4-}) = 1.00 \times 10^{35}$

$Fe^{3+} + 6CN^- = [Fe(CN)_6]^{3-}$ $\quad K_S^{\ominus}([Fe(CN)_6]^{3-}) = 1.00 \times 10^{42}$

$O_2 + 2H_2O + 4e^- = 4OH^-$ $\quad E^{\ominus} = 0.40V$

(1) Find the $E^{\ominus}$ value of $Fe(CN)_6^{3-} + e^- = [Fe(CN)_6]^{4-}$

(2) Discuss the stability of $[Fe(CN)_6]^{4-}$ in air.

14. Ammonia gas is introduced into the mixed solution containing 0.02mol/L NH_4Cl and 0.1mol/L $[Cu(NH_3)_4]^{2+}$, so as to make the concentration of ammonia water reach 1.0mol/L. Please judge whether there is precipitation of $Cu(OH)_2$ by calculation? The known conditions: $K_{sp}^{\ominus}(Cu(OH)_2) = 2.2 \times 10^{-20}$, $K_S^{\ominus}\{[Cu(NH_3)_4]^{2+}\} = 2.09 \times 10^{13}$, $K_b^{\ominus}(NH_3) = 1.74 \times 10^{-5}$

15. Can the following disproportionation reactions occur?

(1) $2Cu^+ \rightleftharpoons Cu + Cu^{2+}$

(2) $2[Cu(NH_3)_2]^+ \rightleftharpoons Cu + [Cu(NH_3)_4]^{2+}$ The known conditions:

$E^{\ominus}(Cu^{2+}/Cu^+) = 0.159V$,

$E^{\ominus}(Cu^+/Cu) = 0.521V$,

$K_S^{\ominus}([Cu(NH_3)_4]^{2+}) = 2.09 \times 10^{13}$,

$K_S^{\ominus}([Cu(NH_3)_2]^+) = 7.20 \times 10^{10}$

16. Mix 0.100 mol/L $AgNO_3$ solution and 6.00 mol/L ammonia water by equal volume, and ask for the concentration of each component in the mixed solution. $K_{dis}^{\ominus}\{[Ag(NH_3)_2]^+\} = 8.91 \times 10^{-8}$

17. The known conditions:

$$Zn^{2+} + 2e = Zn,\ E^{\ominus} = -0.763V$$

$$Zn(CN)_4^{2-} + 2e = 4CN^- + Zn,\ E^{\ominus} = -1.23V$$

Calculate the stability constant of the reaction product:

$Zn_2 + 4CN^- = Zn(CN)_4^{2-}$ (without considering the successive stability constant).

18. 100 ml of 0.40 mol/L ammonia solution was mixed with 0.20 mol/L HCl solution of the same volume at 25°C, and then $[Cu(NH_3)_4]Cl_2$ was added to make the final concentration of 0.20 mol/L $[Cu(NH_3)_4]Cl_2$ solution. Is there any precipitation of $Cu(OH)_2$ in the mixture?

The known conditions: $K_{sp}^{\ominus}(Cu(OH)_2) = 2.2 \times 10^{-20}$, $K_{稳}^{\ominus}\{[Cu(NH_3)_4]^{2+}\} = 2.09 \times 10^{13}$, $K_b^{\ominus}(NH_3) = 1.74 \times 10^{-5}$

目标检测

1. 下列化合物中哪些是复盐？哪些是螯合物？哪些是简单盐？哪些是配合物？

（1）$[CoClNH_3(en)_2]Cl_2$　　（2）$[Cu(NH_3)_4]SO_4$

（3）$CuSO_4 \cdot 5H_2O$　　（4）$(NH_3)_2SO_4 \cdot FeSO_4 \cdot 6H_2O$

（5）$(NH_4)_2[Fe(Br)_5(H_2O)]$　　（6）$KAl(SO_4)_2 \cdot 12H_2O$

2. 命名下列配合物，并指出配合物的中心离子、配位数。

（1）$[Zn(NH_3)_4](OH)_2$　　（2）$Na_3[Ag(S_2O_3)_2]$　　（3）$[CoCl(NH_3)_5]Cl_2$

（4）$H_2[PtCl_6]$　　（5）$[CoCl_2(NH_3)_3(H_2O)]Cl$

3. 何为螯合物和螯合效应？请选择下列化合物中哪些可以作为螯合剂？

$H_2N–NH_2$,　NH_3,　$(CH_3)_2N–NH_2$,　$H_2NCH_2CH_2NH_2$

4. $AgNO_3$ 溶液能从 $Fe(NH_3)_4C1_3$ 溶液中将 1/3 的 Cl^- 沉淀为 AgCl，但在 $Pt(NH_3)_3Cl_4$ 中仅能沉淀 1/4 的氯。试根据这些事实写出这两种配合物的化学式。

5. 根据实验测得的磁矩确定下列配合物是内轨型还是外轨型，并判断几何构型。

（1）$[Fe(CN)_6]^{2-}$　　$\mu = 0$ B.M.

（2）$[Fe(H_2O)_6]^{2+}$　　$\mu = 5.30$ B.M.

（3）$[Co(NH_3)_6]^{3+}$　　$\mu = 0$ B.M.

（4）$[Co(NH_3)_6]^{2+}$　　$\mu = 3.88$ B.M.

（5）$[Ni(CN)_4]^{2-}$　　$\mu = 0$ B.M.

（6）$[Ni(NH_3)_4]^{2+}$　　$\mu = 3.0$ B.M.

6. 根据软硬酸碱规则，比较下列配离子之间的稳定性相对大小。

（1）F^-、SCN 与 Fe^{3+} 形成配离子

（2）F^-、I^- 与 Hg^{2+} 形成配离子

（3）SCN^-、ROH 与 Pd^{2+} 形成配离子

（4）CN^-、NH_3 与 Cd^{2+} 形成配离子

7. 计算下列配离子的晶体场稳定化能。

（1）$[Co(CN)_6]^{3-}$

（2）$[CoF_6]^{3-}$

8. 已知配合物的分裂能 Δ 和成对能 p 之间的关系的关系如下：

（1）$[Co(CN)_6]^{3-}$，$\Delta > p$　　（2）$[Co(NH_3)_6]^{3-}$，$\Delta < p$

请判断配离子是高自旋还是低自旋？各中心离子的未成对电子数，并指出 t_{2g} 和 e_g 轨道的电子数。

9.（1）已知 $K_2[Ni(CN)_4]$ 与 $[Ni(CO)_4]$ 均呈反磁性，请判断这两种配合物的空间构型，以及中心离子的杂化类型。

（2）解释为何大多数过渡元素的配离子有色，而大多数 Zn(Ⅱ) 的配离子无色。

10. 下列说法正确的是

（1）配合物必须同时有内界和外界。

（2）金属离子 A^{3+}、B^{2+} 可分别形成 $[A(NH_3)_6]^{3+}$ 和 $[B(NH_3)_6]^{2+}$，它们的稳定常数依次为 4×10^5

和 2×10^{10}，则相同浓度的 $[A(NH_3)_6]^{3+}$ 和 $[B(NH_3)_6]^{2+}$ 溶液中，A^{3+} 和 B^{2+} 的浓度关系是 $c(A^{3+}) > c(B^{2+})$。

(3) 某配离子的逐级稳定常数分别为 $K_1^\ominus$、$K_2^\ominus$、$K_3^\ominus$、$K_4^\ominus$，则该配离子的不稳定常数 $K_{is}^\ominus = K_1^\ominus \cdot K_2^\ominus \cdot K_3^\ominus \cdot K_4^\ominus$。

(4) 只有金属才能提供空轨道与配体形成配合物。

11. 根据配合物的稳定常数，判断下列反应进行的方向。

（1）$[HgCl_4]^{2-} + 4CN^- \rightleftharpoons [Hg(CN)_4]^{2-} + 4Cl^-$

（2）$[Cu(NH_3)_4]^{2+} + Cd^{2+} \rightleftharpoons [Cd(NH_3)_4]^{2+} + Cu^{2+}$

（3）$[FeF_6]^{3-} + 3C_2O_4{}^{2-} \rightleftharpoons [Fe(C_2O_4)_3]^{3-} + 6F^-$

12. 欲将 0.1mol/L 的 AgCl 溶解在 1L 中，求氨水的初始浓度至少是多少？已知 $K_{sp}^\ominus(AgCl) = 1.77 \times 10^{-10}$, $K_{稳}^\ominus([Ag(NH_3)_2]^+) = 1.12 \times 10^7$

13. 已知下列数据：

$Fe^{3+} + e^- = Fe^{2+}$ $E^\ominus = 0.77V$

$Fe^{2+} + 6CN^- = [Fe(CN)_6]^{4-}$ $K_S^\ominus([Fe(CN)_6]^{4-}) = 1.00 \times 10^{35}$

$Fe^{3+} + 6CN^- = [Fe(CN)_6]^{3-}$ $K_S^\ominus([Fe(CN)_6]^{3-}) = 1.00 \times 10^{42}$

$O_2 + 2H_2O + 4e^- = 4OH^-$ $E^\ominus = 0.40V$

（1）求 $Fe(CN)_6{}^{3-} + e^- = [Fe(CN)_6]^{4-}$ 的 $E^\ominus$ 值

（2）讨论 $[Fe(CN)_6]^{4-}$ 在空气中的稳定性。

14. 往含 0.02mol/L NH_4Cl 和 0.1mol/L $[Cu(NH_3)_4]^{2+}$ 的混合溶液中通入氨气，使氨水的浓度达到 1.0mol/L，通过计算说明有无 $Cu(OH)_2$ 沉淀生成？

$K_{sp}^\ominus(Cu(OH)_2) = 2.2 \times 10^{-20}$, $K_b^\ominus(NH_3 \cdot H_2O) = 1.74 \times 10^{-5}$, $K_S^\ominus([Cu(NH_3)_4]^{2+}\} = 2.1 \times 10^{13}$

15. 判断下列歧化反应能否发生？

（1）$2Cu^+ \rightleftharpoons Cu + Cu^{2+}$

（2）$2[Cu(NH_3)_2]^+ \rightleftharpoons Cu + [Cu(NH_3)_4]^{2+}$ 已知

$E^\ominus(Cu^{2+} / Cu^+) = 0.159V$,

$E^\ominus(Cu^+ / Cu) = 0.521V$,

$K_S^\ominus([Cu(NH_3)_4]^{2+}) = 2.09 \times 10^{13}$,

$K_S^\ominus([Cu(NH_3)_2]^+) = 7.20 \times 10^{10}$

16. 将 0.100mol/L $AgNO_3$ 溶液与 6.00mol/L 氨水等体积混合，求混合溶液中各组份的浓度。已知：$K_{稳}^\ominus\{[Ag(NH_3)_2]^+\} = 8.91 \times 10^{-8}$

17. 已知：

$$Zn^{2+} + 2e = Zn,\ E^\ominus = -0.763V$$

$$Zn(CN)_4^{2-} + 2e = 4CN^- + Zn,\ E^\ominus = -1.23V$$

求：$Zn_2 + 4CN^- = Zn(CN)_4^{2-}$ 反应产物的稳定常数（不考虑逐次稳定常数）。

18. 25℃条件下，将 0.40mol/L 氨水溶液 100ml 与等体积的 0.20mol/L HCl 溶液混合，再加入 $[Cu(NH_3)_4]Cl_2$，使其终浓度为 0.20mol/L $[Cu(NH_3)_4]Cl_2$ 溶液。试问该混合液中有无 $Cu(OH)_2$ 沉淀生成？已知：$K_{sp}^\ominus(Cu(OH)_2) = 2.2 \times 10^{-20}$，$K_{稳}^\ominus\{[Cu(NH_3)_4]^{2+}\} = 2.09 \times 10^{13}$，$K_b^\ominus(NH_3) = 1.74 \times 10^{-5}$

Chapter 10　Main Group Elements and Related Mineral Drugs
第十章　主族元素及相关矿物药

学习目标

知识要求

1. **掌握**　碱金属、碱土金属、Pb、B、Al、Si、As、S、F元素及其重要化合物的基本性质。

2. **熟悉**　碱金属、碱土金属、Pb、B、Al、Si、As、S、F元素的性质与电子层结构的关系。

3. **了解**　碱金属、碱土金属、Pb、B、Al、Si、As、S、F元素及其矿物药在医药中的应用。

能力要求

学会根据元素原子结构的核外电子排布特征分析主族元素的性质，培养提高分析问题的能力。

1. General Properties

According to the arrangement of configuration of extra-nuclear electron, the elements in the periodic table are divided into main group elements (主族元素) and subgroup elements (副族元素). Main group elements fill their *s* and *p* valence orbitals to complete their electron configurations. The metals in the first two groups of the periodic table are characterized as *s* block elements because of their outer shells having one or two electrons in an *s* orbital. The *s* block elements include the elements in group ⅠA and ⅡA of the periodic table. The *p* block elements include the elements in groups Ⅲ - ⅧA of the periodic table. The configuration of extra-nuclear electron of main group elements presents periodic changes. The properties related to structure also show periodic changes, for example atomic radius, ionization energies, electronegativity, etc. Table 10-1 shows the changing law of main group elemental properties.

Table 10-1 The changing law of main group elemental properties

Atomic Radius is Increasing, Ionization Energies and Electronegativity are Decreasing (down the groups ↓)

ⅠA	ⅡA	ⅢA	ⅣA	ⅤA	ⅥA	ⅦA	ⅧA
H							He
Li	Be	B	C	N	O	F	Ne
Na	Mg	Al	Si	P	S	Cl	Ar
K	Ca	Ga	Ge	As	Se	Br	Kr
Rb	Sr	In	Sn	Sb	Te	I	Xe
Cs	Ba	Tl	Pb	Bi	Po	At	Rn
Fr	Ra						

Atomic Radius is Increasing, Ionization Energies and Electronegativity are Decreasing (→)

2. Alkali Metals and Their Compounds

The elements in group ⅠA, the alkali metals (碱金属), are lithium, sodium, potassium, rubidium, caesium and francium. All the elements are metals with valence electron configuration ns^1. Because their hydroxides are soluble in water and strongly alkaline, the elements in group ⅠA are called alkali metals. They conduct electricity and heat, are soft, and have low melting points that decrease down the group. Properties of alkali metals are described in Table 10-2.

Table 10-2 Properties of alkali metals

Properties	Li	Na	K	Rb	Cs
Valence electron configuration	$2s^1$	$3s^1$	$4s^1$	$5s^1$	$6s^1$
Atomic radius (Å)	1.52	1.86	2.27	2.48	2.65
Ionic radius M^+ (Å)	0.68	0.95	1.33	1.48	1.69
First ionization energies (kJ/mol)	521	499	421	405	371
Electronegativity	0.98	0.93	0.82	0.82	0.79
$E^\ominus$ (V)	−3.045	−2.7109	−2.923	−2.925	−2.923
Oxidation number	+1	+1	+1	+1	+1

2.1 Hydrides

Alkali metals form hydrides when they react with hydrogen that (at least formally) contain H^- at high temperature. Because they have the characteristics of ionic compounds (white solids of low volatility), these hydrides are known as the salt like hydrides.

$$2M + H_2 = 2MH \quad (M = \text{alkali metals})$$

The most important characteristic of the ionic hydrides is the very strong reducibility of the hydride ion. Ionic hydrides can remove protons from water to release hydrogen, as illustrated by the following equations.

$$LiH + H_2O = LiOH + H_2\uparrow$$

$$NaH + H_2O = NaOH + H_2\uparrow$$

Ionic hydrides such as NaH can be used as drying agents (desiccant) because they can remove hydrogen from the traces of water present in many solvents.

2.2 Oxides and Hydroxides

Alkali metals all react with vigorous oxygen. Only Li reacts directly with excess oxygen to give "normal" oxides. Sodium reacts with oxygen to give the peroxide. The other alkali metals form the superoxides.

$$4Li + O_2 = 2Li_2O$$

$$2Na + O_2 = Na_2O_2$$

$$K + O_2 = KO_2$$

When the oxygen compounds of alkali metals react with water, strong basic solutions are produced regardless of whether an oxide, peroxide, or superoxide is involved.

$$Li_2O + H_2O = 2LiOH$$

$$Na_2O_2 + 2H_2O = 2NaOH + H_2O_2$$

$$2KO_2 + 2H_2O = 2KOH + H_2O_2 + O_2\uparrow$$

2.3 Halides

The mineral medicine of sodium chloride (NaCl) is called Halitum (大青盐), which is an important salt to maintain the body fluid balance. When the human body lacks sodium chloride, it will cause nausea, vomiting, failure and muscle spasm. Therefore, sodium chloride is often made into physiological saline (0.90%) to supply body fluids for patients with excessive bleeding or water loss. Potassium chloride (KCl) is used for arrhythmia caused by hypokalemia and digitalis poisoning.

2.4 Carbonates

In addition to lithium carbonate, the rest of alkali carbonate is easily soluble in water. $NaHCO_3$ is commonly known as baking soda (小苏打). Its aqueous solution is weakly alkaline, and it is often used to treat excessive gastric acid and acidosis. Because it reacts with potassium hydrogen tartrate in solution to produce CO_2, their mixture is the main component of fermentation powder (发酵粉).

2.5 Sulfates

Alkali metal sulfates are soluble in water, of which sodium sulfate is the most important. $Na_2SO_4 \cdot 10H_2O$ is called mirabilite (Natric Sulfas, 芒硝), which is easily weathered and dehydrated into anhydrous sodium sulfate in the air. Anhydrous sodium sulfate is called Natrii Sulfas Exsiccatus (玄明粉)

in traditional Chinese medicine. It is white powder with deliquescence. It is used as a desiccant for some organic substances in the synthesis of organic drugs. In medicine, mirabilite and Natrii Sulfas Exsiccatus are used as laxatives.

3. Alkaline Earth Metals and Their Compounds

The elements in group ⅡA, the alkaline earth metals (碱土金属), are beryllium, magnesium, calcium, strontium, barium, and radium. All the elements are metals with valence electron configuration ns^2. All the elements are silvery white metals and are denser, harder, and less reactive than the alkali metals, but they are still more reactive than many typical metals. Properties of alkaline earth metals are described in Table 10-3.

Table 10-3 Properties of alkaline earth metals

Properties	Be	Mg	Ca	Sr	Ba
Valence electron configuration	$2s^2$	$3s^2$	$4s^2$	$5s^2$	$6s^2$
Atomic radius (Å)	1.113	1.60	1.973	2.151	2.173
Ionic radius M^{2+} (Å)	0.31	0.65	0.99	1.13	1.35
First ionization energies (kJ/mol)	905	742	593	552	564
Second ionization energies (kJ/mol)	1768	1460	1152	1070	971
Electronegativity	1.57	1.31	1.00	0.95	0.89
$E^{\ominus}$(V)	−1.85	−2.357	−2.76	−2.89	−2.90
Oxidation number	+2	+2	+2	+2	+2

3.1 Hydrides

Calcium, strontium and barium form hydrides with hydrogen that (at least formally) contain H^- at high temperature. Because they have the characteristics of ionic compounds (white solids of low volatility), these hydrides are known as the salt like hydrides.

$$M + H_2 = MH_2 \quad (M = Ca, Sr \text{ and } Ba)$$

The most important characteristic of the ionic hydrides is the very strong reducibility of the hydride ion. Ionic hydrides can remove protons from water to release hydrogen, as illustrated by the following equations.

$$CaH_2 + 2H_2O = Ca(OH)_2 + 2H_2\uparrow$$

3.2 Oxides and Hydroxides

In reactions with oxygen, the lighter members of alkaline earth metals give normal oxides, but barium and radium give peroxides.

$$2Ca + O_2 = 2CaO$$
$$Ba + O_2 = BaO_2$$

The oxides of alkaline earth metals are ionic, so they react with water to produce the hydroxides. However, beryllium hydroxide is quite different, and it exhibits amphoteric behavior. Although other hydroxides of alkaline earth metals are strong bases, the utility of the hydroxides is limited somewhat because they are only slightly soluble (only about 0.12g of $Ca(OH)_2$ dissolves in 100g of water).

3.3 Halides

Calcium chloride ($CaCl_2$) can be used in the treatment of calcium deficiency, as well as in antiallergic and anti-inflammatory drugs. Barium chloride ($BaCl_2$) is an important soluble barium salt, which can be used in medicine and rodenticide. However, barium chloride is highly toxic, so be sure to use it carefully and safely.

3.4 Carbonates

Calcium carbonate ($CaCO_3$) is the most important oxo compound of the element. It occurs widely in nature as limestone, chalk, marble, dolomite, and as coral, pearl, and seashell. Calcium carbonate is an important biomineral and a major constituent of bones and shells. It also has widespread use in construction and road building, and is used as an antacid, an abrasive in toothpaste, in chewing gum, and as a health supplement where it is taken to maintain bone density.

3.5 Sulfates

The dihydrate of calcium sulfate, $CaSO_4 \cdot 2H_2O$, is commonly known as raw gypsum (生石膏). It can dehydrate to form $2CaSO_4 \cdot H_2O$, which is commonly named as mature gypsum (calcined gypsum, burnt gypsum, 熟石膏) by heating. The effect of raw gypsum is to clear away heat and reduce fire. $2CaSO_4 \cdot H_2O$ has the effect of antipyretic and antiphlogistic. It is one of the main drugs of "White tiger soup" for the treatment of epidemic encephalitis B.

Magnesium is also found as Epsom salts (泻盐), $MgSO_4 \cdot 7H_2O$, which is used in solution for medicinal. $BaSO_4$, also known as barite (重晶石), can prevent X-rays from passing through and is non-toxic. It is not absorbed in the gastrointestinal tract, so it is often used for gastrointestinal radiography.

4. Boron and Its Compounds

Boron (硼) is the ⅢA element in the periodic table. The valence electron configuration $2s^22p^1$ suggests that the +3 oxidation state is dominant in the compounds of boron (B). The main use of B is in borosilicate glasses.

4.1 Borane

The binary hydrogen compounds of B are called boranes (硼烷). The simplest member of the series, diborane, B_2H_6 (乙硼烷), is electron deficient. All the boron hydrides are flammable, and several of the lighter ones, including diborane, react spontaneously with air, often with explosive violence and a green lash. The boron hydrides will all burn readily to produce B_2O_3 and water.

$$B_2H_6 + 3O_2 = B_2O_3 + 3H_2O$$

Boranes are readily hydrolysed by water and release a lot of heat to produce boric acid and hydrogen.

$$B_2H_6 + 6H_2O = 2H_3BO_3 + 6H_2\uparrow$$

Borane has strong reducibility and can be oxidized by oxidant such as halogen.

$$B_2H_6 + 6Cl_2 = 2BCl_3 + 6HCl$$

4.2 Boric Acid

Boric acid (硼酸), H_3BO_3, is a white, glossy, scaly crystal. It's slightly soluble in water and greasy. So it can be used as a lubricant.

H_3BO_3 is a weak acid. Its acidity is not due to the ionization, but due to the fact that boric acid is an electron deficient compound, in which the boron atom (with an empty orbit) reacts with the OH^- ion (having the lone electron pair) from the water molecule to release H^+ ion.

$$H_3BO_3 + H_2O = [(OH)_3B \leftarrow OH]^- + H^+$$

Due to weak acidity, H_3BO_3 has been used as an eye wash and a mild antiseptic.

4.3 Borax

The most common minerals containing boron are the tetraborates of sodium or calcium. Borax (硼砂), $Na_2B_4O_7 \cdot 10H_2O$, is the most important source of boron. It is a colorless and translucent crystal, which is easily weathered in dry air. It is not easy to dissolve in water at ordinary temperature, but it is easier to dissolve in boiling water. It is easy to hydrolyze in water. Borax has many domestic uses, for example as a water softener, cleaner, and mild pesticide.

Borax, also known as "basin" sand in traditional Chinese medicine, is used to treat laryngitis, stomatitis, and otitis media. The ingredient of Bingpeng San (冰硼散) and Compound Borax Mouthwash (复方硼砂含嗽剂) is borax.

5. Aluminum and Its Compounds

Aluminum (铝) is the Ⅲ A element in the periodic table. The valence electron configuration $3s^23p^1$ suggests the +3 oxidation state is dominant in the compounds of aluminum (Al). Aluminum is the third most abundant metal element in the earth's crust after oxygen and silicon.

5.1 Oxides

Al_2O_3 is a white amorphous powder. The most stable form of Al_2O_3, α-alumina, is a very hard and refractory material. In its mineral form, it is known as corundum (刚玉) and sapphire or ruby (蓝宝石或红宝石), depending on the metal ion impurities. The blue of sapphire arises from a charge-transfer transition from Fe^{2+} to Ti^{4+} ion impurities. Ruby is α-alumina in which a small fraction of the Al^{3+} ions is replaced by Cr^{3+}. Another form of Al_2O_3, γ-alumina, is a metastable polycrystalline form with a defect spinel structure and a very high surface area. Partly because of its surface acid and base sites, this material is used as a solid phase in chromatography and as a heterogeneous catalyst and catalyst support.

5.2 Hydroxides

Aluminum hydroxide (氢氧化铝) is white gelatinous amorphous precipitates. It can be obtained by adding ammonia or lye in aluminum salt solution.

$$Al_2(SO_4)_3 + 6NH_3 \cdot H_2O = 2Al(OH)_3\downarrow + 3(NH_3)_2SO_4$$

Aluminum hydroxide is a kind of amphoteric hydroxides. Its alkalinity is slightly stronger than its acidity, but it is still a weak base.

$$Al(OH)_3 + 3HNO_3 = Al(NO_3)_3 + 3H_2O$$

$$Al(OH)_3 + KOH = KAlO_2 + 2H_2O$$

Aluminum hydroxide can neutralize gastric acid, protect gastric mucosa, and be used to treat excessive gastric acid and gastric ulcer.

5.3 Sulfates

Aluminum sulfate (硫酸铝) is a white powder which is soluble in water, and its aqueous solution is acidic due to hydrolysis.

Aluminum sulfate is easy to combine with alkali metal or ammonium sulfate to form double salt. For example, $KAl(SO_4)_2 \cdot 12H_2O$, commonly known as alum or white alum (明矾或白矾), is a colorless crystal. Alum has the effect of convergence and detoxification. It is often used as water purifying agent and fire extinguishing agent in industry.

6. Silicon and Its Compounds

Silicon (硅) is the ⅣA element in the periodic table. The valence electron configuration $3s^23p^2$ suggests the +4 oxidation state is dominant in the compounds of silicon (Si).

6.1 Oxides

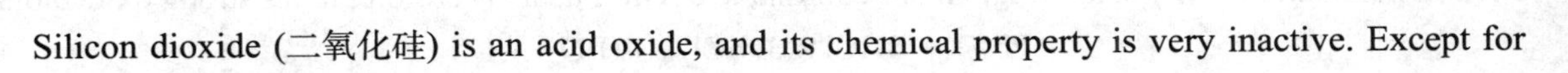

Silicon dioxide (二氧化硅) is an acid oxide, and its chemical property is very inactive. Except for fluorine, hydrogen fluoride and strong base, it cannot react with other substances at room temperature. The following reactions are reaction of silicon dioxide with strong base or molten sodium carbonate.

$$SiO_2 + 2NaOH = Na_2SiO_3 + H_2O$$

$$SiO_2 + Na_2CO_3 = Na_2SiO_3 + CO_2\uparrow$$

Because sodium silicate is soluble in water, glass containing silicon dioxide can be corroded by strong base.

6.2 Oxyacids and Salts of Silicon

Silicic acid (硅酸) has many forms and complex composition, which changes with the formation conditions. It is usually expressed by the general formula $xSiO_2 \cdot yH_2O$. Only orthosilicic acid H_4SiO_4 (正硅酸) and its dehydrated product, metasilicic acid H_2SiO_3 (偏硅酸), exist in the form of simple single acid in silicic acid. H_2SiO_3 is traditionally called silicic acid. Silicic acid is a very weak binary acid. Its solubility is very small and it can be easily replaced by other acids from silicate solution.

$$Na_2SiO_3 + 2HCl = H_2SiO_3 + 2NaCl$$

The silicic acid generated in the above reaction has little solubility in water and does not precipitate immediately. Polysilicic acid sol is formed by the gradual polymerization of polysilicic acid. After drying, it becomes a white transparent porous solid, which is called silica gel. Silica gel has strong adsorption capacity, and it is a good desiccant, adsorbent and carrier, and widely used in chromatography.

There are many kinds of silicates in nature. Except for the silicate of potassium and sodium, they are all insoluble. Soluble sodium silicate is commonly known as water glass (水玻璃). It is a good binder and filler of soap and detergent. Natural zeolite is an aluminosilicate with a porous structure for drying gases and organic solvents.

7. Lead and Its Compounds

Lead (铅) is the ⅣA element in the periodic table. The valence electron configuration $6s^2 6p^2$ suggests that the +2 oxidation state is dominant in the compounds of lead (Pb).

The main oxides of lead are lead oxide (PbO), lead dioxide (PbO_2), lead sesquioxide (Pb_2O_3) and lead tetroxide (Pb_3O_4). Pb_2O_3 and Pb_3O_4 are mixed oxidation state oxides.

PbO is commonly known as Lithargyrum (密陀僧). It is yellow powder, insoluble in water, which is amphoteric and alkalescent. It has the functions of disinfection, insecticidal and antiseptic in medicine.

PbO_2 is a strong oxidant in acid solution, which can oxidize concentrated hydrochloric acid to chlorine. It can also oxidize flesh Mn^{2+} ion to purple MnO_4^- ion.

$$PbO_2 + 4HCl = PbCl_2 + 2H_2O + Cl_2\uparrow$$

$$5PbO_2 + 2Mn^{2+} + 4H^+ = 2MnO_4^- + 5Pb^{2+} + 2H_2O$$

The best known is "red lead" (红铅、铅丹), Pb_3O_4, which contains Pb (Ⅳ) in an octahedral environment and Pb (Ⅱ) in an irregular six-coordinate environment. Therefore, it has strong oxidizability.

$$Pb_3O_4 + 8HCl = 3PbCl_2 + 4H_2O + Cl_2\uparrow$$

8. Arsenic and Its Compounds

Arsenic (砷) is the ⅤA element in the periodic table. The valence electron configuration $4s^24p^3$ suggests the common oxidation values of arsenic are +3 and +5.

8.1 Hydrides

Arsine (AsH_3) is a colorless highly toxic gas with garlic flavor. It burns when heated in the air. Under the condition of anoxia, AsH_3 decomposes into arsenic.

$$2AsH_3 + 3O_2 = 2As_2O_3 + 3H_2O$$

$$2AsH_3 = 2As\downarrow + 3H_2\uparrow$$

The martensitic arsenic test is a sensitive method for arsenic detection (detection limit 0.007mg). The mixture of sample, zinc and hydrochloric acid will lead the generated gas into the hot glass tube, and the arsenic precipitated will gather in the cooling part of the vessel to form a bright black "arsenic mirror".

$$As_2O_3 + 6Zn + 12HCl = 2AsH_3\uparrow + 6ZnCl_2 + 3H_2O$$

Another method to identify arsenic is the ancient's arsenic test, which uses the strong reducibility of arsine to reduce $AgNO_3$ to Ag, so as to achieve the purpose of arsenic detection (detection limit 0.005mg).

$$2AsH_3 + 12AgNO_3 + 3H_2O = As_2O_3 + 12HNO_3 + 12Ag\downarrow$$

8.2 Oxides

There are two kinds of arsenic oxides: As_2O_3 with an oxidation number of +3 and As_2O_5 with an oxidation number of +5.

Commonly known as white arsenic (砒霜), As_2O_3 is a white highly toxic powder with a lethal dose of 0.1g. It is slightly soluble in water to form arsenite (H_3AsO_3). Arsenic compounds with an oxidation number of +3 have both oxidability and reducibility, but are reductive mainly.

$$AsO_3^{3-} + I_2 + 2OH^- = AsO_4^{3-} + H_2O + 2I^-$$

As_2O_5 is soluble in water to form arsenic acid (H_3AsO_4). Arsenic compounds with an oxidation number of + 5 have oxidability.

$$H_3AsO_4 + 2HI = H_3AsO_3 + I_2 + H_2O$$

9. Sulfur and Its Compounds

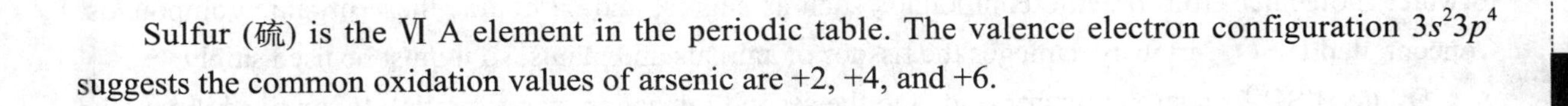

Sulfur (硫) is the Ⅵ A element in the periodic table. The valence electron configuration $3s^23p^4$ suggests the common oxidation values of arsenic are +2, +4, and +6.

9.1 Hydrides

Hydrogen sulfide (硫化氢) is a colorless toxic gas with the smell of rotten eggs, which can be dissolved in water. The water solution of saturated H_2S at room temperature is about 0.1mol/L, and it is called hydrosulfuric acid, which is a weak binary acid.

H_2S has strong reducibility and can be oxidized to sulfur by oxygen in the air. When it reacts with a strong oxidant, it can be oxidized to sulfuric acid.

$$2H_2S + O_2 = 2H_2O + 2S\downarrow$$
$$H_2S + I_2 = 2HI + S\downarrow$$
$$4Cl_2 + H_2S + 4H_2O = H_2SO_4 + 8HCl$$

9.2 Oxyacids and Salts

9.2.1 Sulphite and Its Salts

Sulfur dioxide (SO_2) is soluble in water, and its aqueous solution is called sulphite acid (亚硫酸). It is a binary medium strong acid, and it has the following equilibrium relationship in water.

$$SO_2 + H_2O \rightleftharpoons H_2SO_3 \rightleftharpoons H^+ + HSO_3^- \qquad K_{a1}^{\ominus} = 1.26\times10^{-2}$$
$$HSO_3^- \rightleftharpoons H^+ + SO_3^{2-} \qquad K_{a2}^{\ominus} = 6.31\times10^{-7}$$

Sulfite and salts are both oxidative and reductive. The reducibility is usually main property.

$$Na_2SO_3 + Cl_2 + H_2O = Na_2SO_4 + 2HCl$$

This reaction is widely used in the dechlorination of bleached fabrics in the printing and dyeing industry, and it can be used as a detoxifying agent of halogen poisoning in medicine.

Sulfite salts are more reductive than oxidative. For example, sodium sulfite solution is easily oxidized by oxygen in the air.

$$2Na_2SO_3 + O_2 = 2Na_2SO_4$$

Sodium sulfite is often used as an antioxidant in injection to protect the main components of drugs from oxidation.

9.2.2 Sulfuric Acid and Its Salts

Pure sulfuric acid (硫酸) is a colorless oily liquid, soluble in water, and can be mixed with water in any proportion, releasing a lot of heat when dissolved.

Hot concentrated sulfuric acid has strong oxidizability, which can oxidize many metals and nonmetals. Its reduction product is usually SO_2, but under the action of strong reducing agent, it can be reduced to S or H_2S.

$$2H_2SO_4(\text{浓}) + Cu = CuSO_4 + 2H_2O + SO_2\uparrow$$
$$4H_2SO_4(\text{浓}) + 3Zn = 3ZnSO_4 + 4H_2O + S\downarrow$$

Concentrated H_2SO_4 is a kind of desiccant commonly used in industry and laboratory due to its strong water absorption, such as chlorine, hydrogen and carbon dioxide.

Concentrated H_2SO_4 also has strong dehydration, which can capture hydrogen and oxygen equivalent to water molecules from organic compounds such as sugars, and carbonize these organic compounds. Concentrated H_2SO_4 seriously damages the tissues of animals and plants, so it must be used safely.

Dilute H_2SO_4 is a strong binary acid, and the second ionization step is partial. Its oxidizability is the function of H^+ ion in H_2SO_4, which is different from that of concentrated H_2SO_4.

Sulfuric acid can form two types of salt, normal salt and acid salt. Only alkali metal and alkaline earth metal (and ammonium) can generate acid salt and normal salt, and other metals can only generate normal salt.

9.2.3 Thiosulfuric Acid and Its Salts

Thiosulfuric acid (硫代硫酸, $H_2S_2O_3$) is very unstable, but its salt can exist stably. The most important one is sodium thiosulfate $Na_2S_2O_3 \cdot 5H_2O$ (硫代硫酸钠), commonly known as Hypo. Sodium thiosulfate is a colorless and transparent crystal, which is easily soluble in water. The solution is weakly alkaline due to the hydrolysis of $S_2O_3^{2-}$.

Sodium thiosulfate is stable in neutral and alkaline solution, and it decomposes rapidly in acid solution.

$$Na_2S_2O_3 + 2HCl = 2NaCl + S\downarrow + SO_2\uparrow + H_2O$$

It is used in medicine to treat scabies because of its high bactericidal ability.

$Na_2S_2O_3$ is reductive, and the most important reaction is the reaction between $Na_2S_2O_3$ and iodine to form sodium tetrathionate.

$$2Na_2S_2O_3 + I_2 = Na_2S_4O_6 + 2NaI$$

This reaction is carried out quantitatively and the basis for the determination of substance content by Iodimetry in analytical chemistry.

$S_2O_3^{2-}$ ion has a very strong coordinating ability and is a commonly used coordination agent.

$$2S_2O_3^{2-} + AgBr = [Ag(S_2O_3)_2]^{3-} + Br^-$$

According to reducibility and coordination ability of $Na_2S_2O_3$, it is often used as antioxidant of injection and antidote of halogen and heavy metal ions

9.2.4 Persulfate and Its Salts

Peroxydisulfate (过二硫酸) is a colorless crystal with similar chemical properties to concentrated sulfuric acid. Perbisulphate also has strong water absorption, dehydration and strong oxidation, which can carbonize paper.

The commonly used persulfates are $K_2S_2O_8$ and $(NH_4)_2S_2O_8$. All persulfates are strong oxidants. When the catalysis of Ag^+ ion is introduced, $S_2O_8^{2-}$ can rapidly oxidize colorless Mn^{2+} ions to purple MnO_4^-.

$$2Mn^{2+} + 5S_2O_8^{2-} + 8H_2O = 2MnO_4^- + 10SO_4^{2-} + 16H^+$$

10. Fluorine and Its Compounds

Fluorine (氟) is the ⅦA element in the periodic table. The valence electron configuration $2s^22p^5$ suggests the common oxidation value of arsenic is -1.

Fluorine reacts violently with water to form hydrogen fluoride (氟化氢) and oxygen.

$$2F_2 + 2H_2O = 4HF + O_2\uparrow$$

Hydrofluoric acid is a weak acid because of its special bond energy.

$$HF + 2H_2O \rightleftharpoons F^- + H_3O^+ \qquad K_a^\ominus = 6.31\times10^{-4}$$

Hydrofluoric acid should not be stored in glassware because it can react with SiO_2 or silicate to form gaseous SiF_4. Therefore, it should be placed in plastic containers.

$$SiO_2 + 4HF = SiF_4\uparrow + 2H_2O$$

Using this property of HF, the mark and pattern can be etched on the glass. Hydrofluoric acid has strong corrosiveness, and it has a serious destructive effect on cell tissue and bone. If the skin is found to be stained with hydrofluoric acid, wash it with plenty of water immediately and apply with dilute ammonia water.

The content of fluorine (CaF_2) in human tooth enamel is about 0.5%. The lack of fluoride is one of the causes of caries. The toothpaste made of SnF_2 can enhance the anticorrosion ability of enamel and prevent caries. However, the excessive intake will lead to fluorosis and dental fluorosis.

重点小结

1. *s* 区元素 *s* 区元素是最活泼的金属元素，包括Ⅰ A 族和Ⅱ A 族元素，分别称为碱金属和碱土金属，易形成氢化物、氧化物（过氧化物和超氧化物）、卤化物、碳酸盐和硫酸盐等重要化合物。其中重要的矿物药有大青盐、小苏打、芒硝、玄明粉、生石膏、熟石膏、泻盐、重晶石等，在中医药的临床中具有重要的功效和作用。

2. *p* 区元素 *p* 区元素位于周期表的右侧，包括Ⅲ A 族—Ⅷ A 族，依次为：硼族、碳族、氮族、氧族、卤族和稀有气体元素。本章主要介绍了 Pb、Al 两种金属元素及其重要的化合物。PbO 和 Pb_3O_4 都是强的氧化剂，能把浓盐酸氧化为氯气。PbO（密陀僧）在医药上具有消毒、杀虫、防腐的功效。Pb_3O_4（铅丹或红丹）主要用于配制外用膏药，具有收敛、止痛、消炎和生肌的作用。$Al(OH)_3$ 具有两性，内服能中和胃酸，保护胃黏膜，用于治疗胃酸过多、胃溃疡。$Al_2(SO_4)_3$ 水解显酸性，$Al_2(SO_4)_3 \cdot 12H_2O$ 为中药白矾，内服有祛痰燥湿、敛肺止血的功效，还可用于治疗皮炎和湿疹。除此之外，本章还重点介绍了 B、Si、As、S 和 F 五种非金属元素及其重要的化合物。硼砂（$Na_2B_4O_7 \cdot 10H_2O$）中药上称为盆砂，可治疗咽喉炎、口腔炎、中耳炎。As_2O_3 有剧毒，俗称砒霜，有祛腐拔毒功效，用于慢性皮炎如牛皮癣等。近年来临床用砒霜和亚砷酸内服治疗白血病，取得重大进展。

题库

Exercises

1. Choice question

(1) The valence electron coniguration of the alkaline earth metals is

A. ns^1　　B. ns^2　　C. $(n\text{-}1)d^{10}ns^1$　　D. $(n\text{-}1)d^{10}ns^2$

(2) Among the following compounds, the substance that can be used as both oxidant and reductant is

A. H_2SO_3　　B. H_3BO_3　　C. H_2SO_4　　D. H_2SiO_3

(3) Which of the following elements has maximum electronegativity

A. B　　B. Si　　C. F　　D. S

(4) The chemical composition of Natrii Sulfas Exsiccatus is

A. $CaCO_3$　　B. $Na_2SO_4 \cdot 10H_2O$　　C. Na_2SO_4　　D. $BaSO_4$

(5) The chemical composition of mature gypsum is

A. $MgSO_4 \cdot 7H_2O$　　B. Na_2SO_4　　C. $2CaSO_4 \cdot H_2O$　　D. Na_2SO_3

(6) The acid that can't be put in a glassware is

A. H_2S　　B. HF　　C. $H_2S_2O_8$　　D. H_2SO_4

2. True-false question

(1) H_3BO_3 is a ternary acid containing three ionizable H^+ ions.

(2) HF is a strong acid.

(3) Borax is soluble and hydrolyzable in water.

(4) Al $(OH)_3$ is amphoteric and can react with acid and alkali.

3. Fill in the blanks

(1) The chemical composition of mirabilite is ________.

(2) The chemical composition of borax is ________.

(3) The chemical composition of white arsenic is ________.

(4) The color of Pb_3O_4 is ________, it is commonly known as ________.

4. Complete the following equations

(1) $PbO_2 + HCl \rightarrow$

(2) $PbO_2 + Mn^{2+} + H^+ \rightarrow$

(3) $Al(OH)_3 + KOH \rightarrow$

(4) $Cl_2 + H_2S + H_2O \rightarrow$

(5) $Na_2S_2O_3 + HCl \rightarrow$

(6) $Na_2S_2O_3 + I_2 \rightarrow$

(7) $Mn^{2+} + S_2O_8^{2-} + H_2O \rightarrow$

(8) $SiO_2 + HF \rightarrow$

目 标 检 测

1. 选择题

（1）碱土金属的价电子层结构是

A. ns^1　　B. ns^2　　C. $(n\text{-}1)d^{10}ns^1$　　D. $(n\text{-}1)d^{10}ns^2$

（2）下列含氧的化合物中既可作氧化剂又可作还原剂的物质是

A. H_2SO_3　　B. H_3BO_3　　C. H_2SO_4　　D. H_2SiO_3

（3）下列元素中，电负性最大的是

A. B　　B. Si　　C. F　　D. S

（4）玄明粉的化学成分是

A. $CaCO_3$　　B. $Na_2SO_4 \cdot 10H_2O$　　C. Na_2SO_4　　D. $BaSO_4$

（5）医疗上用作石膏绷带，称之为熟石膏的是

A. $MgSO_4 \cdot 7H_2O$　　B. Na_2SO_4　　C. $2CaSO_4 \cdot H_2O$　　D. Na_2SO_3

（6）不能用玻璃仪器盛放的酸是

A. H_2S　　B. HF　　C. $H_2S_2O_8$　　D. H_2SO_4

2. 判断题

(1) H_3BO_3 是三元酸，在它的分子里含有 3 个可电离的 H^+ 离子。

（2）HF 是强酸。

（3）硼砂易溶于水，易水解。

（4）$Al(OH)_3$ 具有两性，既可以和酸反应，也可以和碱反应。

3. 填空题

（1）芒硝的化学成分是 ________。

（2）硼砂的化学成分是 ________。

（3）砒霜的化学成分是 ________。

（4）Pb_3O_4 呈 ________ 色，俗称 ________。

4. 完成下列反应方程式

（1）$PbO_2 + HCl \rightarrow$

（2）$PbO_2 + Mn^{2+} + H^+ \rightarrow$

（3）$Al(OH)_3 + KOH \rightarrow$

（4）$Cl_2 + H_2S + H_2O \rightarrow$

（5）$Na_2S_2O_3 + HCl \rightarrow$

（6）$Na_2S_2O_3 + I_2 \rightarrow$

（7）$Mn^{2+} + S_2O_8^{2-} + H_2O \rightarrow$

（8）$SiO_2 + HF \rightarrow$

Chapter 11 The Transition Metal Elements and Related Mineral Drugs

第十一章 过渡金属元素及相关矿物药

学习目标

知识要求

1. **掌握** Cr、Mn、Fe、Cu、Zn、Hg元素及其重要化合物的基本性质。
2. **熟悉** Cr、Mn、Fe、Cu、Zn、Hg元素性质与电子层结构的关系。
3. **了解** Cr、Mn、Fe、Cu、Zn、Hg元素及其矿物药在医药中的应用。

能力要求

应用过渡金属元素的性质鉴别含有过渡金属离子的矿物药。

Generally, the subgroup elements are also called a transition elements (过渡元素). Because of all of them are metals, they are also called the transition metals (过渡金属). The term "transition" denotes elements in the middle of the periodic table. They provide a transition between the "base formers" on the left and the "acid formers" on the right. There are more transition elements-members than main group elements (主族元素). Although some of them are rare and of limited use, others play crucial roles in many aspects of modern life. The following are general properties of transition elements.

(1) All are metals.

(2) Most are harder and more brittle than non-transition metals, with higher melting points, boiling points, and heats of vaporization.

(3) Their ions and their compounds are usually colored.

(4) They form many complex ions.

(5) With few exceptions, they exhibit multiple oxidation states.

(6) Many of them are paramagnetic, as are many of their compounds.

(7) Many of the metals and their compounds are effective catalysts.

Some properties of 3*d*-transition metals are listed in Table11-1

Table 11-1 Properties of metals in the first transition series

Properties	Sc	Ti	V	Cr	Mn	Fe	Co	Ni	Cu	Zn
Melting point (℃)	1541	1660	1890	1850	1244	1535	1495	1453	1083	420
Boiling point (℃)	2831	3287	3380	2672	1962	2750	2870	2732	2567	907
Density (g/cm^3)	2.99	4.54	6.11	7.18	7.21	7.87	8.9	8.91	8.96	7.13

(continued)

Properties	Sc	Ti	V	Cr	Mn	Fe	Co	Ni	Cu	Zn
Atomic radius (Å)	1.62	1.47	1.34	1.25	1.29	1.26	1.25	1.24	1.28	1.34
Ionic radius, M^{2+} (Å)	—	0.94	0.88	0.89	0.80	0.74	0.72	0.69	0.70	0.74
Electronegativity	1.36	1.54	1.63	1.66	1.55	1.80	1.88	1.91	1.90	1.65
$E^{\ominus}$ (M^{2+}/M) (V)	−2.08*	−1.63	−1.18	−0.91	−1.18	−0.44	−0.28	−0.25	+0.34	−0.76
IE (kJ/mol) first	631	658	650	652	717	759	758	757	745	906
IE (kJ/mol) second	1235	1310	1414	1592	1509	1561	1646	1753	1958	1733

*For Sc^{3+} (aq) +3e⁻ → Sc (s).

1. Chromium

The valence shell electron configuration (价层电子结构) of chromium atom is $3d^5 4s^1$. Chromium has various oxidation values from + 2 to + 6, the most important of which are + 3 and + 6. Because of its beautiful color and high hardness, chromium is often plated on other metal surfaces for decoration and protection. Chromium can form alloy, and almost all types of stainless steel have a high proportion of chromium (When steel contains about 14% chromium, this is stainless steel).

1.1 Oxyacids and Salts of Chromium

Both H_2CrO_4 and $H_2Cr_2O_7$ are strong acids, which only exist in aqueous solution. $H_2Cr_2O_7$ is more acidic than H_2CrO_4.

Potassium and sodium chromates are yellow crystals, while their dichromates are orange red crystals. $K_2Cr_2O_7$ (commonly known as red alum potassium, 红矾钾) has very little solubility at low temperature, does not contain crystal water, and is not easy to deliquesce. It is often used as a reference substance in quantitative analysis.

When adding acid to the chromic acid solution, the solution changes from yellow to red; otherwise, when adding alkali to the dichromate solution, the solution changes from orange red to yellow. The following equilibrium exists in the aqueous solution:

$$2CrO_4^{2-} + 2H^+ = 2HCrO_4^- = Cr_2O_7^{2-} + H_2O$$

(yellow) (orange red)

The solubility of chromate is generally smaller than that of dichromate, and it can form corresponding precipitates of $BaCrO_4$ (lemon yellow), $PbCrO_4$ (yellow), and Ag_2CrO_4 (brick red) with Ba^{2+}, Pb^{2+}, and Ag^+, which are often used to test the existence of Ba^{2+}, Pb^{2+}, and Ag^+.

Chromium chloride hexahydrate $CrCl_3 \cdot 6H_2O$ (green or purple), $Cr_2(SO_4)_3 \cdot 18H_2O$ (purple) and potassium chromium alum $KCr(SO_4)_2 \cdot 12H_2O$ (blue purple) are easily soluble in water. The salt solution of Cr^{3+} will show different colors under different conditions. In alkaline medium, Cr^{3+} can be oxidized by dilute H_2O_2 solution, and the solution changes from green to yellow. This reaction is often used to identify Cr^{3+}:

$$2\,[Cr(OH)_4]^- + 2OH^- + 3H_2O_2 = 2CrO_4^{2-} + 8H_2O$$

(bright green) (yellow)

$Cr(OH)_3$ is a gray blue precipitate formed by the action of an appropriate amount of alkali on the chromium salt solution (pH value is about 5.3). $Cr(OH)_3$ is an amphoteric hydroxide, which can be dissolved in acid to generate hydrated chromium ion and soluble in alkali to generate tetrahydroxy chromium (III) acid salt. Due to the weak acidity and basicity of $Cr(OH)_3$, both chromium (Ⅲ) salt and tetrahydroxychromium (III) acid salt are hydrolyzed in water.

$$Cr(OH)_3 + 3H^+ = Cr^{3+} + 3H_2O$$

$$Cr(OH)_3 + OH^- = [Cr(OH)_4]^-$$

(gray blue) (bright green)

1.2 Complexes of Chromium

The valence shell electron configuration of Cr^{3+} is $3d^3$, with six empty orbitals, which can form the d^2sp^3 octahedral complex. In these complexes, the d_γ orbitals are all empty, and the *d–d* transition is very easy to occur under visible light irradiation. Therefore, most of the complexes of Cr (Ⅲ) have color. Cr^{3+} can form complexes with H_2O, Cl^-, OH^-, SCN^-, and other ligands.

2. Manganese

Manganese is a fairly abundant element, constituting about 0.1% of Earth's Crust. Its principal ore is Pyrolusite, MnO_2. The electron configuration of Mn is $[Ar]3d^54s^2$ By employing first the two 4*s* electrons and then, consecutively, up to all five of its unpaired 3*d* electrons, manganese exhibits all oxidation states from +2 to +7. The most important reactions of manganese compounds are oxidation-reduction reaction.

2.1 Oxides and Hydroxides of Manganese

Manganese oxide mainly refers to manganese dioxide MnO_2, which is a black solid matter insoluble in water. It is the main component of pyrolusite in nature, and also the main raw material for preparing other manganese compounds and metal manganese. As the main component of Wumingyi (无名异), MnO_2 has the effect of detumescence, removing stasis and relieving pain. Because Mn (Ⅳ) is in the middle oxidation value of Mn, it can be both oxidized and reduced. But it is used mainly as oxidant, especially in acid medium, MnO_2 is a strong oxidant. For example, in the laboratory, MnO_2 can be used to react with concentrated hydrochloric acid to prepare chlorine. MnO_2 can also be used to react with concentrated sulfuric acid to prepare oxygen.

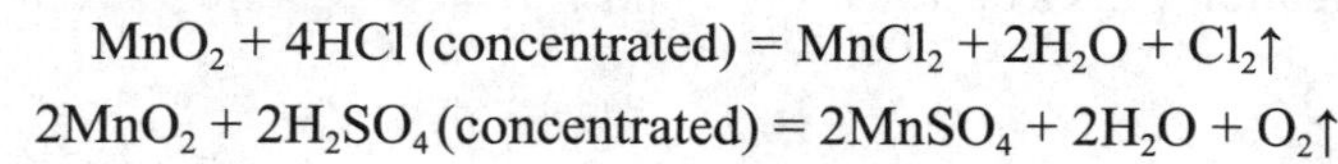

$$MnO_2 + 4HCl\,(concentrated) = MnCl_2 + 2H_2O + Cl_2\uparrow$$

$$2MnO_2 + 2H_2SO_4\,(concentrated) = 2MnSO_4 + 2H_2O + O_2\uparrow$$

The white colloidal $Mn(OH)_2$ precipitate can be obtained by adding alkali into the salt solution of Mn^{2+}. $Mn(OH)_2$ has strong alkalinity and weak acidity, and is easily oxidized to brown $MnO(OH)_2$ by air.

2.2 Salts of Manganese

The salt of Mn^{2+} can be obtained by the reaction of metal manganese with dilute non-oxidizing acid. The strong acid salts of Mn^{2+} are all soluble in water, and only a few weak acid salts, such as $MnCO_3$ and MnS, are difficult to dissolve in water. Manganese salt crystallized from aqueous solution is pink crystal with crystal water. Mn^{2+} is quite stable in acid solution. Only strong oxidants such as $NaBiO_3$, PbO_2, $(NH_4)_2S_2O_8$ can oxidize Mn^{2+} to MnO_4^-, and the color changes from colorless to purple. Therefore, this reaction can be used to identify the existence of Mn^{2+}.

$$5NaBiO_3+2Mn^{2+}+14H^+ = 5Na^++5Bi^{3+}+2MnO_4^-+7H_2O$$

Potassium manganate, a dark green solid, is more important among manganates. It can be produced by congruent melting MnO_2 with KOH.

$$2MnO_2 + 4KOH + O_2 \xrightarrow{\text{melting}} 2K_2MnO_4 + 2H_2O$$

Potassium manganate is relatively stable in strong alkaline solution and is prone to disproportionation in acid solution:

$$3MnO_4^{2-} + 4H^+ = MnO_2 + 2MnO_4^- + 2H_2O$$

The most widely used salt in permanganate is potassium permanganate $KMnO_4$, commonly known as gray manganese potassium (灰锰钾), which is a dark purple crystal. $KMnO_4$ solid will decompose when heated above 200°C, and this reaction can be used to prepare a small amount of oxygen:

$$2KMnO_4 = MnO_2 + K_2MnO_4 + O_2\uparrow$$

$KMnO_4$ decomposes slowly in acid solution and decomposes very slowly in neutral solution, but light and MnO_2 catalyze its decomposition. Therefore, $KMnO_4$ solution should be stored in brown bottle, and after a period of time, MnO_2 should be removed by filtration.

$$4MnO_4^- + 4H^+ = 4MnO_2\downarrow + 3O_2\uparrow + 2H_2O$$

$KMnO_4$ is oxidable in acid, neutral or alkaline solutions, and its reduction products vary with the acidity and alkalinity of solutions, such as colorless Mn^{2+} (acid), Tan MnO_2 precipitation (neutral or weak alkaline), or green MnO_4^{2-} (strongly alkaline). Therefore, potassium permanganate is a commonly used oxidant in chemistry and also used as antiseptic, disinfectant, deodorant and antidote in medicine. $KMnO_4$ dilute solution (0.1%) can be used for disinfection of instruments and equipment, and its 0.5% solution can treat mild scald.

3. Iron

Iron is the ⅧB element in the fourth period. The valence shell electron configuration of iron is $3d^64s^2$. The common oxidation values of iron are + 2 and + 3.

3.1 Oxides and Hydroxides of Iron

The oxides of iron include FeO (ferrous oxide), Fe_2O_3 (ferric oxide) and Fe_3O_4 (ferroferric oxide). Fe_2O_3 is the main component of mineral medicine ochre (赭石或代赭石). It has the functions of calming the liver and suppressing the Yang, cooling blood and hemostasis, and is commonly used in headache, dizziness, palpitation, madness, convulsion, vomiting, etc. Fe_3O_4 is the main component of mineral medicine magnetite (磁石). It has the functions of calming the nerves and quieting the spirit, calming the liver and suppressing the Yang. It can be used to treat palpitation, insomnia, dizziness, tinnitus, deafness, kidney deficiency and asthma.

Adding a strong base to the Fe^{2+} salt solution will produce a white $Fe(OH)_2$ precipitate. $Fe(OH)_2$ is extremely unstable. It will turn dark green soon after contacting with air, and then turn into a reddish brown iron oxide hydrate $Fe_2O_3 \cdot nH_2O$. It is customarily written as $Fe(OH)_3$.

$$4Fe(OH)_2 + O_2 + 2H_2O = 4Fe(OH)_3$$

Both $Fe(OH)_2$ and $Fe(OH)_3$ are insoluble in water. $Fe(OH)_2$ is alkaline and soluble in strong acid to form ferrous salt; $Fe(OH)_3$ is amphoteric, mainly alkaline, soluble in acid to form corresponding ferric salt and soluble in hot concentrated strong alkali solution to form ferric acid salt:

$$Fe(OH)_3 + NaOH = NaFeO_2 + 2H_2O$$

3.2 Sulfides of Iron

FeS_2 is generally called ferrous disulfide, which has a valence of + 2 for Fe and -1 for S. FeS_2 is the main component of the mineral medicine “natural copper” (自然铜), which has the effect of dispersing blood stasis and relieving pain, extending the tendons and connecting the bones. It is used to treat bruise, broken tendons and bones, swelling and pain. FeS_2 is usually transformed into $Fe(Ac)_2$ which is easy to be absorbed by human body after calcined and vinegar quenched.

3.3 Salts of Iron

The most important iron salts are ferrous sulfate and ferric chloride. Ferrous sulfate heptahydrate $FeSO_4 \cdot 7H_2O$, commonly known as green alum (绿矾), is a light green crystal, which is also called soap alum (皂矾) in traditional Chinese medicine. It is used in agriculture for pest control, and in medicine for iron deficiency anemia. Ferrous sulfate will decompose into ferric oxide when exposed to strong heat, and it will gradually be weathered in the air and lose part of water, and the surface is easily oxidized to form yellow brown basic ferric sulfate:

$$2FeSO_4 = Fe_2O_3 + SO_2\uparrow + SO_3\uparrow$$

$$4FeSO_4 + O_2 + 2H_2O = 4Fe(OH)SO_4$$

Ferrous sulfate can form double salt with sulfates of alkali metals and ammonium, such as $(NH_4)_2SO_4 \cdot FeSO_4 \cdot 6H_2O$, which is called Mohr's Salt (摩尔盐). It is more stable and easier to preserve than $FeSO_4$. It is a common reducing agent in analytical chemistry, used to calibrate $K_2Cr_2O_7$ solution or $KMnO_4$ solution. Anhydrous ferric trichloride is a sepia covalent compound, which sublimates easily. It is in a vapor state at 400°C, and it exists as dimer Fe_2Cl_6. $FeCl_3 \cdot 6H_2O$ is generally prepared from the solution, which is dark yellow. Fe^{3+} ions usually exist in the form of lavender $[Fe(H_2O)_6]^{3+}$ in acid aqueous

solution, which is easy to hydrolyze and turn yellow. In addition, Fe^{3+} is a medium strength oxidant in acid solution, which can oxidize I^-, $SnCl_2$, SO_3, H_2S, Fe, Cu, etc., and reduce itself to Fe^{2+}.

3.4 Complexes of Iron

Fe^{2+} and Fe^{3+} are easy to form octahedral complexes with coordination number of 6. The most common ligands are CN^- and SCN^-.

Potassium ferrocyanide $K_4[Fe(CN)_6]$ is a yellow crystal, commonly known as yellow blood salt (黄血盐). Potassium ferricyanide $K_3[Fe(CN)_6]$, a dark red crystal, is commonly known as red blood salt (赤血盐). These two kinds of iron cyanogen complexes are easily soluble in water and are quite stable in water. When $K_3[Fe(CN)_6]$ is added to Fe^{2+} solution or $K_4[Fe(CN)_6]$ is added to Fe^{3+} solution, blue precipitate can be formed. This reaction is often used to identify Fe^{2+} and Fe^{3+}, respectively:

$$K^+ + Fe^{3+} + [Fe(CN)_6]^{4-} = KFe[(CN)_6]\downarrow \quad \text{(Turnbull Blue)}$$

$$K^+ + Fe^{2+} + [Fe(CN)_6]^{3-} = KFe[(CN)_6]\downarrow \quad \text{(Prussian Blue)}$$

Fe^{3+} reacts with SCN^- to form the blood-red $[Fe(SCN)_n]^{3-n}$. The n value depends on the SCN^- concentration and acidity in the solution. Because of its sensitivity, it is often used to identify Fe^{3+}:

$$Fe^{3+} + nSCN^- = [Fe(SCN)_n]^{3-n} \quad (n=1\text{-}6)$$

It is difficult for Fe^{2+} to form stable ammonium compound. Due to the strong hydrolysis, it often forms $Fe(OH)_3$ precipitate when ammonia is added to Fe^{3+} aqueous solution and no ammonium compound is formed.

There are many kinds of cationic compounds of iron element in mineral medicine. Iron powder (铁粉散), Pig iron Luoyin (生铁落饮), Qiwei iron filings pill (七味铁屑丸) and Cizhu pill (磁珠丸) are mainly Fe_3O_4. Gengnian-an (更年安) and Jiangfan pill (绛矾丸) are mainly $FeSO_4 \cdot 7H_2O$. Xuanfu Daizhe Decoction (旋覆代赭汤) is mainly Fe_2O_3. Shehuang pill (蛇黄丸) is mainly $Fe_2O_3 \cdot xH_2O$. Huangfan pill (黄矾丸) is mainly $Fe_2(SO_4)_3 \cdot 10H_2O$. The main components of Shenxiaotaiyi pill (神效太乙丸) and Zhenlingdan (震灵丹) are $Fe_2O_3 \cdot xH_2O$ and FeS_2.

4. Copper

4.1 Oxides and Hydroxides of Copper

Copper oxides are mainly Cu_2O and CuO. Cu_2O is a covalent compound, toxic, insoluble in water, and heat stable. The solid color varies with the particle size from yellow, red, brick-red to dark brown, which can be used as paint pigment. At the same time, it has semiconductor properties, which can be used for making cuprous rectifier. Cu_2O is one component of hematite. The dark red Cu_2O powder can be obtained by reducing CuO with hydrogen:

$$2CuO + H_2 \xrightarrow{150℃} Cu_2O + H_2O\uparrow$$

The black CuO powder can be made by the decomposition of some oxysalts or the heating of copper powder in oxygen:

$$2Cu(NO_3)_2 \xrightarrow{200℃} 2CuO + O_2\uparrow + 4NO_2\uparrow$$

Cu_2O and CuO are basic oxides. CuO is soluble in acid with high thermal stability. When Cu_2O dissolves in dilute acid, the disproportionation reaction takes place, and Cu^{2+} ions and Cu are formed:

$$Cu_2O + 2H^+ = Cu^{2+} + Cu\downarrow + H_2O$$

When strong base is added to Cu^{2+} solution, a blue flocculent $Cu(OH)_2$ precipitate is formed, which is amphoteric and alkalescent. It is soluble in acid, and also in strong base to form bright blue $[Cu(OH)_4]^{2-}$ ion.

Cuprous hydroxide (CuOH) is very unstable and easy to dehydrate to Cu_2O.

4.2 Salts of Copper

4.2.1 $CuSO_4$

Anhydrous copper sulfate is a white powder, which is insoluble in ethanol and ether. It has strong water-absorbing quality, and it is blue after water absorption. However, when crystallized from aqueous solution, blue copper sulfate pentahydrate ($CuSO_4\cdot 5H_2O$), also known as blue vitriol (蓝矾或胆矾) is generated. This property can be used to test or remove a small amount of water in ethanol, ether and other solvents (as desiccant).

When $CuSO_4$ encounters a small amount of aqueous ammonia, it will generate a light blue basic copper sulfate precipitation. The precipitate will dissolve and convert into dark blue tetraammine copper coordination ion when excessive aqueous ammonia is added.

$$2CuSO_4 + 2NH_3\cdot H_2O \rightarrow (NH_4)_2SO_4 + Cu_2(OH)_2SO_4\downarrow \xrightarrow{NH_3} [Cu(NH_3)_4]SO_4$$

The dark blue crystal $[Cu(NH_3)_4]SO_4\cdot H_2O$ is obtained by adding methanol into the solution of tetraammine copper coordination ion. In agriculture, $CuSO_4\cdot 5H_2O$: CaO : H_2O = 1 : 1 : 100 (mas ratio) was mixed to prepare Bordeaux mixture (波尔多液), which was used as pesticides and fungicides in orchard and crops.

4.2.2 $CuCl_2$

$CuCl_2$ is a chain-like covalent molecule, which is liable to deliquesce and easily soluble in water, ethanol, acetone and other solvents. The anhydrous cupric chloride is brownish yellow. When it dissolves in water, yellow $[CuCl_4]^{2-}$ is formed in a very thick solution, while blue $[Cu(H_2O)_4]^{2+}$ is formed in a dilute solution, so $CuCl_2$ solution often shows yellow green or green due to the coexistence of $[CuCl_4]^{2-}$ and $[Cu(H_2O)_4]^{2+}$. $CuCl_2$ decomposes when heated to 773K:

$$2CuCl_2 \xrightarrow{773K} 2CuCl + Cl_2\uparrow$$

4.2.3 CuS

When H_2S is introduced into Cu^{2+} solution, black CuS precipitate is obtained. It is insoluble in water and non-oxidizing acid, but soluble in hot dilute nitric acid (稀硝酸) or concentrated nitric acid (浓硝酸). But Cu_2S only dissolves in concentrated nitric acid or sodium cyanide (氰化钠).

$$3Cu_2S + 16HNO_3\,(concentrated, hot) = 6Cu(NO_3)_2 + 3S\downarrow + 4NO\uparrow + 8H_2O$$

$$Cu_2S + 4CN^- = 2[Cu(CN)_2]^- + S^{2-}$$

4.2.4 $Cu(OH)_2CO_3$

Basic copper carbonate, $Cu(OH)_2CO_3$, is malachite green (孔雀绿). It is a substance produced by the reaction of copper with oxygen, carbon dioxide and water vapor in the air. $Cu(OH)_2CO_3$ is also called malachite (孔雀石) or copper green (铜绿), which can be used for the treatment of pannus, carbuncle, swelling and pain, etc.

4.3 Complexes of Copper

The valence shell electron configuration of Cu^{2+} is $3d^9$, which is easier to form complexes than Cu^+. Cu^{2+} can form planar quadrilateral and deformed octahedral complexes with ligands of 4 and 6, respectively. For example, the coordination numbers of $[Cu(NH_3)_4(H_2O)_2]^{2+}$ and $[CuY]^{2+}$ are both 6. The coordination numbers of $[Cu(NH_3)_4]^{2+}$, $[Cu(H_2O)_4]^{2+}$, $[Cu(OH)_4]^{2-}$, $[CuCl_4]^{2-}$ (light yellow), $[Cu(en)_2]^{2+}$ (dark blue purple), and $Cu(CH_3COO)_2 \cdot H_2O$ are all 4. In addition, tartaric acid (酒石酸) and citric acid (柠檬酸) can also form stable complexes with $Cu(OH)_2$, and their aqueous solutions are called Fehling's solution (斐林试剂) and Banedict's solution (班氏试剂，或本尼迪特试剂), respectively.

The valence shell electron configuration of Cu^+ ion is $3d^{10}$, with 4*s* and 4*p* empty orbitals in the outer shell. It can take *sp*, sp^2 and sp^3 hybridization, forming linear, planar triangular and tetrahedral complexes with ligands of 2, 3 and 4, respectively.

5. Zinc

Zinc (Zn) is located in the group ⅡB of the fourth period in the periodic table. The valence shell electron configuration is $3d^{10}4s^2$, and the common oxidation number is +2. Zinc is a kind of silver white metal. Because 3*d* electrons do not participate in bonding, its melting point and boiling point are low. Zinc mainly exists in nature in the form of sulfide or oxygen-containing compounds, for example, $ZnCO_3$ (smithsonite, 菱锌矿), ZnS (red zinc ore, 红锌矿), etc.

5.1 Oxides and Hydroxides of Zinc

5.1.1 ZnO

Zinc oxide (ZnO), commonly known as zinc white powder (锌白粉), is the main component of traditional Chinese medicine named calcined calamine (煅炉甘石), which has the function of convergence and promoting wound healing. It is used to prepare compound powder (散剂), suspension (混悬剂), ointment (软膏剂) and paste (糊剂) for external use to treat skin dampness and inflammation.

5.1.2 $Zn(OH)_2$

The white precipitate of $Zn(OH)_2$ can be formed by adding an appropriate amount of alkali to the solution containing Zn^{2+}. $Zn(OH)_2$ is acid-base amphoteric. It can be dissolved in acid to form corresponding salt, or in alkali to form $[Zn(OH)_4]^{2-}$.

$$Zn(OH)_2 + 2OH^- = [Zn(OH)_4]^{2-}$$

5.2 Salts of Zinc

5.2.1 $ZnCO_3$

Zinc carbonate ($ZnCO_3$) is a commonly mineral drug known as calamine (炉甘石、甘石、羊甘石), and the original mineral is smithsonite (菱锌矿). $ZnCO_3$ has the functions of detoxification, clearing the eyes and removing the haze, astringent and antipruritic. It is the main component of Chinese patent medicine of Miaohou powder (妙喉散) and Shengji powder (生肌散).

5.2.2 $ZnSO_4$

Zinc sulfate ($ZnSO_4$), the first kind of zinc supplement, is currently substituted by zinc gluconate (葡萄糖酸锌), zinc glycyrrhizinate (甘草酸锌), zinc citrate (枸橼酸锌) and zinc arginine (精氨酸锌), etc. Oral administration is used to treat diseases caused by zinc deficiency. 0.3% ~ 0.5% $ZnSO_4$ can also be used to treat conjunctivitis, and its compound preparation can promote wound healing.

Zinc is an essential trace element for human body. The total content of zinc in normal adults is about 2300mg/70kg. It is mainly distributed in muscle cells and bones, mainly absorbed in intestine and excreted in feces and urine. The daily intake of human body is 12–16mg. Zinc in human body mainly forms complexes with biological macromolecules, such as nucleic acids and proteins, and participates in many physiological and biochemical reactions in the form of enzymes. Now it is known that more than 80 kinds of enzyme's biological activities are related to zinc.

6. Mercury

6.1 Oxides and Hydroxides of Mercury

According to the different preparation methods and conditions, mercuric oxide has two different variants: yellow HgO and red HgO, both of them are difficult to dissolve in water and toxic! The structures of these two variants are the same. The difference in color is due to the size of the particles. The particles of the yellow HgO are small while those of the red HgO are large. Both of them can decompose into black mercury and oxygen at 720K.

$$2HgO \xrightarrow{700K} 2Hg\ (black) + O_2\uparrow$$

When the soluble mercury salt reacts with alkali, the final product is not $Hg(OH)_2$, but yellow HgO precipitation, which is caused by the unstable decomposition of $Hg(OH)_2$.

$$Hg^{2+} + 2OH^- = HgO\downarrow\ (yellow) + H_2O$$

The red HgO is a traditional Chinese medicine named Red powder (红粉), which can be obtained by the thermal decomposition of $Hg(NO_3)_2$ or by heating mercury in oxygen at about 620K. The yellow HgO can be transformed into red HgO when it heated below 570K.

6.2 Sulfides of Mercury

The natural mercuric sulfide mineral is called cinnabar (朱砂、丹砂), which is red and has the effect of calming the nerves and detoxifying. The artificial preparation of mercuric sulfide is obtained by heating and sublimation of mercury and sulfur. HgS is the most insoluble metal sulfide, which is insoluble in hydrochloric acid and nitric acid, but soluble in aqua regia, mixture of hydrochloric acid and KI or excessive Na_2S solution:

$$3HgS + 12Cl^- + 2NO_3^- + 8H^+ = 3[HgCl_4]^{2-} + 3S\downarrow + 2NO\uparrow + 4H_2O$$

$$HgS + 2H^+ + 4I^- = [HgI_4]^{2-} + H_2S$$

$$HgS + Na_2S\ (\text{concentrated}) = Na_2[HgS_2]$$

6.3 Salts of Mercury

6.3.1 $HgCl_2$

$HgCl_2$ is a white solid with low melting point and easy sublimation. Its melting point is 549K, commonly known as Shenggong (升汞). It is soluble in water and highly toxic. Its dilute solution has bactericidal effect and is used as disinfectant in surgery. $HgCl_2$ is a linear covalent molecule, which is difficult to ionize, soluble in organic solvent, slightly hydrolyzed in water, ammonolysis in ammonia.

$HgCl_2$ is oxidable in acid solution. It can be reduced to Hg_2Cl_2 (white precipitate) by reaction with some reducing agents such as $SnCl_2$. When $SnCl_2$ is excessive, it can be further reduced to black mercury. Therefore, the reaction of $SnCl_2$ and $HgCl_2$ can be used to test Hg^{2+} or Sn^{2+} ions.

$$2HgCl_2 + SnCl_2 + 2HCl = Hg_2Cl_2\downarrow\ (\text{white}) + H_2[SnCl_6]$$

$$Hg_2Cl_2 + SnCl_2 + 2HCl = 2Hg\downarrow\ (\text{black}) + H_2[SnCl_6]$$

6.3.2 Hg_2Cl_2

Hg_2Cl_2 is a linear molecule of covalent type (Cl-Hg-Hg-Cl). Hg^+ is hybridized to two *sp* hybrid orbitals and exists as a dimer of Hg_2^{2+} (Hg^+ : Hg^+). Hg_2Cl_2 is a white solid that is difficult to dissolve in water, slightly sweet, commonly known as calomel (甘汞), which is the main component of light powder (轻粉) of traditional Chinese medicine, with a small amount of low toxicity. It is often used to make calomel electrode. Hg_2Cl_2 can be prepared by grinding $HgCl_2$ together with metal Hg. But Hg_2Cl_2 is easy to decompose into $HgCl_2$ and Hg in light, so it is often stored in brown bottle.

$$HgCl_2 + Hg = Hg_2Cl_2\downarrow\ (\text{white})$$

The nitrate of mercury is soluble in water, hydrolyzable, unstable to heat, and easily decomposed into HgO and Hg.

$$2Hg(NO_3)_2 = 2HgO\ (\text{red}) + 4NO_2\uparrow + O_2\uparrow\ (\text{high temperature reaction})$$

$$Hg(NO_3)_2 = Hg\ (\text{black}) + 2NO_2\uparrow + O_2\uparrow\ (\text{low temperature reaction})$$

6.4 Complexes of Mercury

It is difficult for Hg_2^{2+} to form complex. However, Hg^{2+} can form stable complexes with Cl^-, Br^-, I^-, CN^-, SCN^-, S^{2-}, etc. The KOH solution of $K_2[HgI_4]$ is called Nessler reagent (奈斯勒试剂), which is used for the identification of trace ammonia ions.

$$NH_4Cl + 2K_2[HgI_4] + 4KOH = \underset{(\text{reddish-brown})}{[Hg_2ONH_2]I} + KCl + 7KI + 3H_2O$$

重点小结

1. Cr、Mn、Fe、Cu、Zn、Hg 本章介绍了Cr、Mn、Fe元素价电子层构型和元素的通性，各元素重要化合物的结构、性质，以及各化合物之间的相互关系。

2. 重要矿物药 本章介绍了Cr、Mn、Fe、Cu、Zn、Hg元素在医药中的应用，介绍了朱砂、轻粉、代赭石、升汞、炉甘石、摩尔盐、无名异、绿矾、胆矾、铜绿等矿物药的主要化学成分及其功效。

题库

Exercises

1. Fill in the blanks: write the chemical formula of the following substances

(1) cinnabar ________ (2) light powder ________ (3) ochre ________

(4) mercury bichloride ________ (5) calamine ________ (6) Mohr's Salt ________

(7) Wumingyi ________ (8) green alum ________ (9) blue vitriol ________

(10) copper green ________

2. Judgement question

(1) All the atoms containing *d* electrons belong to d-block elements.

(2) Among all metals, the metal with the highest melting point belongs to the subgroup element, and the metal with the lowest melting point also belongs to the subgroup element.

(3) When Pb^{2+} ion is added into $K_2Cr_2O_7$ solution, yellow $PbCrO_4$ precipitate is formed.

(4) In the alkaline medium, $KMnO_4$ will fade when it reacts with Na_2SO_3.

(5) $(NH_4)_2SO_4 \cdot FeSO_4 \cdot 6H_2O$ is not as easy to preserve as $FeSO_4$.

(6) Fe^{3+} can react with ammonia to form $[Fe(NH_3)_6]^{3+}$.

(7) In the presence of air, copper can dissolve in ammonia.

(8) ZnO has astringent effect and can promote wound healing.

(9) $HgCl_2$ is commonly known as calomel.

(10) Red powder in traditional Chinese medicine refers to Fe_2O_3 red brown powder.

3. Complete and balance the following equations

(1) $CrO_2^- + H_2O_2 + OH^- \rightarrow$

(2) $MnO_4^- + SO_3^{2-} + H^+ \rightarrow$

(3) $Mn^{2+} + Na_2BiO_3(s) + H^+ \rightarrow$

4. Explain the following experimental phenomena with chemical reaction equations

(1) Under the condition of absolute absence of oxygen, when NaOH solution is added to the solution containing Fe^{2+}, white precipitate is formed;

(2) After exposed in air, white precipitate gradually transforms into reddish brown precipitate;

(3) After filtration, the precipitate is dissolved in hydrochloric acid converting into yellow solution;

(4) When a few drops of KSCN solution is added to the yellow solution, it turns blood red immediately, and when SO_2 is introduced, the red disappears;

(5) The purple color of $KMnO_4$ solution will fade away when it is added to the above solution;

(6) When yellow blood salt solution is added, blue precipitate is formed.

5. Q & A questions

(1) Why is the molecular formula of mercurous chloride written as Hg_2Cl_2 instead of HgCl?

(2) When NaOH solution is added to the solution of Fe^{2+} salt in the air, what products are obtained respectively? Write the relevant chemical reaction formula?

(3) A is an orange red solid of chromium compound which can be dissolved in water. When A is treated with concentrated HCl, it will produce a yellow green irritant gas B and a dark green solution C. When KOH solution is added to C, it will precipitate D in grayish blue. If excessive KOH solution is added, the precipitate will disappear and become green solution E. When H_2O_2 is added into E, the yellow solution F will be formed after heating. F will be acidified with dilute acid, and then it will become the solution of original compound A.

Q: what substances are A, B, C, D, E, and F, respectively?

目标检测

1．填空题：写出下列物质的化学式

（1）朱砂 ________，（2）轻粉 ________，（3）代赭石 ________，（4）升汞 ________，（5）炉甘石 ________，（6）摩尔盐 ________，（7）无名异 ________，（8）绿矾 ________，（9）胆矾 ________，（10）铜绿 ________。

2．判断题

（1）含有 d 电子的原子都属于 d 区元素。

（2）在所有的金属中，熔点最高的属于副族元素，熔点最低的也属于副族元素。

（3）向 $K_2Cr_2O_7$ 溶液中加入 Pb^{2+} 离子，将生成黄色的 $PbCrO_4$ 沉淀。

（4）在碱性介质中，$KMnO_4$ 与 Na_2SO_3 反应后褪色。

（5）$(NH_4)_2SO_4 \cdot FeSO_4 \cdot 6H_2O$ 不如 $FeSO_4$ 易保存。

（6）Fe^{3+} 能与氨水反应生成 $[Fe(NH_3)_6]^{3+}$。

（7）在空气存在的条件下，铜可以溶于氨水。

（8）ZnO 具有收敛作用，能促进创面愈合。

（9）$HgCl_2$ 俗称甘汞。

（10）中药中的红粉指的是 Fe_2O_3 红棕色粉末。

3．完成并配平下列反应式

（1）$CrO_2^- + H_2O_2 + OH^- \rightarrow$

（2）$MnO_4^- + SO_3^{2-} + H^+ \rightarrow$

（3）$Mn^{2+} + Na_2BiO_3(s) + H^+ \rightarrow$

4．用化学反应方程式说明下列实验现象

（1）在绝对无氧的条件下，向含有 Fe^{2+} 的溶液中加入 NaOH 溶液后，生成白色沉淀。

（2）暴露于空气中后，白色沉淀逐渐转化成红棕色沉淀。

（3）过滤后的沉淀溶于盐酸得到黄色溶液。

（4）向黄色溶液中加几滴 KSCN 溶液，立即变血红色，再通入 SO_2，则红色消失。

（5）向此溶液中滴加 $KMnO_4$ 溶液，$KMnO_4$ 紫色褪去。

（6）最后加入黄血盐溶液时，生成蓝色沉淀。

5. 问答题

（1）为什么氯化亚汞的分子式要写成 Hg_2Cl_2 而不能写成 HgCl?

（2）在 Fe^{2+} 盐溶液中加入 NaOH 溶液，在空气中放置后，得到哪些产物？写出相关的化学反应式?

（3）铬的某化合物 A 是一橙红色溶于水的固体，将 A 用浓 HCl 处理产生黄绿色刺激性气体 B 和暗绿色溶液 C。在 C 中加入 KOH 溶液，先生成灰蓝色沉淀 D，继续加入过量的 KOH 溶液则沉淀消失，变为绿色溶液 E。在 E 中加入 H_2O_2，加热则生成黄色溶液 F，F 用稀酸酸化，又变为原来化合物 A 的溶液。问：A、B、C、D、E、F 各是什么物质?

Appendix

Appendix 1 Physical and Chemical Constants

Name	Symbol	Value and Unit
Molar volume	V_m	22.41410±0.00019 dm^3/mol (273.15K, 101.3kPa)
		22.71108±0.00019 dm^3/mol (273.15K, 100 kPa)
Standard pressure	$p^\ominus$	1bar = 10^5Pa
Molar gas constant	R	8.314510 (70) J/(mol・K)
Boltzman's constant	k	1.380658 (12)×10^{-23}J/K
Avogadro's number	N_A	6.0221367 (36)×10^{23} /mol
Triple point of water	$T_{tp}(H_2O)$	273.16K
Boiling point of water	$t_b(H_2O)$	99.975°C (1990.1.1)
Faraday's constant	F	9.6485309(29)×10^4C/mol
Planck's constant	h	6.6260755(40)×10^{-34}J・s
Speed of light(in a vacuum)	c_0	299792458 m/s
Electron charge	e	1.60217733(49)×10^{-19} C
Electron mass	m_e	9.1093897(54)×10^{-31}kg
Rydbergs constant	R_∞	10973731.534(13) /m
Bohr radius	a_0	5.29177249(24)×10^{-11} m
Bohr magneton	μ_B	9.2740154(31)×10^{-24} J/T
Dielectric constant (in a vacuum)	ε_0	8.854187816×10^{-12} F/m
Atomic unit $\frac{1}{12}m(^{12}C)$	m_u	1.6605402(10)×10^{-27} kg = 1 u

Appendix 2 Thermodynamic Properties of Some Selected Inorganic Compounds

Compound Name	$\Delta_f H_m^\ominus$ (298.15K)kJ/mol	$\Delta_f G_m^\ominus$ (298.15K)kJ/mol	$S_m^\ominus$ (298.15K)J/(K·mol)
Ag(s)	0	0	42.55
Ag_2O(s)	– 31.1	– 11.20	121.3
Al(s)	0	0	28.83
α – Al_2O_3(s)	– 1675.7	– 1582.3	50.92
Br_2(l)	0	0	152.23
Br_2(g)	30.91	3.11	245.46
C(s, graphite)	0	0	5.74
C(s, diamond)	1.90	2.90	2.38
CCl_4(l)	– 135.4	– 65.20	216.4
CO(g)	– 110.52	– 137.17	197.67
CO_2(g)	– 393.51	– 394.36	213.74
Ca(s)	0	0	41.42
CaO(s)	– 635.09	– 604.03	39.75
$Ca(OH)_2$(s)	– 986.09	– 898.49	83.39
$CaCO_3$(s, calcite)	– 1206.92	– 1128.79	92.9
Cl_2(g)	0	0	223.07
Cu(s)	0	0	33.15
Cu_2O(s)	– 168.6	– 146.0	93.14
CuO(s)	– 157.3	– 129.7	42.63
F_2(g)	0	0	202.78
Fe(s)	0	0	27.28
$FeCl_3$(s)	– 399.49	– 334.00	142.3
Fe_2O_3(s, haematite)	– 824.2	– 742.2	87.40
Fe_3O_4(s, magnetite)	– 1118.4	– 1015.4	146.4
H_2(g)	0	0	130.68
HBr(g)	– 36.40	– 53.45	198.70
HCl(g)	– 92.31	– 95.30	186.91
HF(g)	– 271.1	– 273.1	173.78
HI(g)	26.48	1.70	206.59

(continued)

Compound Name	$\Delta_f H_m^\ominus$ (298.15K)kJ/mol	$\Delta_f G_m^\ominus$ (298.15K)kJ/mol	$S_m^\ominus$ (298.15K)J/(K·mol)
HNO_3(l)	– 174.10	– 80.71	155.60
HNO_3(g)	– 135.06	– 74.72	266.38
H_3PO_4(s)	– 1279.0	– 1119.1	110.50
H_2S(g)	– 20.63	– 33.56	205.79
H_2O(l)	– 285.83	– 237.13	69.91
H_2O(g)	– 241.82	– 228.57	188.82
I_2(s)	0	0	116.14
I_2(g)	62.44	19.33	260.69
K(s)	0	0	64.18
KCl(s)	– 436.75	– 409.14	82.59
Mg(s)	0	0	32.68
$MgCl_2$(s)	– 641.32	– 591.79	89.62
MgO(s)	-601.70	– 569.43	26.94
$Mg(OH)_2$(s)	– 924.54	– 833.51	63.18
Na(s)	0	0	51.21
NaCl(s)	– 411.15	– 384.14	72.13
Na_2O(s)	– 414.22	– 375.46	75.06
NaOH(s)	– 425.61	– 379.49	64.46
Na_2CO_3(s)	– 1130.68	– 1044.44	134.98
Na_2SO_4(s)	-1387.08	– 1270.16	149.58
N_2(g)	0	0	191.61
NH_3(g)	– 46.11	– 16.45	192.45
NO(g)	90.25	86.55	210.76
NO_2(g)	33.18	51.31	240.06
N_2O(g)	82.05	104.20	219.85
N_2O_4(g)	9.16	97.89	304.29
N_2O_5(g)	11.3	115.1	355.7
O_3(g)	142.7	163.2	238.93
O_2(g)	0	0	205.14
P(s,white)	0.0	0.0	41.09
P(s,red)	– 17.6	– 12.1	22.80
S(s, rhombic)	0	0	31.80
S_8(g)	102.3	49.63	430.98
SO_2(g)	– 296.83	– 300.19	248.22
SO_3(g)	– 395.72	– 371.06	256.76

(continued)

Compound Name	$\Delta_f H_m^\ominus$ (298.15K)kJ/mol	$\Delta_f G_m^\ominus$ (298.15K)kJ/mol	$S_m^\ominus$ (298.15K)J/(K·mol)
Si(s)	0.0	0.0	18.83
SiO_2(s, quartz)	– 910.94	– 856.64	41.84
SiO_2(s, amorphous)	– 903.49	–850.70	46.9
$SiCl_4$(l)	– 687.0	– 619.83	240
$SiCl_4$(g)	– 657.01	– 616.98	330.7
Zn(s)	0	0	41.63
ZnO(s)	– 348.28	– 318.30	43.64
$ZnCl_2$(s)	– 415.05	– 369.40	111.46
$ZnCO_3$(s)	– 812.78	– 731.52	82.4
CH_4(g)	– 74.81	– 50.72	186.26
C_2H_2(g)	226.73	209.20	200.94
C_2H_4(g)	52.26	68.15	219.56
C_2H_6(g)	– 84.68	– 32.82	229.60
C_3H_8(g)	– 103.85	– 23.37	270.02
C_6H_6(g)	82.93	129.73	269.31
C_6H_6(l)	49.04	– 124.45	173.26
CH_3OH(l)	– 238.66	– 166.27	126.8
C_2H_5OH(l)	– 277.69	– 174.78	160.7
HCOOH(l)	– 424.72	– 361.35	128.95
CH_3COOH(l)	– 484.5	– 389.9	159.8
$(BH_2)_2CO$(s)	–332.9	– 196.7	104.6

Wagman D.D et al., The NBS tables of Chemical Thermodynamic Properties, Standards Press of China, 1998.

Appendix 3 Dissociation Constants of Weak Acids and Bases

Weak electrolyte	Molecular formula	Step	$K_a^\ominus$(or $K_b^\ominus$)	$pK_a^\ominus$(or $pK_b^\ominus$)
Arsenic acid	H_3AsO_4	1	5.50×10^{-3}	2.26
		2	1.74×10^{-7}	6.76
		3	5.13×10^{-12}	11.29
Arsenious acid	H_3AsO_3	1	5.13×10^{-10}	9.29
Boric acid	H_3BO_3	1	5.37×10^{-10}	9.27

(continued)

Weak electrolyte	Molecular formula	Step	$K_a^\ominus$ (or $K_b^\ominus$)	$pK_a^\ominus$ (or $pK_b^\ominus$)
Carbonic acid	H_2CO_3	1	4.47×10^{-7}	6.35
		2	4.68×10^{-11}	10.33
Hydrocyanic acid	HCN		6.17×10^{-10}	9.21
Chromic acid	H_2CrO_4	1	1.82×10^{-1}	0.74
		2	3.24×10^{-7}	6.49
Hydrofluoric acid	HF		6.31×10^{-4}	3.20
Nitrous acid	HNO_2		5.62×10^{-4}	3.25
Hydrogen peroxide	H_2O_2	1	2.40×10^{-12}	11.62
Phosphoric acid	H_3PO_4	1	6.92×10^{-3}	2.16
		2	6.17×10^{-8}	7.21
		3	4.79×10^{-13}	12.32
Phosphorous acid	H_3PO_3	1	5.01×10^{-2}	1.30
		2	2.00×10^{-7}	6.70
Hydrogen sulphuric acid	H_2S	1	8.91×10^{-8}	7.05
		2	1.00×10^{-19}	19.00
Sulphuric acid	H_2SO_4	2	1.02×10^{-2}	1.99
Sulphurous acid	H_2SO_3	1	1.41×10^{-2}	1.85
		2	6.31×10^{-8}	7.20
Thiocyanic acid	HSCN		0.141	0.85
Metasilicic acid	H_2SiO_3	1	1.70×10^{-10}	9.77
		2	1.60×10^{-12}	11.80
Hypochloric acid	HClO		3.98×10^{-8}	7.40
Hypobromous acid	HBrO		2.82×10^{-9}	8.55
Hypoiodous acid	HIO		3.16×10^{-11}	10.50
Thiosulfuric acid	$H_2S_2O_3$	1	2.52×10^{-1}	0.60
		2	1.90×10^{-2}	1.72
Formic acid	HCOOH		1.78×10^{-4}	3.75
Acetic acid	HAc		1.75×10^{-5}	4.756
Oxalic acid	$H_2C_2O_4$	1	5.62×10^{-2}	1.27
		2	1.55×10^{-4}	3.81
Ammonia	$NH_3\cdot H_2O$		1.74×10^{-5}	4.76
Hydroxylamine	$NH_2OH\cdot H_2O$		9.12×10^{-9}	8.04
Calcium hydroxide	$Ca(OH)_2$	1	3.72×10^{-3}	2.43
		2	3.98×10^{-2}	1.40
Silver hydroxide	AgOH		1.10×10^{-4}	3.96
Zinc hydroxide	$Zn(OH)_2$		9.55×10^{-4}	3.02

W. M. Haynes, CRC Handbook of Chemistry and Physics, 94th ed., 2013-2014

Appendix 4 Solubility Product Constants of Selected Slightly Electrolytes (291—298K)

Slightly electrolyte	$K_{sp}^{\ominus}$	Slightly electrolyte	$K_{sp}^{\ominus}$	Slightly electrolyte	$K_{sp}^{\ominus}$
Halide		As_2S_3	2.1×10^{-22}	CuSCN	1.77×10^{-13}
AgCl	1.77×10^{-10}	Ag_2S	6.3×10^{-50}	$Hg_2(CN)_2$	5×10^{-40}
AgBr	5.35×10^{-13}	Bi_2S_3	1.0×10^{-97}	$Hg_2(SCN)_2$	2.0×10^{-20}
AgI	8.52×10^{-17}	CuS	6.3×10^{-36}	Sulfate	
BiI_3	7.71×10^{-19}	Cu_2S	2.5×10^{-48}	Ag_2SO_4	1.20×10^{-5}
BaF_2	1.84×10^{-7}	α–CoS	4.0×10^{-21}	$BaSO_4$	1.08×10^{-10}
CuBr	6.27×10^{-9}	β–CoS	2.0×10^{-25}	$CaSO_4$	4.93×10^{-5}
CaF_2	3.45×10^{-11}	FeS	6.3×10^{-18}	Hg_2SO_4	6.5×10^{-7}
CuI	1.27×10^{-12}	Hg_2S	1.0×10^{-47}	$PbSO_4$	2.53×10^{-8}
CuCl	1.72×10^{-7}	HgS red	4×10^{-53}	$SrSO_4$	3.44×10^{-7}
Hg_2Cl_2	1.43×10^{-18}	HgS black	1.6×10^{-52}	Oxalate	
Hg_2I_2	5.2×10^{-29}	MnS crystal form	2.5×10^{-13}	$Ag_2C_2O_4$	5.40×10^{-12}
MgF_2	5.16×10^{-11}	MnS samorphous	2.5×10^{-10}	$BaC_2O_4\cdot H_2O$	2.3×10^{-8}
$PbBr_2$	6.6×10^{-6}	α-NiS	3.2×10^{-19}	BaC_2O_4	1.6×10^{-7}
PbI_2	9.8×10^{-9}	β-NiS	1.0×10^{-24}	$CaC_2O_4\cdot H_2O$	2.32×10^{-9}
PbF_2	3.3×10^{-8}	γ-NiS	2.0×10^{-26}	$CdC_2O_4\cdot 3H_2O$	1.42×10^{-8}
$PbCl_2$	1.7×10^{-5}	PbS	8.0×10^{-28}	$MgC_2O_4\cdot 2H_2O$	4.83×10^{-6}
SrF_2	4.33×10^{-9}	α-ZnS	1.6×10^{-24}	$MnC_2O_4\cdot 2H_2O$	1.7×10^{-7}
MgF_2	5.16×10^{-11}	β-ZnS	2.5×10^{-22}	$ZnC_2O_4\cdot 2H_2O$	1.38×10^{-9}
PbI_2	9.8×10^{-9}	Sb_2S_3	1.5×10^{-93}	Phosphate	
Hydroxid		CdS	8.0×10^{-27}	Ag_3PO_4	8.89×10^{-17}
AgOH	2.0×10^{-8}	Carbonate		$AlPO_4$	9.84×10^{-21}
$Al(OH)_3$	1.3×10^{-33}	Ag_2CO_3	8.46×10^{-12}	$Ba_3(PO_4)_2$	3.4×10^{-23}

(continued)

Slightly electrolyte	$K_{sp}^{\ominus}$	Slightly electrolyte	$K_{sp}^{\ominus}$	Slightly electrolyte	$K_{sp}^{\ominus}$
$Bi(OH)_3$	6.0×10^{-31}	$BaCO_3$	2.58×10^{-9}	$BiPO_4$	1.3×10^{-23}
$Co(OH)_2$ new	5.92×10^{-15}	$CaCO_3$	3.36×10^{-9}	BaP_2O_7	3.2×10^{-11}
CuOH	1×10^{-14}	$CoCO_3$	1.4×10^{-13}	$Ca_3(PO_4)_2$	2.07×10^{-29}
$Cu(OH)_2$	2.2×10^{-20}	$CuCO_3$	1.4×10^{-10}	$CaHPO_4$	1.0×10^{-7}
$Cr(OH)_3$	6.3×10^{-31}	$FeCO_3$	3.13×10^{-11}	$Co_3(PO_4)_2$	2.05×10^{-35}
$Ca(OH)_2$	5.5×10^{-6}	Hg_2CO_3	3.6×10^{-17}	$CoHPO_4$	2.0×10^{-7}
$Cd(OH)_2$ new	7.2×10^{-15}	$MnCO_3$	2.34×10^{-11}	$Cu_3(PO_4)_2$	1.40×10^{-37}
$Co(OH)_3$	1.6×10^{-44}	$MgCO_3$	6.82×10^{-6}	$FePO_4\cdot2H_2O$	9.91×10^{-16}
$Fe(OH)_3$	2.79×10^{-39}	$NiCO_3$	1.42×10^{-7}	$MgNH_4PO_4$	2.5×10^{-13}
$Fe(OH)_2$	4.87×10^{-17}	$PbCO_3$	7.4×10^{-14}	$Mg_3(PO_4)_2$	1.04×10^{-24}
$Hg(OH)_2$	3.2×10^{-26}	$SrCO_3$	5.6×10^{-10}	$Ni_3(PO_4)_2$	4.74×10^{-32}
$Hg_2(OH)_2$	2.0×10^{-24}	$ZnCO_3$	1.46×10^{-10}	$Pb_3(PO_4)_2$	8.0×10^{-43}
$Mg(OH)_2$	5.61×10^{-12}	Chromate		$PbHPO_4$	1.3×10^{-10}
$Mn(OH)_2$	1.9×10^{-13}	Ag_2CrO_4	1.12×10^{-12}	$Sr_3(PO_4)_2$	4.0×10^{-28}
$Ni(OH)_2$ new	5.48×10^{-16}	$BaCrO_4$	1.17×10^{-10}	$Zn_3(PO_4)_2$	9.0×10^{-33}
$Pb(OH)_2$	1.43×10^{-15}	$Ag_2Cr_2O_7$	2.0×10^{-7}	Other	
$Pb(OH)_4$	3.2×10^{-66}	$CaCrO_4$	7.1×10^{-4}	AgAc	1.94×10^{-3}
$Sn(OH)_2$	5.45×10^{-28}	$PbCrO_4$	2.8×10^{-13}	BiOCl	1.8×10^{-31}
$Sn(OH)_4$	1×10^{-56}	$SrCrO_4$	2.2×10^{-5}	$K[B(C_6H_5)_4]$	2.2×10^{-8}
$Zn(OH)_2$	3×10^{-17}	Cyanide and thiocyanide		$K_2[PtCl_6]$	7.48×10^{-6}
$Zn(OH)_2$ old	1.2×10^{-17}	AgSCN	1.03×10^{-12}	$KClO_4$	1.05×10^{-2}
$Ti(OH)_3$	1.68×10^{-44}	AgCN	5.97×10^{-17}	$Zn_2[Fe(CN)_6]$	4.0×10^{-16}
Sulfide		CuCN	3.47×10^{-20}		

J. G. Speight, Lange's "Handbook of Chemistry" 16th ed., 2005

Appendix 5　Electronegativities of Selected Elements

Atom	χ_{spec}	χ_p	$\chi_{A\&R}$	Atom	χ_{spec}	χ_p	$\chi_{A\&R}$
H	2.300	2.20	2.20	Na	0.869	0.93	1.01
Li	0.912	0.98	0.97	Mg	1.293	1.31	1.23
Be	1.576	1.57	1.47	Al	1.613	1.61	1.47
B	2.051	2.04	2.01	Si	1.916	1.90	1.74
C	2.544	2.55	2.50	P	2.253	2.19	2.06
N	3.066	3.04	3.07	S	2.589	2.58	2.44
O	3.610	3.44	3.50	Cl	2.869	3.16	2.83
F	4.193	3.98	4.10	Ar	3.242		
K	0.734	0.82	0.91	Rb	0.706	0.82	0.89
Ca	1.034	1.00	1.04	Sr	0.963	0.95	0.99
Ga	1.756	1.81	1.82	In	1.656	1.78	1.49
Ge	1.994	2.01	2.02	Sn	1.824	1.96	1.72
As	2.211	2.18	2.20	Sb	1.984	2.05	1.82
Se	2.424	2.55	2.48	Te	2.158	2.10	2.01
Br	2.685	2.96	2.74	I	2.359	2.66	2.21
Kr	2.966			Xe	2.582		
Ne	4.787						

Appendix 6　Standard Half-Cell Reduction Potentials of Selected Elements (291—298K)

1. In acid solution

Electrode reaction	$E_A^\ominus$/V
$Li^+ + e^- = Li$	−3.0401
$K^+ + e^- = K$	−2.931
$Ba^{2+} + 2e^- = Ba$	−2.912
$Sr^{2+} + 2e^- = Sr$	−2.899

(continued)

Electrode reaction	$E_A^{\ominus}$/V
$Ca^{2+} + 2e^- = Ca$	–2.868
$Na^+ + e^- = Na$	–2.71
$Mg^{2+} + 2e^- = Mg$	–2.372
$Al^{3+} + 3e^- = Al$	–1.676
$Mn^{2+} + 2e^- = Mn$	–1.185
$Se + 2e^- = Se^{2-}$	–0.924
$Cr^{2+} + 2e^- = Cr$	–0.913
$Zn^{2+} + 2e^- = Zn$	–0.7618
$Cr^{3+} + 3e^- = Cr$	–0.744
$Ag_2S(s) + 2e^- = 2Ag + S^{2-}$	–0.691
$As + 3H^+ + 3e^- = AsH_3$	–0.608
$Ga^{3+} + 3e^- = Ga$	–0.549
$H_3PO_3 + 2H^+ + 2e^- = H_3PO_2 + H_2O$	–0.499
$2CO_2 + 2H^+ + 2e^- = H_2C_2O_4$	–0.481
$S + 2e^- = S^{2-}$	–0.47627
$Fe^{2+} + 2e^- = Fe$	–0.447
$Cr^{3+} + e^- = Cr^{2+}$	–0.407
$Cd^{2+} + 2e^- = Cd$	–0.403
$Se + 2H^+ + 2e^- = H_2Se$	–0.399
$PbSO_4(s) + 2e^- = Pb + SO_4^{2-}$	–0.3588
$In^{3+} + 3e^- = In$	–0.3382
$Tl^+ + e^- = Tl$	–0.336
$Co^{2+} + 2e^- = Co$	–0.280
$H_3PO_4 + 2H^+ + 2e^- = H_3PO_3 + H_2O$	–0.276
$Ni^{2+} + 2e^- = Ni$	–0.257
$CuI(s) + e^- = Cu(s) + I^-$	–0.1858
$AgI(s) + e^- = Ag + I^-$	–0.15224
$Sn^{2+} + 2e^- = Sn$	–0.1375
$Pb^{2+} + 2e^- = Pb$	–0.1262
$Fe^{3+} + 3e^- = Fe$	–0.037
$2H^+ + 2e^- = H_2$	0.000
$AgBr(s) + e^- = Ag + Br^-$	+0.07133
$S_4O_6^{2-} + 2e^- = 2S_2O_3^{2-}$	+0.08

(continued)

Electrode reaction	$E_A^{\ominus}$/V
$TiO^{2+}+2H^{+}+e^{-} = Ti^{3+}+H_2O$	+0.1
$S+2H^{+}+2e^{-} = H_2S(g)$	+0.142
$Sn^{4+}+2e^{-} = Sn^{2+}$	+0.151
$Cu^{2+}+e^{-} = Cu^{+}$	+0.153
$SO_4^{2-}+4H^{+}+2e^{-}=H_2SO_3+H_2O$	+0.2172
$SbO^{+}+2H^{+}+3e^{-} = Sb+H_2O$	+0.212
$AgCl(s)+e^{-} = Ag +Cl^{-}$	+0.22233
$HAsO_2+3H^{+}+3e^{-} = As+2H_2O$	+0.248
$IO_3^{-}+3H_2O+6e^{-} = I^{-}+6OH^{-}$	+0.26
$Hg_2Cl_2(s)+2e^{-} = 2Hg+2Cl^{-}$	+0.26808
$BiO^{+}+2H^{+}+3e^{-} = Bi+H_2O$	+0.302
$VO^{2+}+2H^{+}+e^{-} = V^{3+}+H_2O$	+0.337
$Cu^{2+}+2e^{-} = Cu$	+0.3419
$Fe(CN)_6^{3-}+e^{-} = Fe(CN)_6^{4-}$	+0.358
$2H_2SO_3+2H^{+}+4e^{-} = S_2O_3^{2-}+H_2O$	+0.40
$SO_3^{2-}+6H_2O +4e^{-} = S+6OH^{-}$	+0.45
$S_2O_3^{2-}+6H^{+} +4e^{-} = 2S+3H_2O$	+0.5
$4H_2SO_3+4H^{+}+6e^{-} = S_4O_6^{2-}+6H_2O$	+0.51
$Cu^{+}+e^{-} = Cu$	+0.521
$I_2(s)+2e^{-} = 2I^{-}$	+0.5355
$MnO_4^{-}+e^{-} = MnO_4^{2-}$	+0.558
$H_3AsO_4+2H^{+}+2e^{-} = H_3AsO_3+H_2O$	+0.560
$2HgCl_2+2e^{-}=Hg_2Cl_2(s)+2Cl^{-}$	+0.63
$O_2(g)+2H^{+}+2e^{-} = H_2O_2$	+0.695
$Fe^{3+}+e^{-} = Fe^{2+}$	+0.771
$Hg_2^{2+}+2e^{-} = 2Hg$	+0.7973
$Ag^{+}+e^{-} = Ag$	+0.7996
$AuBr_4^{-}+2e^{-} = AuBr_2+2Br^{-}$	+0.802
$Hg^{2+}+2e^{-} = Hg$	+0.851
$AuBr_4^{-}+3e^{-} = Au+4Br^{-}$	+0.854
$Cu^{+}+I^{-}+e^{-} = CuI(s)$	+0.86
$2Hg^{2+}+2e^{-} = Hg_2^{2+}$	+0.920
$NO_3^{-}+3H^{+}+2e^{-} = HNO_2+H_2O$	+0.934
$AuBr_2^{-}+e^{-} = Au+2Br^{-}$	+0.959

(continued)

Electrode reaction	$E^{\ominus}_{A}$/V
$HNO_2+H^++e^- = NO(g)+H_2O$	+0.983
$HIO+H^++2e^- = I^-+H_2O$	+0.987
$VO_2^++2H^++e^- = VO^{2+}+H_2O$	+0.991
$AuCl_4^-+3e^- = Au+4Cl^-$	+1.002
$Br_2(l)+2e^- = 2Br^-$	+1.066
$Br_2(H_2O)+2e^- = 2Br^-$	+1.0873
$IO_3^-+5H^++4e^- = HIO+2H_2O$	+1.14
$ClO_3^-+3H^++2e^- = HClO_2+H_2O$	+1.181
$ClO_4^-+2H^++2e^- = ClO_3^-+H_2O$	+1.189
$IO_3^-+6H^++5e^- = 1/2I_2+3H_2O$	+1.195
$MnO_2(s)+4H^++2e^- = Mn^{2+}+2H_2O$	+1.224
$O_2(g)+4H^++4e^- = 2H_2O$	+1.229
$Cl_2(g)+2e^- = 2Cl^-$	+1.3583
$Cr_2O_7^{2-}+14H^++6e^- = 2Cr^{3+}+7H_2O$	+1.36
$ClO_4^-+8H^++8e^- = Cl^-+4H_2O$	+1.389
$ClO_4^-+8H^++7e^- = 1/2Cl_2+4H_2O$	+1.39
$BrO_3^-+6H^++6e^- = Br^-+3H_2O$	+1.423
$HIO+H^++e^- = 1/2I_2+H_2O$	+1.439
$HBrO+H^++2e^- = Br^-+H_2O$	+1.444
$ClO_3^-+6H^++6e^- =Cl^-+3H_2O$	+1.451
$PbO_2(s)+4H^++2e^- = Pb^{2+}+2H_2O$	+1.455
$ClO_3^-+6H^++5e^- = 1/2Cl_2+2H_2O$	+1.47
$HClO+H^++2e^- = Cl^-+H_2O$	+1.482
$BrO_3^-+6H^++5e^- =1/2\ Br_2+3H_2O$	+1.482
$Au^{3+}+3e^- = Au$	+1.498
$MnO_4^-+8H^++5e^- = Mn^{2+}+4H_2O$	+1.507
$Mn^{3+}+ e^- = Mn^{2+}(7.5mol.L^{-1}H_2SO_4)$	+1.5415
$HBrO+H^++e^- = 1/2Br_2(H_2O)+H_2O$	+1.574
$HBrO+H^++e^- = 1/2Br_2(l)+H_2O$	+1.596
$H_5IO_6+H^++2e^- = IO_3^-+3H_2O$	+1.601
$HClO+H^++e^- = 1/2Cl_2+H_2O$	+1.611
$HClO_2+2H^++2e^- = HClO+H_2O$	+1.645
$MnO_4^-+4H^++3e^- = MnO_2+4H_2O$	+1.679
$PbO_2(s)+SO_4^{2-}+4H^++2e^- = PbSO_4(s)+2H_2O$	+1.6913

(continued)

Electrode reaction	$E_A^\ominus$/V
$Au^+ + e^- = Au$	+1.692
$Ce^{4+} + e^- = Ce^{3+}$	+1.72
$H_2O_2 + 2H^+ + 2e^- = 2H_2O$	+1.776
$BrO_4^- + 2H^+ + 2e^- = BrO_3^- + H_2O$	+1.853
$Co^{3+} + e^- = Co^{2+}$	+1.92
$S_2O_8^{2-} + 2e^- = 2SO_4^{2-}$	+2.010
$O_3 + 2H^+ + 2e^- = O_2 + H_2O$	+2.076
$S_2O_8^{2-} + 2H^+ + 2e^- = 2HSO_4^-$	+2.123
$FeO_4^{2-} + 8H^+ + 3e^- = Fe^{3+} + 4H_2O$	+2.20
$F_2(g) + 2e = 2F^-$	+2.866
$F_2(g) + 2H^+ + 2e^- = 2HF$	+3.053

2. In alkaline solution

Electrode reaction	$E_B^\ominus$/V
$Ca(OH)_2 + 2e^- = Ca + 2OH^-$	−3.02
$Ba(OH)_2 + 2e^- = Ba + 2OH^-$	−2.99
$La(OH)_3 + 3e^- = La + 3OH^-$	−2.90
$Mg(OH)_2 + 2e^- = Mg + 2OH^-$	−2.69
$H_2BO_3^- + H_2O + 3e^- = B + 4OH^-$	−2.5
$SiO_3^{2-} + 3H_2O + 4e^- = Si + 6OH^-$	−1.697
$HPO_3^{2-} + 3H_2O + 2e^- = H_2PO_2^- + 3OH^-$	−1.65
$Mn(OH)_2 + 2e^- = Mn + 2OH^-$	−1.56
$Cr(OH)_3 + 3e^- = Cr + 3OH^-$	−1.48
$As + 3H_2O + 3e^- = AsH_3 + 3OH^-$	−1.37
$Zn(CN)_4^{2-} + 2e^- = Zn + 4CN^-$	−1.34
$ZnO_2^{2-} + 2H_2O + 2e^- = Zn + 4OH^-$	−1.215
$CrO_2^- + 2H_2O + 3e^- = Cr + 4OH^-$	−1.2
$2SO_3^{2-} + 2H_2O + 2e^- = S_2O_4^{2-} + 4OH^-$	−1.12
$PO_4^{3-} + 2H_2O + 2e^- = HPO_3^{2-} + 3OH^-$	−1.05
$Zn(NH_3)_4^{2+} + 2e^- = Zn + 4NH_3$	−1.04
$SO_4^{2-} + H_2O + 2e^- = SO_3^{2-} + 2OH^-$	−0.93
$P + 3H_2O + 3e^- = PH_3$ (气) $+ 3OH^-$	−0.87
$2NO_3^- + 2H_2O + 2e^- = N_2O_4 + 3OH^-$	−0.85
$Co(OH)_2 + 2e^- = Co + 2OH^-$	−0.73
$SO_3^{2-} + 3H_2O + 4e^- = S + 6OH^-$	−0.59

(continued)

Electrode reaction	$E_B^{\ominus}$/V
$PbO + H_2O + 2e^- = Pb + 2OH^-$	−0.580
$2SO_3^{2-} + 3H_2O + 4e^- = S_2O_3^{2-} + 6OH^-$	−0.571
$Fe(OH)_3+e^-= Fe(OH)_2+OH^-$	−0.56
$S+2e^- =S^{2-}$	−0.4763
$NO_2^-+H_2O+e^- =NO+2OH^-$	−0.46
$Cu(OH)_2+2e^- =Cu+2OH^-$	−0.222
$CrO_4^{2-} + 2H_2O +3e^- = Cr(OH)_3 + 5OH^-$	−0.13
$O_2 + H_2O + 2e^- = HO_2^- + OH^-$	−0.076
$HgO+H_2O+2e^-=Hg+2OH^-$	+0.0977
$[Co(NH_3)_6]^{3+}+e^- = [Co(NH_3)_6]^{2+}$	+0.108
$IO_3^-+2H_2O+4e^- =IO^-+4OH^-$	+0.15
$IO_3^-+3H_2O+6e^- =I^-+6OH^-$	+0.26
$O_2+2H_2O+4e^- =4OH^-$	+0.401
$IO^-+H_2O+2e^- =I^-+2OH^-$	+0.485
$MnO_4^-+2H_2O+3e^- =MnO_2+4OH^-$	+0.595
$MnO_4^{2-}+2H_2O+2e^- =MnO_2+4OH^-$	+0.60
$ClO_3^-+3H_2O+6e^- =Cl^-+6OH^-$	+0.62
$ClO^-+H_2O+2e^- =Cl^-+2OH^-$	+0.81
$O_3+H_2O+2e^- =O_2+2OH^-$	+1.24
$Cl_2(g) +2e^- =2Cl^-$	+1.3583

Appendix 7 Formation Constants of Some Selected Complex Ions in Aqueous Solution (293—298K, I=0)

Ligand	Metal ion	Ligand number n	$K_S^{\ominus}$	$lgK_S^{\ominus}$
Cl^-	Ag^+	2	1.1×10^5	5.04
	Cd^{2+}	4	6.31×10^2	2.80
	Co^{3+}	1	2.630	1.42
	Cu^+	3	5.01×10^5	5.7
	Hg^{2+}	4	1.17×10^{15}	15.07
	Pt^{2+}	4	1.0×10^{16}	16.0

(continued)

Ligand	Metal ion	Ligand number n	$K_s^\ominus$	$lgK_s^\ominus$
Cl^-	Sb^{3+}	6	1.29×10^4	4.11
	Sn^{2+}	4	3.02	1.48
	Tl^{3+}	4	1.00×10^{18}	18.00
	Zn^{2+}	4	1.58	0.20
Br^-	Ag^+	4	5.37×10^8	8.73
	Bi^{3+}	4	1.99×10^7	7.30
	Bi^{3+}	6	5.01×10^9	9.70
	Cd^{2+}	4	5.01×10^3	3.70
NH_3	Ag^+	2	1.12×10^7	7.05
	Cd^{2+}	4	1.32×10^7	7.12
	Cd^{2+}	6	1.38×10^5	5.14
	Co^{2+}	6	1.29×10^5	5.11
	Co^{3+}	6	1.58×10^{35}	35.2
	Cu^+	2	7.24×10^{10}	10.86
	Cu^{2+}	4	2.09×10^{13}	13.32
	Fe^{2+}	2	1.58×10^2	2.2
	Hg^{2+}	4	1.9×10^{19}	19.28
	Ni^{2+}	4	9.12×10^7	7.96
	Ni^{2+}	6	5.5×10^8	8.74
	Pt^{2+}	6	2.00×10^{35}	35.3
	Zn^{2+}	4	2.88×10^9	9.46
CN^-	Ag^+	2	1.26×10^{21}	21.1
	Au^+	2	2.00×10^{38}	38.3
	Cd^{2+}	4	6.02×10^{18}	18.78
	Cu^+	2	1.0×10^{24}	24.0
	Cu^+	4	2.00×10^{30}	30.30
	Fe^{2+}	6	1.0×10^{35}	35
	Fe^{3+}	6	1.0×10^{42}	42
	Hg^{2+}	4	2.51×10^{41}	41.4
	Ni^{2+}	4	2.0×10^{31}	31.3
	Zn^{2+}	4	5.01×10^{16}	16.70
F^-	Al^{3+}	6	6.92×10^{19}	19.84
	Fe^{2+}	1	6.3	0.8
	Fe^{3+}	1	1.9×10^5	5.28
	Fe^{3+}	2	2.0×10^9	9.30
	Fe^{3+}	3	1.15×10^{12}	12.06
	Fe^{3+}	5	5.89×10^{15}	15.77
	Sb^{3+}	4	7.94×10^{10}	10.9
	Sn^{2+}	3	3.16×10^9	9.50

(continued)

Ligand	Metal ion	Ligand number n	$K_s^\ominus$	$lgK_s^\ominus$
I^-	Ag^{2+}	2	5.5×10^{11}	11.74
	Ag^{2+}	3	4.79×10^{13}	13.68
	Bi^{3+}	6	6.31×10^{18}	18.80
	Cd^{2+}	4	2.57×10^{5}	5.41
	Cu^{+}	2	7.08×10^{8}	8.85
	Hg^{2+}	2	6.61×10^{23}	23.82
	Hg^{2+}	4	6.76×10^{29}	29.83
	Pb^{2+}	4	2.95×10^{4}	4.47
SCN^-	Ag^{+}	2	3.72×10^{7}	7.57
	Ag^{+}	4	1.20×10^{10}	10.08
	Cu^{+}	2	1.51×10^{5}	5.18
	Cd^{2+}	4	4.0×10^{3}	3.6
	Fe^{3+}	3	1.00×10^{5}	5.00
	Fe^{3+}	6	1.26×10^{6}	6.10
	Hg^{2+}	4	1.70×10^{21}	21.23
$S_2O_3^{2-}$	Ag^{+}	2	2.88×10^{13}	13.46
	Cd^{2+}	2	$2..75\times10^{6}$	6.44
	Cu^{+}	2	1.66×10^{12}	12.22
	Hg^{2+}	4	1.74×10^{33}	33.24
EDTA(Y^{4-})	Al^{3+}	1	1.35×10^{16}	16.13
	Bi^{3+}	1	6.31×10^{22}	22.8
	Ca^{2+}	1	1.0×10^{11}	11.0
	Cd^{2+}	1	2.51×10^{16}	16.4
	Co^{2+}	1	2.04×10^{16}	16.31
	Co^{3+}	1	1.00×10^{36}	36
	Cr^{3+}	1	1.0×10^{23}	23.0
	Cu^{2+}	1	5.01×10^{18}	18.7
	Fe^{2+}	1	2.14×10^{14}	14.33
	Hg^{2+}	1	6.31×10^{21}	21.80
	Mg^{2+}	1	4.36×10^{8}	8.64
	Ni^{2+}	1	3.63×10^{18}	18.56
	Pb^{2+}	1	2.00×10^{18}	18.3
	Sn^{2+}	1	1.26×10^{22}	22.1
	Zn^{2+}	1	2.51×10^{16}	16.4
en	Ag^{+}	2	5.01×10^{7}	7.70
	Cd^{2+}	3	1.23×10^{12}	12.09
	Co^{2+}	3	8.71×10^{13}	13.94
	Co^{3+}	3	4.90×10^{48}	48.69
	Cu^{+}	2	6.31×10^{10}	10.80

(continued)

Ligand	Metal ion	Ligand number n	$K_s^\ominus$	$lgK_s^\ominus$
en	Cu^{2+}	2	1.00×10^{20}	20.00
	Fe^{2+}	3	5.01×10^{9}	9.70
	Hg^{2+}	2	2.00×10^{23}	23.3
	Mn^{2+}	3	4.68×10^{5}	5.67
	Ni^{2+}	3	2.14×10^{18}	18.33
	Zn^{2+}	3	1.29×10^{14}	14.11
$C_2O_4^{2-}$	Co^{2+}	3	5.01×10^{9}	9.7
	Cu^{2+}	2	3.16×10^{8}	8.5
	Fe^{2+}	3	1.66×10^{5}	5.22
	Fe^{3+}	3	1.58×10^{20}	20.2
	Mn^{2+}	2	6.31×10^{5}	5.80
	Mn^{3+}	3	2.63×10^{19}	19.42
	Ni^{2+}	3	3.16×10^{8}	-8.5

Appendix 8 Selected Bond Energies

Hydrogen					
H—H	435.8	H—O	429.9	H—S	353.6
H—C	338.4	H—N	338.9	H—B	345.2
H—F	569.7	H—Cl	431.4	H—Br	366.2
H—I	298.3				
Group 13					
B—C	448	B—N	377.9	B—O	809
B—F	732	B—Cl	427	B—Br	390.9
B—I	361				
Group 14					
C—C	618.3	C—O	1076.4	C—N	750.0
C— F	513.8	C—Cl	394.9	C—Br	318.0
C—I	253.1	C—P	507.5		
Si—Si	310	Si—F	576.4	Si—Cl	416.7
Si—Br	358.2	Si—I	243.1	Si—O	799.6

(continued)

Group 15					
N—N	944.8	N—O	631.6	N—F	349
N—Cl	333.9	N—Br	280.8	N—I	159
P—P	489.1	P—F	405	P—Cl	376
P—Br	329				
Group 16					
O—O	498.4	O—F	220	O—Cl	267.5
O—Br	237.6	O—I	240	O—S	517.9
S—S	425.3	S—F	343.5	S—Cl	241.8
S—Br	218				
Group 17					
F—F	158.7	F—Cl	260.8	F—Br	280
F—I	271.5	F—Xe	14.18		
Cl—Cl	242.4	Cl—Br	219.3	Cl—I	211.3
Cl—Xe	7.08				
Br—Br	193.9	Br—I	179.1	Br—Xe	5.94
I—I	152.3	Br—Xe	6.48		
Group 18					
Xe—Xe	6.023				

Appendix 9 Charge Densities of Selected Ions

Cation	Charge density	Cation	Charge density	Cation	Charge density
Ac^{3+}	57	At^{7+}	609	Bk^{3+}	86
Ag^{+}	15	Au^{+}	11	Br^{7+}	1796
Ag^{2+}	60	Au^{3+}	118	C^{4+}	6265(T)
Ag^{3+}	163	B^{3+}	7334(T)	Ca^{2+}	52
Al^{3+}	770(T)	B^{3+}	1663	Cd^{2+}	59
Al^{3+}	364	Ba^{2+}	23	Ce^{3+}	75
Am^{3+}	82	Be^{2+}	1108(T)	Ce^{4+}	148
As^{3+}	307	Bi^{3+}	72	Cf^{3+}	88

(continued)

Cation	Charge density	Cation	Charge density	Cation	Charge density
As^{5+}	884	Bi^{5+}	262	Cl^{7+}	3880
Cm^{3+}	84	K^{+}	11	Pu^{4+}	153
Co^{2+}	155(LS)	La^{3+}	72	Ra^{2+}	18
Co^{2+}	108(HS)	Li^{+}	98(T)	Rb^{+}	8
Co^{3+}	349(LS)	Li^{+}	52	Rc^{7+}	889
Co^{3+}	272(HS)	Lu^{3+}	115	Rh^{3+}	224
Cr^{2+}	116(LS)	Mn^{2+}	144(LS)	S^{4+}	1 152
Cr^{2+}	92(HS)	Mn^{2+}	84(HS)	S^{6+}	2 883
Cr^{3+}	261	Mn^{3+}	307(LS)	Sb^{3+}	157
Cr^{4+}	465	Mn^{3+}	232(HS)	Sb^{5+}	471
Cr^{5+}	764	Mn^{4+}	508	Sc^{3+}	163
Cr^{6+}	1175	Mn^{7+}	1238	Sc^{4+}	583
Cs^{+}	6	Mo^{3+}	200	Se^{6+}	1305
Cu^{+}	51	Mo^{6+}	589	Si^{4+}	970
Cu^{2+}	116	NH_4^{+}	11	Sm^{3+}	86
Dy^{2+}	43	Na^{+}	24	Sn^{2+}	54
Dy^{3+}	99	Nb^{3+}	180	Sn^{4+}	267
Er^{3+}	105	Nb^{5+}	402	Sr^{2+}	33
Eu^{2+}	34	Nd^{3+}	82	Ta^{3+}	180
Eu^{3+}	88	Ni^{2+}	134	Ta^{5+}	402
F^{7+}	25 110	No^{2+}	40	Tb^{3+}	96
Fe^{2+}	181 (LS)	Np^{5+}	271	Tc^{4+}	310
Fe^{2+}	98(HS)	Os^{4+}	335	Tc^{7+}	780
Fe^{3+}	349(LS)	Os^{6+}	698	Te^{4+}	112
Fe^{3+}	232(HS)	Os^{8+}	2 053	Te^{6+}	668
Fe^{6+}	3 864	P^{3+}	587	Th^{4+}	121
Fr^{+}	5	P^{6+}	1358	Ti^{2+}	76
Ga^{3+}	261	Pa^{5+}	245	Ti^{3+}	216
Gd^{3+}	91	Pb^{2+}	32	Ti^{4+}	362
Ge^{2+}	116	Pb^{4+}	196	Ti^{+}	9
Ge^{4+}	508	Pd^{2+}	76	Tl^{3+}	105
Hf^{4+}	409	Pd^{4+}	348	Tm^{2+}	48
Hg^{+}	16	Pm^{3+}	84	Tm^{3+}	108
Hg^{2+}	49	Po^{4+}	121	U^{4+}	140
Ho^{3+}	102	Po^{6+}	431	U^{6+}	348
I^{7+}	889	Pr^{3+}	79	V^{2+}	95
In^{3+}	138	Pr^{4+}	157	V^{3+}	241

(continued)

Cation	Charge density	Cation	Charge density	Cation	Charge density
Ir^{3+}	208	Pt^{2+}	92	V^{4+}	409
Ir^{5+}	534	Pt^{4+}	335	V^{5+}	607
W^{4+}	298	Y^{3+}	102	Zn^{2+}	112
W^{6+}	566	Yb^{3+}	111	Zr^{4+}	240

HS, high spin; LS, low spin; T, four-coordinate tetrahedral ions.

Anion	Charge density	Anion	Charge density	Anion	Charge density
As^{3-}	12	I^-	4	O^{2-}	19
Br^-	6	MnO_4^-	4	OH^-	23
CN^-	7	N^{3-}	50	p3⁻	14
CO_3^{2-}	17	N_3^-	6	S^{2-}	16
C 1⁻	8	NO_3^-	9	SO_4^{2-}	5
ClO_4^-	3	O^{2-}	^{d}46	Se^{2-}	12
F^-	24	O_2^-	13	Te^{2-}	9

Appendix 10 Elemental Symbols and English Names

Element	Chinese	English	Element	Chinese	English	Element	Chinese	English
H	氢	Hydrogen	Sr	锶	Strontium	Re	铼	Rhenium
He	氦	Helium	Y	钇	Yttrium	Os	锇	Osmium
Li	锂	Lithium	Zr	锆	Zirconium	Ir	铱	Iridium
Be	铍	Beryllium	Nb	铌	Niobium	Pt	铂	Platinum
B	硼	Boron	Mo	钼	Molybdenum	Au	金	Gold
C	碳	Carbon	Tc	锝	Technetium	Hg	汞	Mercury
N	氮	Nitrogen	Ru	钌	Ruthenium	Tl	铊	Thallium
O	氧	Oxygen	Rh	铑	Rhodium	Pb	铅	Lead
F	氟	Fluorine	Pd	钯	Palladium	Bi	铋	Bismuth
Ne	氖	Neon	Ag	银	Silver	Po	钋	Polonium
Na	钠	Sodium	Cd	镉	Cadmium	At	砹	Astatine
Mg	镁	Magnesium	In	铟	Indium	Rn	氡	Radon

(continued)

Element	Chinese	English	Element	Chinese	English	Element	Chinese	English
A l	铝	Aluminum	Sn	锡	Tin	Fr	钫	Francium
Si	硅	Silicon	Sb	锑	Antimony	Ra	镭	Radium
P	磷	Phosphorus	Te	碲	Tellurium	Ac	锕	Actinium
S	硫	Sulfur	I	碘	Iodine	Th	钍	T horium
Cl	氯	Chlorine	Xe	氙	Xeno n	Pa	镤	Protactinium
Ar	氩	Argon	Cs	铯	Cesium	U	铀	Uranium
K	钾	Potassium	Ba	钡	Barium	Np	镎	Neptunium
Ca	钙	Calcium	La	镧	lanthanum	Pu	钚	Plutonium
Sc	钪	Scandium	Ce	铈	Cerium	Am	镅	Americium
Ti	钛	Titanium	Pr	镨	Praseodymium	Cm	锔	Curium
V	钒	Vanadium	Nd	钕	Neodymium	Bk	锫	Berkelium
Cr	铬	Chromium	Pm	钷	Promethium	Cf	锎	Californium
Mn	锰	Manganese	Sm	钐	Samarium	Es	锿	Einsteinium
Fe	铁	Iron	Eu	铕	Europium	Fm	镄	Fermium
Co	钴	Cobalt	Gd	钆	Gadolinium	Md	钔	Mendelevium
Ni	镍	Nickel	Tb	铽	Terbium	No	锘	Nobelium
Cu	铜	Copper	Dy	镝	Dysprosium	Lw	铹	Lawrencium
Zn	锌	Zinc	Ho	钬	Holmium	Rf	钅卢	unnilquadium
Ga	镓	Gallium	Er	铒	Erbium	Db	钅杜	dubnium
Ge	锗	Germanium	Tm	铥	Thulium	Sg	𬭳	Seaborgium
As	砷	Arsenic	Yb	镱	Ytterbium	Bh	𬭛	Bohrium
Se	硒	Selenium	Lu	镥	Lutetium	Hs	𬭶	Hassium
Br	溴	Bromine	Hf	铪	Hafnium	Mt	鿏	Meitneriumconti.
Kr	氪	Krypton	Ta	钽	Tantalum	Ds	𫟼	Darmstadtium
Rb	铷	Rubidium	W	钨	Tungsten	Rg	𬬭	Roentgenium

Appendix 11 Greek Alphabet Symbols and Chinese Transliteration

Upper case	Lower case	English	Chinese transliteration
Α	α	Alpha	阿尔法
Β	β	Beta	贝塔
Γ	γ	Gamma	伽马
Δ	δ	Delta	德尔塔
Ε	ε	Epsilon	艾普西龙
Ζ	ζ	Zeta	截塔
Η	η	Eta	艾塔
Θ	θ	Theta	西塔
Ι	ι	Iota	约塔
Κ	κ	Kappa	卡帕
Λ	λ	Lambda	兰姆达
Μ	μ	Mu	米尤
Ν	ν	Nu	纽
Ξ	ξ	Xi	克西
Ο	ο	Omicron	奥米克龙
Π	π	Pi	派
Ρ	ρ	Rho	洛
Σ	σ	Sigma	西格马
Τ	τ	Tau	套乌
Υ	υ	Upsilon	宇普西隆
Φ	φ	Phi	斐
Χ	χ	Chi	喜
Ψ	ψ	Psi	普西
Ω	ω	Omega	奥米伽

Answer to Exercises

Chapter 2

1. Choice question

(1) A (2) A (3) B (4) C (5) D (6) C (7) B (8) B (9) B

(10) C (11) A (12) A (13) C (14) D

2. Judgment Question

(1) T (2) T (3) F (4) F (5) F

Chapter 3

1. α = 66.7%
2. $p(CO) = 274kPa$, $p(H_2) = 2p(CO) = 548kPa$
3. α = 80%, $[CO]_{ep} = (1 - \alpha)/2 = 0.1mol/L$
4. 11.97
5. 0.20

Chapter 4

1. H_3O^+, $H_2PO_4^-$, H_3PO_4, HCO_3^-, H_2S.
2. OH^-, HCO_3^-, CN^-, NH_3, CH_3COO^-.
3. (1) 1.30 (2) 1.35
4. (1) 10^{-6} (2) 1%
5. pH = 10.97; $\alpha = 1.87\%$
6. 0.044mol/L
7. $[HCO_3^-] \approx [H^+] = 1.23\times10^{-4}mol/L$; pH = 3.91; $[CO_3^{2-}] = K_{a2}^{\ominus} = 4.68\times10^{-11}$ mol/L
8. $[H^+] = 8.8\times10^{-6}mol/L$; $\alpha = \dfrac{[H^+]}{0.10}\times100\% = 0.0088\%$
9. 9.4g
10. V=2.12 ml
11. $K_a^{\ominus} = 7.83\times10^{-6}$
12. pH = 6.00
13. (1) 9.24 (2) 5.12 (3) 1.40

Chapter 5

1. True or False

(1) ×, (2) ×, (3) ×, (4) √

2. One-Choice Questions

(1) D, (2) A, (3) B, (4) A, (5) C, (6) C, (7) C, (8) A, (9) C, (10) B

3. Calculation Question

(1) ① Answer: we assume the solubility of Ag_2CrO_4 is x mol/L in the 0.10mol/L $AgNO_3$ solution.

$$Ag_2CrO_4(s) \rightleftharpoons 2Ag^+ \quad + CrO_4^{2-}$$

equilibrium concentrations /(mol/L) $\quad 0.10 + 2x \approx 0.10 \quad x$

$K_{sp}^{\ominus}(Ag_2CrO_4)=[Ag^+]^2[CrO_4^{2-}]$;

$1.12\times10^{-12}=(0.10)^2 x$;

$\therefore x=1.12\times10^{-10}$mol/L

② Answer: we assume the solubility of Ag_2CrO_4 is y mol/L in the 0.10mol/L Na_2CrO_4 solution.

$$Ag_2CrO_4(s) \rightleftharpoons 2Ag^+ \quad + CrO_4^{2-}$$

equilibrium concentrations /(mol/L) $\quad 2y \quad 0.10 + y \approx 0.10$

$K_{sp}^{\ominus}(Ag_2CrO_4) = [Ag^+]^2[CrO_4^{2-}]$;

$1.12\times10^{-12} = (2y)^2 \times 0.10$;

$\therefore y = 1.67\times10^{-6}$mol/L

(2) Answer: ① we assume the solubility of $Zn(OH)_2$ is $s_1$mol/L in water.

$Zn(OH)_2(s) \rightleftharpoons Zn^{2+}+2OH^-$

$K_{sp}^{\ominus} = s_1 \cdot (2s_1)^2$

$$s_1 = \sqrt[3]{\frac{K_{sp}^{\ominus}[Zn(OH)_2]}{4}} = \sqrt[3]{\frac{3\times10^{-17}}{4}} = 2\times10^{-6}\text{mol/L}$$

② In the $Zn(OH)_2$ saturated solution:

$[Zn^{2+}] = 2\times10^{-6}$mol/L; $[OH^-] = 4\times10^{-6}$mol/L

pH = 14 − pOH = 14 − 5.4 = 8.6

③ We assume the solubility of $Zn(OH)_2$ is $s_2$mol/L in the solution of 0.10mol/L NaOH solution.

$$Zn(OH)_2(s) \rightleftharpoons Zn^{2+} + 2OH^-$$

equilibrium concentrations / mol/L $\quad s_2 \quad 2s_2+0.10$

Because of common-ion effect, s_2 is small and $2s_2 + 0.10 \approx 0.10$

$K_{sp}^{\ominus} = s_2(0.10)^2$

$$s_2 = \frac{K_{sp}^{\ominus}[Zn(OH)_2]}{0.01} = \frac{3\times10^{-17}}{0.01} = 3\times10^{-15}\text{mol/L}$$

④ We assume the solubility of $Zn(OH)_2$ is $s_3$mol/L in the solution of 0.10mol/L $ZnSO_4$ solution.

$$Zn(OH)_2(s) \rightleftharpoons Zn^{2+} + 2OH^-$$

equilibrium concentrations / mol/L $\quad s_3+0.10 \quad 2s_3$

Because of common-ion effect, s_3 is small and $s_3 + 0.10 \approx 0.10$

$K_{sp}^{\ominus} = 0.10 \cdot (2s_3)^2$

$$s_3 = \sqrt{\frac{K_{sp}^{\ominus}[Zn(OH)_2]}{4\times0.1}} = \sqrt{\frac{3\times10^{-17}}{0.4}} = 9\times10^{-9}\,mol/L$$

Chapter 6

1. +5; +2; +6; +5; −1; −1; +4; +3

2. (1) $14H^+ + Cr_2O_7^{2-} + 6Cl^- \rightleftharpoons 2Cr^{3+} + 3Cl_2 + 7H_2O$

(2) $3As_2S_3 + 14ClO_3^- + 18H_2O \rightleftharpoons 6H_3AsO_4 + 14Cl^- + 18H^+ + 9SO_4^{2-}$

(3) $6H^+ + 2MnO_4^- + 5SO_3^{2-} \rightleftharpoons 2Mn^{2+} + 5SO_4^{2-} + 3H_2O$

(4) $PbO_2 + 2Cl^- + 4H^+ \rightleftharpoons Pb^{2+} + Cl_2 + 2H_2O$

(5) $HgS + 2NO_3^- + 4Cl^- + 4H^+ \rightleftharpoons HgCl_4^{2-} + 2NO_2 + S + 2H_2O$

(6) $2CrO_4^{2-} + 3HSnO_2^- + H_2O \rightleftharpoons 3HSnO_3^- + 2CrO_2^- + 2OH^-$

3. (1) (−) Zn| $Zn^{2+}(c_1)$ || $Ag^+(c_2)$ | Ag (+)

Cathode: $Ag^+ + e^- \rightleftharpoons Ag$ Anode: $Zn^{2+} + 2e^- \rightleftharpoons Zn$

(2) (−) Pt| $Sn^{2+}(c_1), Sn^{4+}(c_2)$ || $Fe^{2+}(c_3), Fe^{3+}(c_4)$ | Pt (+)

Cathode: $Fe^{3+} + e^- \rightleftharpoons Fe^{2+}$ Anode: $Sn^{4+} + 2e^- \rightleftharpoons Sn^{2+}$

(3) (−) Pt| $Fe^{2+}(c_1), Fe^{3+}(c_2)$ || $Cl^-(c_3)$| $Cl_2(p)$ | Pt (+)

Cathode: $Cl_2 + e^- \rightleftharpoons 2Cl^-$ Anode: $Fe^{3+} + e^- \rightleftharpoons Fe^{2+}$

(4) (−) Zn | $Zn^{2+}(c_1)$ || $H^+(c_2)$| $H_2(p)$ | Pt (+)

Cathode: $2H^+ + 2e^- \rightleftharpoons H$ Anode: $Zn^{2+} + 2e^- \rightleftharpoons Zn$

4. (1) To prevent the oxidation of Sn^{2+} into Sn^{4+} by the dissolved oxygen in the solution; (2) Na_2SO_3 or $FeSO_4$ in solution is easily oxidized by the dissolved oxygen to Na_2SO_4 or $Fe_2(SO_4)_3$ and then deteriorated.

5. $E^{\ominus}(Zn^{2+}/Zn)$=−0.7618V, $E^{\ominus}(Fe^{2+}/Fe)$=−0.447V, $E^{\ominus}(Ce^{4+}/Ce^{3+})$=1.71V, $E^{\ominus}(Ag^+/Ag)$=0.7996V, $E^{\ominus}(MnO_4^-/MnO_2)$=1.679V

(1) The strongest reducing agent is Zn^{2+}, and the strongest oxidizing agent is Ce^{4+};

(2) Zn;

(3) Ce^{4+} or MnO_4^-.

6. (1) The cell reaction is $Ni(s) + Sn^{2+} \rightleftharpoons Ni^{2+} + Sn(s)$, $E_{MF} = 0.1491V$

(2) The cell reaction is $MnO_2 + 4H^+ + 2Cl^- \rightleftharpoons Cl_2 + Mn^{2+} + 2H_2O$, $E_{MF} = 0.0433V$

7. $[Zn^{2+}] = 0.02mol/L$

8. $[H^+] = 2\times10^{-4}mol/L$

9. pH = 2.33

10. $K_{sp}^{\ominus} = 1.38 \times 10^{-8}$
11. $E^{\ominus}(Ag^+/Ag)$=0.799V
12. $E^{\ominus}(Ag_2C_2O_4/Ag)$=0.4896V
13. (1) $E_{MF}^{\ominus} = 0.1848V$; (2) $K_{sp}^{\ominus} = 5.7\times10^{-7}$
14. (1) $E(Cu^{2+}/Cu)$=0.3123V; $E(Ag^+/Ag)$=0.7399V

The galvanic cell symbol is $(-)\ Cu(s)|Cu^{2+}(0.1mol/L)\ ||\ Ag^+(0.1mol/L)|Ag(s)\ (+)$

(2) (+) $Ag^+ + e^- \rightleftharpoons Ag(s)$; (−) $Cu^{2+} + 2e^- \rightleftharpoons Cu(s)$

The galvanic cell symbol is $2Ag^+ + Cu \rightleftharpoons 2Ag(s) + Cu^{2+}$

(3) $K^{\ominus} = 2.8 \times 10^{15}$

15. (1) $E_{MF}^{\ominus} = -0.134V < 0$, reverse spontaneous process;
(2) $E_{MF} = 0.058V > 0$, forward spontaneous process.
16. $K^{\ominus} = 4.68\times10^{39}$
17. $E^{\ominus}(ClO_3^- / ClO_2^-)$=0.34V
18. (1) H_2O_2 is a strong oxidant in acidic medium and a medium-strength reducing agent in alkaline media;
(2) H_2O_2 is unstable in both acidic and alkaline media, which is prone to disproportionation.

Chapter 7

1. Judgement question

(1)- (8): T F T T T F F T

2. Single-Choice question

(1)- (5): A D D D C

(6)- (11): D E E C D C

3. Multiple - choice question

(1)- (5): ADE DEDE BDE BCD

(6)- (9): ACD ABC ABCD BCD

4. Quesqion

(1) n; n^2; $2l$+1

(2) When solving Schrödinger equation, three parameters (n, l, m) must be introduced for getting a reasonable solution ψ. The three quantum numbers n, l, and m describe the size, shape, and orientation of the orbital, respectively.

(3) F: $1s^22s^22p^5$ Mg: $1s^22s^22p^63s^2$ Cu: $1s^22s^22p^63s^23p^64d^{10}4s^1$

Ar: $1s^22s^22p^63s^23p^6$ Co: $1s^22s^22p^63s^23p^63d^74s^2$

(4) Period 4, Group IIIB, d block, Sc.

(5) Be lost its $2s$ electronics. B lost its $2p$ electronics electron The penetration effect of $2p$ is weaker than $2s$ electron. The shielding effect of inner electrons it feels is stronger than that of $2s$ electrons. So its energy is higher than $2s$ and it is easy to lose. In addition, the electron configuration of B is $2s^22p^1$. When the $2p$ electron is lost, it becomes $2s^22p^0$. That is a stable structure with $2p$ full-empty. So the ionization energy of B is lower than Be.

The 2*p* orbit of N is half-filled. O have a fourth 2*p* electron. The fourth 2*p* electron is easier to remove because doing so relieves the repulsions and leaves a half-filled (stable) 2*p* sublevel.

Chapter 8

1. Indicate the electron shell configurations of the following ions.

Be^{2+}, Fe^{2+}, Ca^{2+}, F^{-}, Ag^{+}, Pb^{2+}, S^{2-}, Na^{+}, Cl^{-}, Zn^{2+}, Sn^{2+}

2 electronic configuration: Be^{2+}

8 electronic configuration: Na^{+}, F^{-}, Ca^{2+}, S^{2-}, Cl^{-}

18 electronic configuration: Ag^{+}, Zn^{2+}

18+2 electronic configuration: Sn^{2+}, Pb^{2+}

9-17 electronic configuration: Fe^{2+}

2. Indicate which of the following combinations can form a molecular orbital and the type of orbital formed.

p_x-s: σ molecular orbital

p_x-p_x: σ molecular orbital

p_y-p_y: π molecular orbital

p_y-p_z: no molecular orbital can be formed due to the mismatching of symmetry

s-s: σ molecular orbital

p_y-s: no molecular orbital can be formed due to the mismatching of symmetry

3. Compare the sizes of the following ionic.

(1) $Na^{+} > Mg^{2+} > Al^{3+}$

(2) $F^{-} < Cl^{-} < Br^{-} < I^{-}$

(3) $Fe^{3+} < Fe^{2+}$

4. Determine the spatial configuration of the following molecules by VSEPR theory.

BF_3: $vp = \frac{1}{2}(3+1\times3) = 3$, number of bonding pairs = 3, trigonal planar

SiF_4: $vp = \frac{1}{2}(4+1\times4) = 4$, number of bonding pairs = 4, tetrahedral

CCl_4: $vp = \frac{1}{2}(4+1\times4) = 4$, Number of bonding pairs = 4, tetrahedral

XeF_4: $vp = \frac{1}{2}(8+1\times4) = 6$, number of bonding pairs=4, square planar

ClF_3: $vp = \frac{1}{2}(7+1\times3) = 5$, number of bonding pairs = 3, T-shaped

PCl_5: $vp = \frac{1}{2}(5+1\times5) = 5$, number of bonding pairs = 5, trigonal bipyramidal

5. Determine the type of hybridization of BF_3 and $HgCl_2$.

BF_3: The outer electron configuration of B is $2s^22p^1$. One electron in 2*s* orbital is excited to an empty 2*p* orbital. One 2*s* orbital and two 2*p* orbitals combine to form three sp^2 hybrid orbitals.

$HgCl_2$: One electron in the 6*s* orbital of Hg is excited to 2*p* orbital. One 6*s* orbital and one 6*p* orbital combine to form two *sp* hybrid orbitals.

6. Write the molecular orbital electron arrangements for the following molecules and ions and

indicate the stability of the compound.

H_2^+: $(\sigma_{1s})^1$ Bond order = 0.5

O_2: $(\sigma_{1s})^2(\sigma_{1s}^*)^2(\sigma_{2s})^2(\sigma_{2s}^*)^2(\sigma_{2p_x})^2(\pi_{2p_y}, \pi_{2p_z})^4(\pi_{2p_y}^*, \pi_{2p_z}^*)^2$ Bond order = 2

N_2: $(\sigma_{1s})^2(\sigma_{1s}^*)^2(\sigma_{2s})^2(\sigma_{2s}^*)^2(\pi_{2p_y}, \pi_{2p_z})^4(\sigma_{2p_x})^2$ Bond order = 3

O_2^+: $(\sigma_{1s})^2(\sigma_{1s}^*)^2(\sigma_{2s})^2(\sigma_{2s}^*)^2(\sigma_{2p_x})^2(\pi_{2p_y}, \pi_{2p_z})^4(\pi_{2p_y}^*, \pi_{2p_z}^*)^1$ Bond order = 2.5

7. Which of the following molecules have hydrogen bonds in their structures?

Toluene: no hydrogen bond,

HF: intermolecular hydrogen bond,

CH_3COOH: intermolecular hydrogen bond,

HNO_3: intramolecular hydrogen bond,

O-nitrophenol: intramolecular hydrogen bond

8. Indicate the type of van der Waals force between the following substances.

I_2 and H_2O: induced force and dispersion force,

C_6H_6 and CCl_4: dispersion force,

CH_3OH and H_2O: orientation force, induced force and dispersion force,

HI and HBr: orientation force, induced force and dispersion force.

9. Compare the melting points of the following substances.

NaF < MgO, CaO < MgO, $AlCl_3$ < NaCl, NaCl > NaI

10. Explain the following phenomena by ionic polarization theory.

(1) The color of AgCl, AgBr and AgI are white, light yellow and yellow respectively, the stronger the ionic polarization, the darker the material color. With the same cation of the three substances, the radius of halogen as anion gradually increases, the polarization gradually increases, so the color gradually deepens.

(2) The solubility of AgCl, AgBr and AgI decreased in turn. From AgCl to AgI, the cations of the three substances are the same, the anionic radius increases in turn, the covalent bond component in the bond increases gradually, the polarity of the bond decreases, so the solubility decreases.

11. Are the following statements true or false? Explain why?

(1) Incorrect. NH_3 is hybridized by sp^3 and BF_3 by sp^2.

(2) Incorrect. Non homonuclear diatomic molecules, such as HF, are polar linear molecules; CH_4 is tetrahedral, however the molecule is nonpolar.

(3) Incorrect. The hybrid orbital of dsp^2 is composed of one d orbital in the inner shell, and one *s* and two *p* orbitals in the outer shell.

(4) Correct

(5) Incorrect, the formation of intramolecular hydrogen bonds will reduce the melting point and boiling point of the substance.

12. Compare the cationic polarizability of the following compounds.

(1) $ZnCl_2$, $FeCl_2$, $CaCl_2$, KCl

$Zn^{2+} > Fe^{2+} > Ca^{2+} > K^+$

(2) $SiCl_4$, $AlCl_3$, $MgCl_2$, NaCl

$Si^{4+} > Al^{3+} > Mg^{2+} > Na^+$

13. Compare the anionic deformability of the following compounds.

(1) KF, KCl, KBr, KI

$F^- < Cl^- < Br^- < I^-$

(2) Na_2O, Na_2S, NaF

$F^- < O^{2-} < S^{2-}$

14. Indicate whether the following molecules are polar or not.

CCl_4, BCl_3, NH_3, SO_2, CO_2, $HgCl_2$

CCl_4 (Nonpolar molecule), BCl_3 (Nonpolar), NH_3 (Polar molecule), SO_2 (Polar molecule), CO_2 (Nonpolar molecule), $HgCl_2$ (Nonpolar molecule)

15. Which of the following compounds do not have lone electron pairs?

H_2O, NH_4^+, NH_3, H_2S, ClF_3

NH^{4+} does not have lone electron pairs

16. Please illustrate the difference between ionic bond and covalent bond.

The essence of covalent bond is the overlapping of atomic orbitals. The larger the probability density of electrons between two nuclei is, the stronger the covalent bond is. Covalent bond has two characteristics: saturation and directivity. The essence of ion bond is the electrostatic attraction between positive and negative ions. The higher the charge of ion is, the stronger the ion bond is. The ionic bond has neither saturation nor directionality.

17. Why covalent bond has both saturation and directionality?

The number of single electrons an atom can provide is limited, so the covalent bond is saturated. When the covalent bond is formed, the maximum overlap can be formed only in a certain direction, so the covalent bond is saturated.

18. Explain the following phenomena by molecular orbital theory.

(1) Expression of He_2 molecular orbital: $(\sigma_{1s})^2(\sigma_{1s}^*)^2$, bond level = 0. According to the molecular orbital theory, two of the four electrons in He_2 are in the bonding molecular orbital with reduced energy and two are in the antibonding molecular orbital with increased energy. After the formation of molecules, the overall energy does not decrease and has no contribution to bonding. Therefore, He_2 cannot exist stably.

(2) Expression of O_2 molecular orbital:

$(\sigma_{1s})^2(\sigma_{1s}^*)^2(\sigma_{2s})^2(\sigma_{2s}^*)^2(\sigma_{2p_x})^2(\pi_{2p_y}, \pi_{2p_z})^4(\pi_{2p_y}^{*1}, \pi_{2p_z}^{*1})$

There is a single electron on each and . So O2 is a paramagnetic molecule.

Chapter 9

1. Answer: double salt: (4) (6); coordination compound: (1) (2) (5); chelate: (1); simple salt: (3)

2. (1) $[Zn(NH_3)_4](OH)_2$ Tetrazine hydroxide (II); Zn^{2+}; 4;

(2) $Na_3[Ag(S_2O_3)_2]$ Sodium silver (I) dithiosulfate; Ag^+; 2;

(3) $[CoCl(NH_3)_5]Cl_2$ Monochlorohydrin pentahydrate cobalt (III); Co^{2+}; 6;

(4) $H_2[PtCl_6]$ Hexachloroplatinum (IV) acid; Pt^{4+}; 6;

(5) $[Cocl_2(NH_3)_3(H_2O)]Cl$ Dichlorochloride · triamine · cobalt monohydrate (III); Co^{3+}; 6

3. To be omitted.

4. Answer: the chemical formulas of the two are $[FeC1_2(NH_3)_4]$ C1 and $[PtCl_3(NH_3)_3]Cl$ respectively.

5. Answer: The inner-orbital coordination compounds include (1), (3) and (5). The outer-orbital

coordination compounds include (2), (4) and (6). The geometry of (1), (2), (3) and (4) is octahedron. (5) is a plane quadrilateral. And (6) is tetrahedral.

6. Answer: (3) and (4), the former is more stable than the latter. For (2), the former is less stable than the latter.

7. Answer:

(1) $\text{CFSE} = x\,E\,(t_{2g}) + y\,E\,(e_g) + (n_2 - n_1)\,P$

$= 6\,E\,(t_{2g}) + 0\,E\,(e_g) + (3-1)\,P$

$= 6\times(-4\ \text{Dq}) + 2P$

$= -24\ \text{Dq} + 2P < -4\ \text{Dq}\ (\Delta_o > P)$

(2) $\text{CFSE} = x\,E\,(t_{2g}) + y\,E\,(e_g) + (n_2 - n_1)\,P$

$= 4\,E\,(t_{2g}) + 2\,E\,(e_g) + (1-0)\,P$

$= 4\times(-4\ \text{Dq}) + 12\ \text{Dq} + P$

$= -4\ \text{Dq} + P < 6\ \text{Dq}\ (\Delta_o < P)$

8. Answer: (1) low-spin coordination compounds; 0; There are six electrons in the t_{2g} orbital and zero in the e_g orbital.

(2) high-spin coordination compounds; 4; There are four electrons in the t_{2g} orbital and two electrons in the e_g orbital.

9. To be omitted.

10. Answer: (1), (3) and (4) are wrong. (2) is right.

Chapter 10

1. Choice question

(1) B (2) A (3) C (4) C (5) C (6) B

2. True-false question

(1) × (2) × (3) √ (4) √

Chapter 11

1. Fill in the blanks: write the chemical formula of the following substances

(1) HgS, (2) Hg_2Cl_2, (3) Fe_2O_3, (4) $HgCl_2$, (5) $ZnCO_3$, (6) $(NH_4)_2SO_4\cdot FeSO_4\cdot 6H_2O$, (7) MnO_2, (8) $FeSO_4\cdot 7H_2O$, (9) $CuSO_4\cdot 5H_2O$, (10) $Cu_2(OH)_2CO_3$

2. Judgement question

(1) ×, (2) √, (3) √, (4) ×, (5) ×, (6) ×, (7) √, (8) √,(9) ×, (10) ×

3. Complete and balance the following equations

(1) $2CrO_2^- + 3H_2O_2 + 2OH^- = 4H_2O + 2CrO_4^{2-}$

(2) $2MnO_4^- + 5SO_3^{2-} + 6H^+ = 2Mn^{2+} + 5SO_4^{2-} + 3H_2O$

(3) $2Mn^{2+} + 5NaBiO_3(s) + 14H^+ = 2MnO_4^- + 5Bi^{3+}\ 5Na^+ + 7H_2O$

4. Explain the following experimental phenomena with chemical reaction equations

(1) $Fe^{2+} + 2OH^- = Fe(OH)_2\downarrow$（白色）

(2) $2H_2O + 4Fe(OH)_2 + O_2 = 4Fe(OH)_3\downarrow$（红棕色）

(3) $Fe(OH)_3 + 3H^+ = Fe^{3+} + 3H_2O$

(4) $Fe^{3+} + SCN^- = Fe(SCN)^{2+}$（血红色）; $2Fe(SCN)^{2+} + SO_2 + H_2O = 2Fe^{2+} + SO_4^{2-} + 4H^+ + 2SCN^-$

(5) $5Fe^{2+} + MnO_4^- + 8H^+ = Mn^{2+} + 5Fe^{3+} + 4H_2O$

(6) $Fe^{3+} + Fe(CN)_6^{4-} + K^+ = KFe[Fe(CN)_6]\downarrow$（蓝色）

5. Q & A questions

(1) Hg^+ is hybridized into two *sp* hybrid orbitals and exists as a dimer of Hg_2^{2+} ($Hg^+ : Hg^+$).

(2) $Fe(OH)_2$ can be slowly oxidized to brownish red $Fe(OH)_3$ by oxygen in the air.

The equation is as follow: $2H_2O + 4Fe(OH)_2 + O_2 = 4Fe(OH)_3\downarrow$（红棕色）

(3) A: $K_2Cr_2O_7$; B: Cl_2; C: $CrCl_3$; D: $Cr(OH)_3$; E: $KCrO_2$; F: K_2CrO_4

参考文献

[1] 铁步荣，杨怀霞. 无机化学 [M]. 北京：中国中医药出版社，2016.
[2] 刘幸平，吴巧凤. 无机化学 [M]. 2 版. 北京：人民卫生出版社，2016.
[3] 北京师范大学，华中师范大学，南京师范大学. 无机化学 [M]. 4 版. 北京：高等教育出版社，2010.
[4] 大连理工大学无机化学教研室编. 无机化学 [M]. 5 版. 北京：高等教育出版社，2006.
[5] 张祖德，无机化学 [M]. 2 版. 合肥：中国科学技术大学出版社，2014.
[6] 张天蓝，无机化学 [M]. 7 版. 北京：人民卫生出版社，2016.
[7] 武汉大学，吉林大学. 无机化学 [M]. 北京：高等教育出版社，2010.
[8] 宋天佑，无机化学 [M]. 2 版. 北京：高等教育出版社，2015.
[9] 天津大学无机化学教研室编. 无机化学 [M]. 4 版，北京：高等教育出版社，2010.
[10] 曹凤歧. 无机化学 [M]. 南京：东南大学出版社，2010.
[11] 章伟光. 无机化学 [M]. 2 版. 北京：科学出版社，2017.
[12] 杨宏孝. 无机化学 [M]. 北京：高等教育出版社，2010.
[13] 王书民. 无机化学 [M]. 北京：科学出版社，2017.
[14] 司学芝，刘捷，展海军. 无机化学 [M]. 北京：化学工业出版社，2009.
[15] 杨怀霞，吴培云. 无机化学 [M]. 2 版. 北京：中国医药科技出版社，2018.
[16] 徐家宁，无机化学核心教程 [M]. 北京：科学出版社，2017.
[17] Gary L. Miessler, Donald A. Tarr. 无机化学 [M]. 3 版. 影印本. 北京 : 高等教育出版社，2006.
[18] Ralph H. Petrucci, William S. Harwood, F. Geoffrey Herring. 普通化学原理与应用. [M]. 8 版. 影印本. 北京 : 高等教育出版社 , 2004.
[19] Kenneth W. Whitten, Raymond E. Davis, M. Larry Peck, George G. Stanley. Chemistry[M]. 9rd Ed. USA. BROOKS/COLE, 2010.
[20] Theodore L. Brown, H. Eugene LeMay Jr., Bruce E. Bursten, Catherine J. Murphy, Patrick M. Woodward, Matthew W. Stoltzfus. Chemistry The Central Science[M]. 13rd Ed. England. PEARSON, 2015.